# Standardization and Digital Enclosure:

## The Privatization of Standards, Knowledge, and Policy in the Age of Global Information Technology

Timothy Schoechle
*University of Colorado, USA*

A volume in the Advances in IT Standards and Standardization Research (AITSSR) Book Series

Director of Editorial Content: Kristin Klinger
Senior Managing Editor: Jamie Snavely
Managing Editor: Jeff Ash
Assistant Managing Editor: Carole Coulson
Typesetter: Chris Hrobak
Cover Design: Lisa Tosheff

Published in the United States of America by
Information Science Reference (an imprint of IGI Global)
701 E. Chocolate Avenue
Hershey PA 17033
Tel: 717-533-8845
Fax: 717-533-8661
E-mail: cust@igi-global.com
Web site: http://www.igi-global.com



Library of Congress Cataloging-in-Publication Data

Schoechle, Timothy D. (Timothy Duncan)
Standardization and digital enclosure : the privatization of standards, knowledge, and policy in the age of global information technology / by Timothy Schoechle.
p. cm.
Includes bibliographical references and index.
Summary: "This study of the privatization of standards and standardization seeks to establish a framework of analysis for public policy discussion and debate with such related issues as technical innovation, access to information and media, intellectual property, and emerging industrial economic development"--Provided by publisher.
ISBN 978-1-60566-334-0 (hardcover) -- ISBN 978-1-60566-335-7 (ebook) 1. Information technology--Standards. 2. Privatization. I. Title.
T58.5.S316 2009
389'.6--dc22
2008047748

This book is published in the IGI Global book series Advances in IT Standards and Standardization Research (AITSSR) Book Series (ISSN: 1935-3391; eISSN: 1935-3405)

British Cataloguing in Publication Data
A Cataloguing in Publication record for this book is available from the British Library.

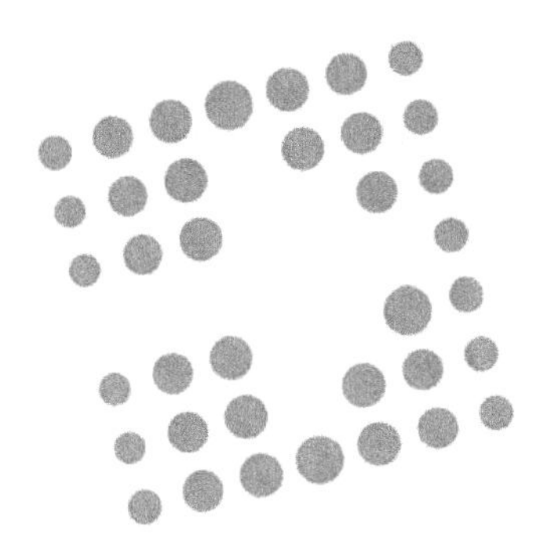

# Advances in IT Standards and Standardization Research (AITSSR) Book Series

*Kai Jakobs (Aachen University, Germany)*

ISSN: 1935-3391
EISSN: 1935-3405

## Mission

IT standards and standardization are a necessary part of effectively delivering IT and IT services to organizations and individuals as well as streamlining IT processes and minimizing organizational cost. In implementing IT standards, it is necessary to take into account not only the technical aspects, but also the characteristics of the specific environment where these standards will have to function.

The **Advances in IT Standards and Standardization Research (AITSSR)** book series seeks to advance the available literature on the use and value of IT standards and standardization. This research provides insight into the use of standards for the improvement of organizational processes and development in both private and public sectors.

## Coverage

- Analyses of standards-setting processes, products, and organization
- Descriptive theory of standardization
- Emerging roles of formal standards organizations and consortia
- Intellectual property rights
- Management of standards
- National, regional, international and corporate standards strategies
- Open Source and standardization
- Risks of standardization
- Technological innovation and standardization
- User-related issues

IGI Global is currently accepting manuscripts for publication within this series. To submit a proposal for a volume in this series, please contact our Acquisition Editors at Acquisitions@igi-global.com or visit: http://www.igi-global.com/publish/.

The Advances in IT Standards and Standardization Research (AITSSR) Book Series (ISSN 1935-3391) is published by IGI Global, 701 E. Chocolate Avenue, Hershey, PA 17033-1240, USA, www.igi-global.com. This series is composed of titles available for purchase individually; each title is edited to be contextually exclusive from any other title within the series. For pricing and ordering information please visit http://www.igi-global.com/book-series/advances-standards-standardization-research-aitssr/37142. Postmaster: Send all address changes to above address. 

## Titles in this Series

*For a list of additional titles in this series, please visit: www.igi-global.com*

*Evolution and Standardization of Mobile Communications Technology*
DongBack Seo (University of Groningen, The Netherlands)
Information Science Reference • copyright 2013 • 328pp • H/C (ISBN: 9781466640740) • US $195.00 (our price)

*Information Technology for Intellectual Property Protection Interdisciplinary Advancements*
Hideyasu Sasaki (Ritsumeikan University, Japan)
Information Science Reference • copyright 2012 • 412pp • H/C (ISBN: 9781613501351) • US $195.00 (our price)

*Frameworks for ICT Policy Government, Social and Legal Issues*
Esharenana E. Adomi (Delta State University, Nigeria)
Information Science Reference • copyright 2011 • 350pp • H/C (ISBN: 9781616920128) • US $180.00 (our price)

*Data-Exchange Standards and International Organizations Adoption and Diffusion*
Josephine Wapakabulo Thomas (Rolls-Royce, UK)
Information Science Reference • copyright 2010 • 336pp • H/C (ISBN: 9781605668321) • US $180.00 (our price)

*Toward Corporate IT Standardization Management Frameworks and Solutions*
Robert van Wessel (Tilburg University, Netherlands)
Information Science Reference • copyright 2010 • 305pp • H/C (ISBN: 9781615207596) • US $180.00 (our price)

*Standardization and Digital Enclosure The Privatization of Standards, Knowledge, and Policy in the Age of Global Information Technology*
Timothy Schoechle (University of Colorado, USA)
Information Science Reference • copyright 2009 • 384pp • H/C (ISBN: 9781605663340) • US $165.00 (our price)

www.igi-global.com

701 E. Chocolate Ave., Hershey, PA 17033
Order online at www.igi-global.com or call 717-533-8845 x100
To place a standing order for titles released in this series,
contact: cust@igi-global.com
Mon-Fri 8:00 am - 5:00 pm (est) or fax 24 hours a day 717-533-8661

# Table of Contents

# Abstract

This is a study of the standardization process and establishment of technical standards that define virtually every artifact of the modern world. In the information and communication technology field, such standards specify everything from the prongs on plugs to software protocols that make the Internet work. Historically, standards have been set largely by volunteers in committees operating within a range of environments, institutional rules and social practice, but they generally have espoused traditional principles of accessibility, democratic deliberation, public accountability, and balanced stakeholder representation.

A recent trend that has prompted much discussion is the increasing privatization of standardization activities under various corporations, trade associations, and consortia. This trend is far removed from the traditional, and, as claimed, more "open," democratic, and inclusive, practices of voluntary consensus committees. Because standards play a powerful role in shaping technologies and their diffusion into society, the trend raises significant public policy issues about how the public interest may be represented and served in today's digital information age that is increasingly dependent on technical standards.

Within fields of law and public policy, an oppositional discourse has emerged concerning the privatization or "enclosure" of ideas—analogous to the land enclosure movement in 16th century England—and on the expansion of intellectual property rights, resulting in the "fencing off" of the intellectual commons. Since standardization practice can be viewed as a collaborative effort in idea formation—an intellectual commons—this study uses the enclosure discourse as a framework for examining the debate over the privatization of standards.

This study, which draws upon a theoretical perspective of political economy and theories of the public sphere, applies methods of discourse analysis and historiography in examining the quest for legitimization by consortia and arguments for their inclusion in the international standards system. Furthermore, the study seeks to clarify the discourse on standards and standardization by showing how the ideas and arguments that form, apply, and justify policy decisions rely on symbols, beliefs, and ideologies that are rhetorically constituted. It is further argued that key terms used in the discourse, that is, open, public, and private, are often ill defined and conflicting.

Keywords: commons, consensus, democracy, discourse, enclosure, open, policy, private, privatization, public, public sphere, rhetoric, standards, standardization.

*Timothy D. Schoechle*
*University of Colorado, USA*
*2004*

# Preface

*He who controls trade controls the world's wealth, and therefore the world itself.*
*—Sir Walter Raleigh, c. 1600*

## STANDARDS AND STANDARDIZATION

The modern world is in large part defined by technical standards. In everyday life, standards define such diverse things as screw threads, paint colors, wine glass dimensions, and film speeds. In the information and communication technology field, such standards are documents that specify everything from the prongs on plugs to the software protocols that make the Internet work. Today, the essential purpose of international standards is to facilitate trade and commerce, although at times in the past standards have been established by nations or regions for defensive purposes, to impede invasion (either military or commercial), and to protect markets. Historically, standards have been set largely by volunteers[1] in committees operating within a range of environments, institutional rules, and social practices, but they generally have espoused traditional principles of accessibility, democratic deliberation, public accountability, and balanced stakeholder representation. With the development of global markets, standards are increasingly being set at the international level, through institutions that operate by employing various forms of nation-based representation, participation, and voting. This book is about the process by which these standards are set and the forces that are re-shaping that process.

### What this Book is and why it Matters

On the surface, this book is about standards and how they are made. But, more fundamentally, it is about the privatization—or what I call the enclosure—of that standards-making process. It is about closely related issues of democratic deliberation in societal decision-making and about the institutionalization of that process. This topic is increasingly important—as dynamic political, economic, social, and other forces are dramatically re-shaping our technological society and our global economy—and as new actors are entering the traditional nation state based system of standards-making institutions.

Two key questions that are addressed here include, what is meant by open standards and what is the public? All actors seem to claim the high grounds of open standards and of acting in the public interest. These claims deserve critical examination and thought—given the policy decisions that governments, standards organizations, and private parties are now being called upon to make in re-shaping or re-directing the course of national, regional, and global policies and institutions. Technical standards will continue to not only define and limit the future of technology—but of society itself. They will determine what the technologies of the future are to be, and, in great part, who owns them, who gets to use them, and whose interests are served by them.

## Standards Setting Organizations

The term Standards Setting Organization (SSO) refers to a vast array of organizations that set standards under a wide range of institutional rules and practices. This range of rules and practices involves, in part, the issue of how "public" or "private" the standards-making process is. Some, known as private "consortia," have restricted or well-defined "members," upon which strict rules can be imposed. Others, of initial interest here, are the more formal traditional public bodies that include three major international SSOs: (1) the International Organization for Standardization (ISO); (2) the International Electrotechnical Commission (IEC); and (3) the International Telecommunications Union (ITU). Also of interest here are the various national bodies that send delegates and technical experts to the international meetings. The tension and dynamic between these formal traditional bodies and the many newer emerging industry consortia is the primary focus of this book.

The oldest of these traditional organizations is the ITU, which was established in 1886 in Geneva, Switzerland, and is now part of the United Nations. The ITU, perhaps the first international intergovernmental organization, was founded as a treaty organization. Its initial purpose was primarily related to interconnection of telephone networks and radio communications—new technologies at the time. Until the mid-1980s, the telephone industry was dominated by nationally owned (or regulated, in the case of the United States) monopolies, and technical standardization served a quasi-governmental function, with participants acting as national government representatives. In the 1992-1994 timeframe, the CCITT (*Comité Consultatif International Telegraphique et Telephonique*), a part of the ITU, was renamed the ITU-T (telephone) sector. Standards produced by the ITU are called "Recommendations."

The two other institutions that are also important features on the international landscape of standardization are the IEC, founded in 1906 with the development of electricity industries, and the ISO, formed in 1947, as the post-World War II successor to the International Association for Standardization (ISA), which was founded in 1926. The IEC standardizes electrical and electronic-related topics and the ISO standardizes everything else. With the development of computers and the need to standardize an increasing variety of hardware interfaces and software functions, the ISO and IEC formed a joint organization in 1988 for information technology standards known as Joint Technical Committee 1 (JTC1). JTC1 took over management of all computer-related ISO (*e.g.,* software) and IEC (*e.g.*, hardware) standards work. Now, with the convergence of telecommunication and computers, the complexity of coordinating work among the various standards-making organizations—ISO, IEC, JTC1, and ITU-T—has become increasingly more challenging. All four organizations operate under different organizational structures, different "memberships," and different governing directives.

## Traditional System Structure and Practices

The ISO, IEC, and ITU-T share certain structural similarities. They are hierarchical, with layers of governing committees, representing a taxonomy of technical topics. Within the ISO and IEC structures are committees called "Technical Committees" (TC) and "Sub-Committees" (SC). In the ITU-T, such committees are referred to as a "Study Group" (SG) and "Working Party" (WP). The TCs and SCs of the ISO and IEC, along with the SGs and WPs of the ITU-T, are all located at the top of the hierarchical organization chart. The actual detailed technical work—including the drafting of technical standards documents—takes place at the bottom level of the hierarchy.

In the ISO and IEC, a bottom level group is usually known as a "Working Group" (WG). In the ITU-T it is usually referred to as a "Rapporteur Group" identified by a "Question" the group addresses. Committee meetings above the bottom level are often referred to as "plenary" meetings because they deal with broader management, policy, coordination issues, or oversight of the technical work. No technical work is performed in plenary meetings. The ITU-T's SGs (thirteen exist today) are renewed or disbanded, and sometimes re-named, routinely on a four-year cycle known as a "Study Period." In contrast, the ISO's over 200 TCs and the IEC's over 100 TCs remain in existence until specifically changed. As an indicator of the scope of its worldwide standards activity, the ISO estimates that 30,000 individual technical experts participate in its work in 2,850 various technical committees, subcommittees, and working groups, and that "... there are, on average, a dozen ISO meetings taking place somewhere in the world every working day of the year."[2]

### Emerging Consortia

Beginning roughly during the late 1980s, due to a variety of forces discussed later in this book, a new breed of SSO began to emerge, known generically as the *consortium* or *forum*. These groups tended to be more focused on specialized technical areas and to have a more limited scope of membership—usually with a particular economic or business interest around a specific technical standard or industry—rather than the nation-based membership of the traditional system. Often, consortia followed a different business model, deriving their income from membership or licensing income rather than from publishing, as did the traditional bodies. These groups have proliferated, although some have proven somewhat ephemeral, depending on the commercial fortunes of their technical foci. The term *consortium* has come to refer to roughly to any of the scores or hundreds of SSOs that are not part of the traditional system. Generally, although there are major exceptions, consortia represent a privatized process clamoring for status and for access to governmental and institutional procurement markets that traditionally have favored formal national, regional, or international standards.

## CONTEXT AND SIGNIFICANCE OF THIS STUDY

This study of the privatization of standards and standardization seeks to establish a framework of analysis for public policy discussion and debate. This book and its topic fit within the larger issues of globalization and trade, as well as with such related issues as technical innovation, access to information and media, intellectual property, and emerging industrial economic development. These are not transitory issues. They are deeply problematic, and how they are resolved will have much to do with determining the shape of the world we will be living in the coming decades—and handing on to future generations.

### Globalization

The noted economist Joseph Stiglitz[3] defines globalization in terms of the continuing decline in transportation and communication costs, and the reduction of man-made barriers to the flow of goods, services, and capital. He comments that:

*...we have a process of "globalization" analogous to the earlier processes in which national economies were formed. Unfortunately, we have no world government, accountable to the people of every country, to oversee the globalization process in a fashion comparable to the way national governments guided the nationalization*

*process. Instead, we have a system that might be called global governance without global government, one in which a few institutions...and a few players...dominate the scene, but in which many of those affected by their decisions are left almost voiceless.* (Stiglitz, 2003, pp. 21–22)

The principle institutions that Stiglitz is referring to include the International Monetary Fund (IMF), the World Bank, and the World Trade Organization (WTO), although the lesser known ISO, IEC, ITU, as well as the numerous other SSOs, could also fit well within the point he is making. The other "players" he is referring to include government and private commercial interests that guide or otherwise influence these institutions.

Stiglitz' primary concern is the failure of global institutions to meet the world's needs in our current age of globalization, and he attributes this failure to the ascendancy during the 1980s of a "free market ideology," championed by the political régimes of President Ronald Reagan and Prime Minister Margaret Thatcher, in the United States and in the United Kingdom respectively. This ideology became hegemonic and carried with it a global wave of deregulation and privatization of a vast array of formerly public functions and enterprises, enforced by the aforementioned institutions and the economic interests that controlled them. Stiglitz optimistically sees much of this free market ideology as having run its course, tempered now by the financial and economic debacles of the late 1990s and early 2000s and a general backlash against globalization. The 2008 U.S. and global economic collapse and government interventions may have finally brought the era of "free market" ideology to a close, globally if not entirely in the United States. Stiglitz finds it essential to establish a new or reformed set of global institutions responsive to public needs. This book addresses how these influences of privatization and deregulation have affected the international standardization system—a crucial element of international trade and development. This book also seeks to help establish a framework for policy discourse around these issues, which will facilitate or guide such a reformation. This book is particularly timely in this respect, given the dramatic changes that have occurred in the economic and political landscape since 2008.

## Audience

This Book was intended to address three audiences: (1) academic—to provide a basis and framework for further academic research, (2) policy makers—as an exercise in critical thinking[4] about policy proposals and basis for analysis of the terms of discourse that are used in policy arguments, and (3) practitioners—to offer historical and structural (organizational) contexts to their work and to encourage them to question the hidden, taken-for-granted, and potentially misleading assumptions behind the everyday terms used in their committee work (*e.g.*, "open," "public," "private," etc.). It is intended that the latter two audiences should find this book to be of extremely practical value.

## ENCLOSURE THEORY—A FRAMEWORK OF ANALYSIS

We take the concept of private property and the notion of ownership as fundamental to our society. It is interesting to note, however, that it is largely a modern concept tracing from the Enlightenment.[5] Our assumptions about property are so pervasive that we seldom consider that it is largely alien to most older cultures—at least in its present form—and does not translate into many developing societies. Most of human history was experienced in tribal societies wherein modern conceptions of property or ownership have little meaning.

The term "enclosure" was used beginning in 15th century Britain (*i.e.,* the Enclosure Movement) to describe the Parliamentary laws and the practices that were used by a rising commercial merchant class to "fence off" (*i.e.,* privatize) what had traditionally been a commons[6] (also known as "open fields") inhabited by rural peasants in an agrarian subsistence economy. The underlying theory was that land is *more productive when held in private hands than when held in common.* Thus, the English Parliament justified the enclosure of the commons and the subsequent evictions on the basis that privatization would lead to more productive use of the land (*i.e.,* greater agricultural production) and thus to the greater good of society. Although some economic research challenges this base conception that privatization is an unambiguous economic good (McClosky,1972; Humphries, 1990), it represents a firmly entrenched belief system that has had an enormous formative influence on British and American culture—and subsequently on global economics and politics. A corollary to this theory that has grown into an ideological belief and has played a predominant and recurring role in U.S. public policy formation could be stated as *the public interest is best benefited by benefiting private interests*—otherwise more recently known as "trickle down" economics.[7]

Recent legal discourse has borrowed the terminology of enclosure, characterized as the *Second Enclosure Movement*—enclosure in the age of digital information—as the enclosure of information and ideas rather than of land (Boyle, 2001). This notion of enclosure is logically an extension of the notion of "intellectual property"—another Enlightenment concept. The Enclosure Discourse is explored in detail later in this book. Here the focus of interest is to use *Enclosure Theory* to examine the privatization of the standards *process* and its implications for society and for technical innovation—rather than on the enclosure of the *content* of standards. The full impact of intellectual property rights (IPR) on standardization and on technical innovation and economic development is a vast and rapidly growing topic, but unfortunately a thorough treatment of the topic is beyond the scope of this book and must be left for future examination.

## SCOPE OF THE STUDY

### Abstract Approach

This work approaches the subjects of standards and standardization, Enclosure Theory, rhetoric, and theories of democratic governance at a rather abstract and theoretical level. This approach was considered necessary in order to build a rigorous basis for the analysis of actual social practices and political economic discourse—and for those researchers that may wish to build upon this foundation. Some individual readers may find such theoretical rigor tedious or unnecessary, in which case they can simply skip over parts of it and still benefit from the analysis that follows. Likewise, historical background is included because it is this author's belief that entirely too much policy discourse today is essentially ahistorical in nature, and that historical context explains much of why things are the way they are. How we got here is important.

### Interdisciplinary Approach

This book seeks to bring together two worlds that ordinarily have little to do with one another—the technical world of standards geeks or engineers from the specialized and often Balkanized world of standards bodies and the specialized world of policy wonks, economists, and political scientists. In this respect, sthis book is a highly interdisciplinary undertaking, as must be the case for any compelling discussion of standardization. No single discipline provides all of the tools needed to understand such a multidimensional field. This author came from over three decades of computer and communication engineering to try to find out how our society made its decisions about what technologies got invented and how and why they were applied to what. This quest required reaching beyond engineering methodology and technology into political philosophy and into perhaps the oldest academic "discipline" of all—rhetoric.

### Standards Research

Another purpose of this book is to contribute to the basis for "Standards Research" which began to emerge as a distinct field of academic study in the late 1990s. Initially prompted by a couple of books, government studies, and university courses, this fledgling field has, over less than a decade, led to the development of a set of periodic conferences, academic journals, and the beginnings of a literature tradition of its own. Advancing and improving the standardization process is an important undertaking because technical standards define the modern world around us. Dr. Kai Jakobs of the University of Aachen observes that "standards setting is big business today…" with the cost of developing a single ICT standard "easily running into seven- or eight-digit numbers"—not to mention the possibly life-or-death strategic role of standards for companies large and small. Yet, he further observes,

*Knowledge about the issues surrounding IT standards and standardization is still rather underdeveloped. Curricula are few and far between, and the number of researchers and scholars with an interest in the subject, albeit increasing, is still fairly small. Possibly even worse, relevant knowledge also seems to be scarce within industry.* (Jakobs, 2005, p. viii)

Establishing a legitimate field of research with a body of literature is important because universities will not establish related curricula without a publishing path for their faculty and career path for their students.

One of the most challenging problems with a new field such as standards research is its inherently interdisciplinary nature. Standardization cannot be well understood without examining it from multiple viewpoints. Some of these perspectives include economics, political science, behavioral sciences, and business, as well as engineering and computer science—and none of these provides a complete picture in itself. It is hoped that Enclosure Theory, and this book as a whole, will provide a useful interdisciplinary framework of analysis for further study.

## ORGANIZATION OF THE BOOK

This book is organized into eight chapters. A brief description of each of the chapters follows:

### Summary

Chapter I includes the Introduction which identifies the research problem—that of the increasing privatization of the traditional formal international standardization system. The chapter sets the scene for the study and the rest of the book by defining research questions and the rationale for its approach to these questions. It then summarizes the discourses on Enclosure and on Standards that will be examined, and describes the methodological approaches and the scope of the study. The chapter closes with comments on the intended audience and the perspective of the author.

The study itself will begin in Chapter II, providing a broad historical background on standards and standardization practice. This history will begin with a broad overview based on historical practice, provide a definition of standards, offer a description of the structure of the international system and how it emerged, and finally focus on the U.S. system and its significant structural and ideological divergence from the European and international standardization system.

Chapter III reviews the literature and discourses that inform the study and are pertinent to its central questions, including the relevant discourses on Enclosure and on standards and standardization, and drawing primarily from the disciplines of economics, law, social studies, and political philosophy. These discourses establish the key concepts, terms, and arguments that will be explored later in the book.

Chapter IV examines the theoretical and methodological approaches taken in the study. This examination will include discourse analysis, as it is to be applied here, the relationship between theory and method, relevant discourses and social practices, theoretical perspectives on discourse and on the public sphere, theories of rhetoric and discourse, and political economy. The basic research questions are then established.

Chapter V situates the discourse of standardization and identifies and attempts to establish meanings for certain essential terms of the discourse—namely, *public*, *private* and *open*. From the perspective provided by examining such terms and their meanings, the book, in Chapter VI, reviews the institutional history of the principal international organizations that form the basis of the traditional global standardization system: the ITU, ISO, and IEC.

Chapter VII proceeds to an analysis of certain actual discourses and argumentation surrounding standardization and consortia—drawing on influential public policy documents (including U.S. Congressional testimony, E.U. policy studies, and rebuttal papers). Finally, Chapter VIII states the results, draws conclusions, makes recommendations, identifies the study's limitations, suggests areas and paths for further research, and makes several concluding observations.

## Advice to the Reader

The heart of the book is in Chapters IV, V, and VII—the analysis and the framework defined for this analysis. Chapters IV and V provide a methodical and terminological framework for how the policy discussion of standards development and role of the consortium is being analyzed. Chapter VII is the analysis itself and the essence of the book. The early chapters and Chapter VI set the stage, providing the antecedent background that provides the contextual framework on which the book relies in terms of history, language, and literature. Readers already familiar with this background might skip over this material and refer back to it as needed. In particular, Chapter III provides an extensive review of a range of applicable literature that should be of interest to academic readers, and it is hoped this material will provide a useful survey and resource to future researchers. There are few published works (relatively speaking) in the area of standards and standardization and it is hoped that this work will further the understanding of standards theory and development.

## CONCLUSION

One purpose of the book is to critically examine the dominant discourse of standardization, which lies within a larger discourse on globalization, geopolitics, trade, and foreign policy. The dominant discourse for thirty years in this arena has been a celebration the many successes of globalization. That dominant discourse has become severely challenged in recent years for reasons detailed by Stiglitz (2003), Gray (1998), Kupchan (2002, 2002a), Khanna (2008), and others. It isn't "globalization" itself (simply a consequence of technology) that is suffering a backlash, but the institutions and the ideology that rode the wave of globalization. Perhaps this backlash is most recently evidenced by the 2008 collapse of the Doha round of WTO talks, as well as the dramatic 2008 collapse of major national and global financial institutions. The enclosure of standards and standardization has been an aspect of this global wave, and as the wave subsides, this analysis can help re-frame the policy discourse around standardization for a newly emerging global order.

By laying the groundwork for Enclosure Theory as a framework of analysis, it is hoped that further studies by this author and by others can build upon it. This book begins the process by examining the enclosure of the institutions and products of standardization and the underlying economic and political logic. The pressures of globalization and its emerging political backlash make resolution of current global trade and economic development issues urgent. For example, the global standards system is challenged today by the emergence of new economic and industrial players—particularly China.

As the standards system became increasingly privatized by economic players (*e.g.,* consortia and patent pools or license authorities formed around technical standards and dominated by Western or Japanese corporations), China found itself to be an outsider with little or no patent position and having to pay royalties even though it manufactures much or most of actual end products. As a result, China has moved forward recently with its own initiatives, both domestically and within ISO, IEC, and ITU, to standardize competitive versions of technologies that have served as the underpinning much of the nation's production (as well as new technologies).[8] China has been able to take these initiatives because it has a domestic market sufficiently large to make its own standards and technologies economically viable, regardless of international markets (Stuttmeier, 2006). China has also made a substantial effort to participate in and understand the formal international standards system (with significant government backing), attending meetings in substantial numbers, hosting meetings, and undertaking the burden of committee secretariats whenever they become available.

During the 1990s, Korea became an important global technology player and moved well up the ladder of economic development, but its influence in the international standards system was limited because of its relatively small domestic market. With China's rather sudden emergence and with Korea's experience and its cultural and economic ties with China, the calculus of influence is changing dramatically in Asia and globally.[9]

China and Korea are but two examples illustrating how the locus of technical innovation, knowledge production, and global markets are changing. There are others. Technical standards will be a crucial element of trade and economic development in the coming decades. It is important to understand the factors that shape the standardization process if it is to be responsive to the needs of commerce and of our emerging global society. It is hoped that this book can advance that understanding.

## REFERENCES

Boyle, James (2001). "The Second Enclosure Movement and the Construction of the Public Domain," *Conference on the Public Domain*. Duke Law School, Durham, NC, November 9-11, 2001.

Farrell, Joseph (1996). "Choosing the Rules for Formal Standardization." Berkeley: UC Berkeley.

Gray, John (1998). *False Dawn: The Delusions of Global Capitalism.* London: Granta Books. 262 pp.

Humphries, Jane (1990, March). "Enclosures, Common Rights, and Women: The Proletarianization of Families in the Late Eighteenth and Early Nineteenth Centuries," *Journal of Economic History, L*(1), The Economic History Association, pp. 17–42.

Jakobs, Kai (2005). *Advanced Topics in Information Technology Standards and Standardization Research: Volume 1.* Hershey, PA: Idea Group Publishing. pp. 348.

Khanna, Parag (2008). *The Second World: Empires and Influence in the New Global Order.* New York: Random House. 466 pp.

Kupchan, Charles A. (2002). "The End of the West." *The Atlantic Monthly*. vol. 290, no. 4. November 2002. pp. 42-44.

Kupchan, Charles A. (2002a). *The End of the American Era: U.S. Foreign Policy and the Geopolitics of the Twenty-first Century*. New York: Alfred A. Knopf. 391 pp.

McClosky, Donald M. (1972, March). "The Enclosure of Open Fields: Preface to a Study of Its Impact on the Efficiency of English Agriculture in the Eighteenth Century," *Journal of Economic History, 32*, The Economic History Association, pp. 15-35.

MIC (2004). *IT839 Strategy: The Road to $20,000 GDP/Capita*. Ministry of Information and Communication, Republic of Korea. <http://www.mic.go.kr>.

Stiglitz, Joseph E. (2003). *Globalization and its Discontents*. New York: W.W. Norton, 288 pp.

Stuttmeier, Richard P., Xiangkui Yao, and Alex Zixiang Tan (2006). *Standards of Power? Technologies, Institutions and Politics in the Development of China's National Standards Strategy*. Seattle: The National Bureau of Asian Research. 52 pp.

## ENDNOTES

1. The term *volunteers* used in this context refers to the participants who are either individuals, consultants acting as individuals or on behalf of voluntary clients, or as employees sent by organizations acting on a voluntary basis *(i.e.,* the sponsoring standards body in not paying the participants to make standards). "The active participants are 'volunteers' willing to spend substantial time and travel money" (Farrell, 1996, p. 2).
2. <http://www.iso.ch>.
3. Columbia University Professor of Economics, of International and Public Affairs, and of Business, and recipient of the 2001 Nobel Prize in Economics.
4. By *critical thinking*, it is meant the discipline or habit of routinely identifying and questioning the unstated assumptions that underlie one's own, as well as others, ideas, narratives, and truth claims.
5. The Enlightenment was a broad social and political movement in the west during the 16th through the 18th centuries that represented a critical questioning of traditional institutions, customs, and morals, and its values centered on principles of freedom, democracy, rights, and reason. The Enlightenment resulted in many of the political and economic ideas that shaped the modern world, known as classical liberalism, including individual rights, the rule of

law, private property rights, science and rationality, the public, democratic governance, socialism, and capitalism.

6 The term "commons" in this context does not mean "owned" in common (as we take the idea of "public" ownership today), but rather the notion of ownership or property is entirely absent.

7 Virtually all public policy proposals are couched in terms of the "public interest."

8 Examples include DVD formats, MPEG compression, 3G Wireless telephony, WiFi, RFID, and home networks.

9 For example, see Korea's Ministry of Information and Communication IT839 Strategy (MIC, 2004).

# Acknowledgment

*Reading maketh a full man; conference a ready man; and writing an exact man.*
—Sir Francis Bacon, 1625 essay, Of Studies

Putting pen to paper for any research/analytical/theoretical work is essentially a solitary undertaking and its ultimate completion is a test of the commitment and perseverance of the author—no one else really having such a vital stake. At the same time, such an effort would be impossible to accomplish without the support of many others who provide a wide range of "conference," from guidance and information to less tangible but nevertheless critically important inspiration, validation and encouragement.

Since the foundations of this book originated as a PhD dissertation, I would be remiss if I failed to acknowledge those who were influential early in my academic life. I would like to recognize Dr. Andrew Calabrese, who first introduced me to the field of communication policy and to many of the basic concepts that underlie this work. I also hold great appreciation for Dr. Willard Rowland and Dr. Gerard Hauser whose writings, teachings, and mentoring provided the principal theoretical and methodological underpinnings of this research. In addition, I am grateful to Dr. Robert Trager and Dr. Stewart Hoover, for their insightful critiques of my original research proposal and their overall encouragement. Sincerest thanks to all.

For much of the substance of this research, I am indebted to Carl Cargill, formerly of Sun Microsystems, who stands above all for his goading me forward with spirited debate and stimulating argument. Carl's commitment to the international standardization system and to crafting education and research in this new field is without parallel. I must also thank Professor Dr. Tineke Egyedi of the Technical University of Delft, The Netherlands, for providing many challenging arguments, questions, and, above all, for leading the way as a role model with her own pioneering dissertation in the field of standards research.

I must also thank Gary Robinson of EMC Corporation, a long-time standards warrior, for wading through manuscripts and offering the insightful critiques that allowed me to move an academic dissertation into a book that could guide both standards practitioners and policy-makers. I thank my colleague, Dr. Kai Jakobs of the Technical University of Aachen for his encouragement in this project and his assistance in bringing it to a publisher.

In engaging in this field of research in the first place, I must thank Professors Randall Bloomfield and Dr. Frank Barnes of the CU College of Engineering and the Interdisciplinary Telecommunication Program (ITP), for their vision in establishing the International Center for Standards Research at the University of Colorado. The Center provided the intellectual platform for launching this work. I am also indebted to my students in TLEN 5190 "Telecom Standards: Current Issues" over several semesters for their curiosity and challenging questions.

I express my gratitude to Professor Lawrence Lessig of Stanford University for his writings that provided validation of my vision of technical standardization as policy-making, and specifically for his charging me with the mission to relate standardization practice to his own advocacy of the commons of ideas. I also thank his colleagues, Professors Mark Lemley of U.C. Berkeley and Molly Van Houweiling of the University of Michigan for providing key papers and ideas.

It is important for me to also thank my friends and colleagues; Philipp Braun of RegTP in Bonn, Germany, for providing translations of key policy documents; Dr. Douglas Wagner for reading, commenting, and editing not only my early drafts, but those right up through the final edit, making it possible to meet my publisher's deadline, and for his own guiding research/writings in communication policy and discourse analysis; and Dr. Mark Andrejevic for introducing me to the "digital enclosure" metaphor. I am grateful to Hans Meierhofer of RegTP for calling my attention to certain key DIN documents and for arranging for me to present this work to his colleagues. Among those, I specifically thank Volker Gebauer of RegTP for his thoughtful comments and access to pertinent documents.

I thank my colleagues in the JTC1 SC25/WG1 U.S. Technical Advisory Group for their ongoing encouragement, and specifically Frank Farance of Farance, Inc., and Dr. Kenneth Wacks of MIT for sharing their detailed knowledge, experience, and references regarding ISO, IEC and JTC1 committee affairs.

I am grateful to my sister, Susan Canterbury for her hospitality and sustenance, and a mountain cabin among the Aspens to retreat to over many weeks when focus and clearing my mind were crucial to bringing the project to completion. I am thankful for my loyal Afghan hounds, Lucy, Liza, Polar Bear, and Sabrina who kept me company through endless evenings by the fireside, pouring through books and papers, and I only regret that they were unable to see me bring it all to a conclusion.

And finally, I give gratitude to my contemporary feline companions, Leo, Wilson, Calliope, Aruba, and Holly (affectionately known as the Office Manager), for "help" with paperwork and for their antics, which never failed to lighten the mood and bring a smile to my face while working to meet the myriad of deadlines.

*Timothy D. Schoechle*

# Chapter I
# Introduction to the Study

*The Congress shall have the Power...To coin Money, regulate the Value, and of foreign Coin, and fix the Standard of Weights and Measures.*

—U.S. Constitution, Article 1, Section 8

## INTRODUCTION

This book is a study of the process of standardization—the process of establishing the technical standards that define nearly every artifact of the modern world. In the field of Information and Communication Technology (ICT)[1] such standards are documents that specify everything from the prongs on plugs and cables to the software protocols that make the Internet work. Technical standards and standardization play a vital role in trade and commerce, and increasingly in economic and cultural globalization. The aim of this study was to setup a research project to explore the discourse around standardization and to analyze it to provide a better understanding of the underlying issues.

Historically, these standards have been set largely by volunteer[2] participants in committees that operate within a wide range of environments, institutional rules and social practices. In general, these groups have espoused a traditional commitment to general principles of democratic deliberation, public accessibility, and balanced stakeholder representation. The historical practice is now being challenged by a variety of forces, including newer, more private organizations that do not necessarily have the same commitment to these principles.

## PURPOSE

The purpose of this study is to establish an overall framework, that of *Digital Enclosure*—Enclosure in the Digital Age—a concept borrowed from legal, economic and public policy discourse for understanding important, contentious and interwoven issues in the current global standardization system. These issues include the rise of new standards-setting "consortia" and the challenge they pose to the traditional standardization process, intellectual property rights (IPRs), competition, anti-trust policy, business and commercial strategies, the Open Source movement, geopolitics, and technical innovation.

This study will begin by problematizing the Enclosure of standards and its relationship to notions of intellectual property, the concept of the "public," and public goods. It will then suggest some forms of Enclosure and some of the counter movements and institutional responses and adaptations to Enclosure. In particular, the current work will seek to explore and understand the consortia phenomenon by analyzing the discourse surrounding it to ascertain the cultural meanings that drive related policy decisions by corporations, governments and institutions. Finally, it will consider Enclosure with respect to technical innovation and competition. Primarily, the purpose here is to establish the terms of the discourse surrounding standardization and consortia, the meanings of such terms, and to provide a framework for further study, analysis, and informed debate.

## THE PROBLEM

The central focus of this study is on the practice of technical standardization as a form of public discourse and idea production within a technical culture. In particular, the study focuses on the discourse surrounding a recent trend toward the privatization or "enclosure" of standardization activities under various corporations, trade associations, and consortia, and away from the more traditional, and possibly, as often claimed, more "open," more democratic, and more inclusive voluntary consensus[3] committees. The distinction between the traditional formal system[4] and the newer consortia is one of legitimacy based on accreditation, on adherence to certain principles and rules of process, and on custom. The public or open nature of the traditional bodies was already somewhat problematic in various respects, given their institutional history, their economic importance, and the variety of interests they have served. Now, it can be argued that parts of the system are being further enclosed. The central issue that emerges in this study from the standards discourse is that of *legitimacy*—the real or perceived legitimacy of the standards and of the process by which they are derived.

These standards are technical specifications that play a powerful role shaping technologies and influencing the ways in which they are diffused into society. Often such standards are incorporated into law or form the basis for administrative policy and rulemaking. Enclosure raises significant issues of public policy concern and questions about how the public interest may be represented and served in a digital information age that is increasingly dependent on technical standards and on the experts that create them. The basic question is: how can decisions made in private serve the public? In any case, such decisions affect the public interest and their basis of legitimacy is a matter of valid public concern.

Technical standards are documents comprising agreements that establish a vast array of processes, practices and procedures vital to the functioning of a modern society and a market economy. These standards address diverse needs such as communication protocols and systems; languages and data structures; accounting procedures and practices; manufacturing processes and quality; and product and process conformity assessment and certification. Many of these standards cross over from the technical realm into the public policy arena by setting specifications that impact public issues and concerns. A sampling of such issues includes privacy, security, access to networks and information, the "digital divide," social equity, societal cohesion, patents, trademarks, copyrights and fair use, freedom of speech, technical innovation, regulation, industrial policy, monopoly, and competitive markets. For example, the technical architectures of the Internet and of the conventional public switched telephone network (PSTN) are very different—with vastly different implications about who manages them, who can use them, and for what purposes.

Originating within the field of law and public policy, an oppositional discourse has recently emerged around the theme of the *privatization* or *enclosure* of ideas[5]—analogous to the *land* enclosure movement in 16th century Britain[6]—and addresses the expansion of intellectual property rights and a resulting general "fencing off" of the intellectual *commons*. Since open standards and standardization practices can be viewed as a form of group collaborative idea generation—an *intellectual commons*—this study begins by using the *Enclosure discourse* as a framework within which to examine the privatization of standards. The Enclosure discourse (including its focus on intellectual property) is not only relevant but also central to the study of standards and standardization because the *commons* historically has been an essential concept of standardization.

## RATIONALE AND APPROACH

This study approaches the problem of Enclosure of standardization by applying the following rationale. The Enclosure discourse is about ideas and the public domain—and about the intellectual commons. Standards and standardization are public goods and have always inhabited the public domain (at least partially), and they have been seen as constituting an intellectual commons (including its essential publicness, however flawed or limited). Therefore, the general discourse on Enclosure may be relevant to, and applied to, the specific problem of standardization. Since the Enclosure of standards and standardization is discursively constructed, the terms of that discourse are therefore important and determinative of material consequences.

The problem is therefore established and approached using the following logic:

- An oppositional[7] enclosure discourse asserts that an enclosure of the intellectual "commons" is increasingly occurring that deeply affects the future of ideas.
- This oppositional discourse deals principally with real and potential negative material outcomes for society—about technical innovation, wealth generation and equity—and therefore it is important.
- This oppositional discourse emerges from a perspective of law, policy and economics, but does not explicitly address the general field of technical standards and standardization in any detail.
- Standardization is a form of collaborative group idea generation and deals with the same sort of ideas as does the Enclosure discourse (*e.g.*, "code,"[8] architecture, intellectual property, *etc.*).
- A discourse on standards and standardization argues for and against the ascendancy of alternative enclosed standards bodies and their legitimation and incorporation into the global standards system.
- Therefore, it is proposed here to logically extend and apply these oppositional enclosure arguments to the discourse on standards and standardization, including concepts of production, innovation,"code," architecture and intellectual property rights, to see how they might fit.

Enclosure may take a wide variety of forms. For instance, standards may be viewed as a form of *intellectual property*,[9] or as being made into intellectual property in a number of ways: (1) standards are often viewed as the intellectual property of the sponsoring institution of the authoring committee or body, to sell, publish, or otherwise distribute and control (*e.g.*, open source GPL or General Public License, consortia, traditional bodies, *etc.*); (2) standards may be viewed as the partial intellectual property of inventors (intellectual property incorporated into a standard—deliberately and knowingly by the standards body) under a RAND (Reasonable And Non Discriminatory) licensing policy; and (3) standards may be seen as targets of strategic manipulation or stealth practices and subject to intellectual property claims or control (*i.e.*, intellectual property incorporated into a standard—unwittingly and unknowingly by the standards body)[10].

From the perspective of the oppositional enclosure discourse, the general hypothetical problem may be stated as follows:

*The public nature of standards as a commons is being challenged. Privatization for the benefit of certain specific interests may have deleterious consequences on society as a whole, including the impairment of access to information and communication networks, of competition, and of technical innovation.*

## RESEARCH QUESTION

This study proceeds by establishing a research question and a set of corollary questions. It is important to understand the explicit and implied meanings of the terms that serve as the basis for standards discourse. A rhetorical analysis of these terms raises questions such as the following. What is meant by "public" and "private"? What is the meaning of "open" and how is its meaning constructed and applied? Can a standard developed in a closed committee be an open standard? What are the views on the effects of open/closed processes of standards making? What is the basis of legitimacy of such standards and related decisions made on behalf of the public?

## THE DISCOURSE ON ENCLOSURE

A body of literature has emerged about the nature of technical innovation and markets, and about the relationship between intellectual property rights (IPRs) and the intellectual "commons." Lessig (1999) maintains that network architectures and software "code" (which, it is asserted here, are largely defined by technical standards) represent *de facto* policy decisions about who is to benefit from ICTs, and emphasizes the derivative nature of intellectual creation, which he calls *ideas,* and the importance of preserving a commons for innovation. (Lessig, 2001). Boyle (2001) likens the increasing emphasis on IPR and the privatization of ideas in the present digital economy to the land enclosure movement of the 16th century. This enclosure discourse is centered on a classic debate of market economics—whether the generation of wealth in society is better served by holding property publicly (in the commons) or in private hands (Boyle, 1996, Lessig, 1999). Benkler (2001; 2002) describes alternative modes of production outside of the market model, including "peer" production and collaborative authorship in the digital information economy, including the Open Source movement. Litman (1990) describes the derivative nature of creation and the process of authorship (standards committees are authors). Van Houweiling (2002) describes forms of enclosure applied to standards and proposes extension of land ownership models into the realm of IPR.

The terms of this discourse include such words as *free*, *open*, *public* and *private*—terms that are deeply problematic and loaded with rhetorical and ideological significance. The Enclosure discourse focuses on how ideas and IPRs are formed and dealt with in society in general, but this discourse does not apply enclosure directly to the problem and practice of standardization, except by implication. It is here proposed that standardization is an important form of social idea generation, and perhaps a form of IPRs creation (although the term *IPR* is also itself somewhat problematic). The first step here will be to relate the Enclosure discourse to the discourse on standards.

## THE DISCOURSE ON STANDARDS

The emerging field of *standards research* is beginning to produce a significant body of literature on standards and on standardization processes. This literature seeks to establish the theoretical underpinnings for the longstanding practice of standards making, and it takes two basic approaches: (1) economic and social theory applied to standards, and (2) case studies of standardization projects and related controversies. This literature will be described in more detail in Chapter III.

Some of the standards theory and case study literature deals with the relative contrasting roles of traditional Standards Development Organizations (*SDOs*) and newer industry private *consortia*. A highly polarized debate has formed around the recent ascendancy of these consortia, pro and con, and it is couched in terms of the aforementioned discourse, as well as that of *market vs. democracy*. Cargill (2001; 2001a) and Egyedi (2001) advocate recognition of consortia as legitimate standards bodies.[11] Sherif (2001) and the *Deutsche Institute für Normung* (DIN, 2002) argue against such recognition. The discourse is largely centered on language and claims for legitimation and market relevance *vs.* language and claims of openness and democracy. Both sides claim to best serve the public interest. I will seek to pry these arguments open by critically examining the associated discursive practice as well as its discourse, language, and embedded meanings.

## METHODOLOGICAL APPROACHES

The primary theoretical framework for this study is *public sphere* theory and the related branch of communication studies known as *political economy.* This study applies methods of discourse analysis and historiography, and draws upon a perspective derived from political economy and theories of the public sphere. The aforementioned methods are then applied to an examination of the quest for legitimation by consortia and the arguments for their incorporation into the international standards system. It seeks to clarify this standards discourse by arguing that many of the ideas and arguments that form, apply, and justify related policy decisions rely on symbols, signs, beliefs and ideologies that are rhetorically constituted, and that the terms of the discourse are often problematic, ill-defined, and conflicting, and sometimes perhaps purposefully so.

The primary method of inquiry employed here will be discourse analysis. The problem of Enclosure is complex, and the approach taken here is only one of many possible approaches. No single approach is, or can be, in itself complete. Discourse reveals prevailing ideological commitments that lead people to interpret data in conflicting ways. Analysis of the discourse will identify the variety of meanings or interpretations found in the various embedded symbols, myths, and ideologies. It is hoped that such an analysis may lead to a better understanding of policy implications and responses. It will be argued that reliance on uncritical, incomplete, inconsistent, and contradictory notions concerning the practice of standardization and its products constrains possible policy options.

## SCOPE OF THIS STUDY

As previously stated, the purpose of this study is to establish an overall discursive framework—Digital Enclosure—within which the evolution of the global standardization system can be analyzed and clarified. This is not intended to be a history, but the study must rely on historical perspectives in order to understand the basis of the present system and its conflicts. Much cultural meaning is embedded in the institutions and practices of the standardization system, and therefore, this study visits enough history to establish the meaning of the important terms of discourse.

## Process Enclosure—Making Standards

The various forms of Enclosure are many, and they are only briefly surveyed here in order to put the topic in perspective and map the general landscape of standardization and of its Enclosure. One particular form of Enclosure—that of the emergence of private consortia—is the primary focus of this study. This is a qualitative study, which attempts to identify representative sample documents and discourses for detailed analysis, guided by the experience of the researcher. Such discourses are chosen because of the influential nature of their authors and/or their audiences and because they address key issues. They are not chosen because of any statistical or proportional representation of opinion in the field. This study does not seek to provide final answers or policies, but to be exploratory and guiding in its conclusions, and to identify issues and improve clarity for those concerned with policy-making in the standards arena.

## Content Enclosure—Intellectual Property and Antitrust Issues

As previously noted, a thorough treatment of the important and lively topic of IPR *vs.* standards—another related form of Digital Enclosure—is beyond the scope of this study. Unfortunately, the IPR/Antitrust topic is too vast to treat thoroughly in this book and not loose the more fundamental focus of the book on the basic terms of discourse and Enclosure Theory. The first stage of enclosure is "privatization" of the standards process—the topic of this book. A further stage is the enclosure of the standards' content, a topic on which this author has previously written (Schoechle, 2001; 2003). Nevertheless, a brief explanation is necessary here, even at the risk of some digression, because of the sheer importance and impact of IPR/antitrust on standards and standardization, and to provide a more complete context for the rest of this book.

Patents may be seen as representing private invention and standardization as representing collaborative public invention. The prevailing view in the literature on standardization seems to be that the problem is one of balancing the private property interests of IPR holders with the common public interests of standardization (Iversen, Bekkers & Blind, 2006; David, 2006). For example, Iversen, Bekkers and Blind (2006) characterize the problem as one of "...balancing the collective gains to be reaped from the elaboration of a common standard against the individual gains to be allocated to relevant individual rights-holders" (p. 1).

The view taken here is different. It is that the *common* or the *public* is the prevailing interest and that private interests that seek to enclose the commons must bear the burden of proof. This principle is not unlike the traditional concept of eminent domain. Antitrust law tradition privileges the public interest, and patent monopoly rights were originally granted as a limited exception—not as a true "property right". As will be discussed later in the book, IPRs have come to be extended well beyond their original Jeffersonian conception, and their contribution to innovation is problematic. David (2006) and others propose that standards *vs.* patent conflicts may be mitigated by the (problematic) notion of "club goods" or limited (dimished) or quasi-privatized "commons" that can somehow live in a world of inflated IPRs—an argument relying on a perverse distinction between the "common" and the "public domain."

Much recent academic and legal discourse proceeds toward the idea of solving the problem by "fixing" the standards process by expanding and "strengthening" the IPR policies of standards bodies, allowing, or even mandating, *ex ante* license negotiation in committees (Skitol, 2005; Lemley, 2002; Majoris, 2005). This remedy fixes the wrong problem and is generally proposed by lawyers, economists, or academics that have no actual experience "on the ground" in a standards committee. Those who actually work in such committees, being acquainted with the cultural environments and social practices, understand the impracticality of such a notion, aside from the thorny legal thicket into which it leads.[12] Not unexpectedly, the formal standards system has not—and probably cannot, and will not, be drawn in such a direction. Another widely proposed solution is the patent pool, a remedy that risks morphing the standards committee into a cartel—a risk more willingly assumed by consortia than by formal bodies.

The importance of the standards *vs.* patents issue is highlighted by an emerging and troubling trend to make the embedding of patents within standards a national industrial strategy by both developed and developing nations. In an important report, *Standards of Power: Technology, Institutions, and Politics in the Development of China's National Standards Strategy*, Stuttmeier, Yao, and Tan (2006) lay out the means by which China is adapting to the established IPR/Standards régime, traditionally dominated by Western and Japanese corporations, by targeting certain industries to establish its own IPR/standards régime and move it into the international system. A similar but somewhat less ambitious strategy was described in the Korean *IT839 Strategy* (MIC, 2004).

In yet another important paper, Yamada (2006) describes a national strategy for Japan as "Underscoring patent exploitation through the *Intellectual Property Strategic Program*" (p. 19). He continues,

> *Japan has been paying growing attention to standardization activities and patent pools as approaches to the effective utilization of intellectual property. This is the right direction for a country that aims to become an* ***intellectual property-based nation*** (p. 19). [emphasis added]

This strategy is based on "taking advantage of international standardization activities as political negotiations," and it is recommended that Japanese companies "move rapidly beyond mere participation in committees" and "have their patented technologies incorporated into standards," form patent pools, provide employees with educational opportunities to improve negotiation skills, and "delegate skilled negotiators to participate in international committees" (p. 20). In other words, enclosing standards is seen as a means by which a mature developed nation can make its future living by "rent" collecting when it reaches a stage at which it no longer manufactures anything. Two problems with such strategy are (1) its assumption that any nation can continue to "corner" technical innovation, and that developing nations that may currently be manufacturing will not soon learn the design/innovation/patent/standards game also, and (2) its dependency on an IPR régime that is resulting in a "feeding frenzy" that is unsustainable. It is likely that such strategies as aforedescribed will simply fragment the international standards system and defeat its basic purposes—an example of the "tragedy of the commons."[13]

Ultimately, the standards *vs.* patents issue will likely be dealt through higher court decisions and/or legislation throttling the out-of-control patent régime or changing antitrust practices, and not by "fixing" the policies and procedures of standards bodies. To this point, the debate surrounding an initiative in the European Parliament, adopted by the E.U. Council of Ministers—known as the proposed *Directive on the Patentability of Computer-Implemented Inventions* (CII), or simply the *CII Patents Directive*—has focused attention on the pros and cons of software patents and the interests of the public and of small and medium enterprises (SMEs).

A further and longer-term response to the patent problem was released in 2007 by the European Patent Office (EPO) under the title, *EPO Scenarios for the Future Project* (EPO 2007). This 128 page document, accompanied by a CD ROM, reported findings from over 100 interviews conducted over several years with various stakeholders who had explored the questions, "How might IP régimes evolve by 2025?" and "What global legitimacy might such régimes have?" Four possible future scenarios for the global patent system were distilled from the study. These four non-exclusive scenarios, titled, *Market Rules*, *Whose Game*, *Trees of Knowledge*, and *Blue Skies*, do not attempt to predict, but simply explore, possible futures as a first step in informing the emerging policy debate.

## TARGET AUDIENCE

This Book addresses three audiences: (1) academics—to provide a basis and framework for further academic research, (2) policy makers—as an exercise in critical thinking about policy proposals and basis for analysis of the terms of discourse that are used in policy arguments, and (3) standards practitioners—to offer historical and structural (organizational) contexts to their work and to encourage them to question the hidden, taken-for-granted, and potentially misleading assumptions behind the everyday terms used in their committee work (*e.g.*, "open," "public," "private," *etc.*). It is intended that the latter two audiences should find this book to be of highly practical value.

This work is intended to help elucidate the debate over the role of standards institutions and their reform. By examining the discourse surrounding enclosure and standardization, it may be possible to obtain a deeper understanding of the arguments and their terminology. The essential analysis is in Chapter VII, and the necessary framework of analysis is in Chapters IV and V. Other chapters provide the historical, methodological, and terminological background on which the analysis is based. Hopefully, this analysis will be useful to those engaged in standardization practice and in public policy discourse, including policy makers, industry leaders, and academic researchers. Also, it is hoped that this book can make a contribution to further establishing and expanding the field of standards research. In particular, this study seeks to establish the terms *enclosure* and Enclosure Theory as a framework for further study and analysis in this field.

## PERSPECTIVE OF THE RESEARCHER

This researcher has undertaken this study from a perspective of twenty-five years of active personal involvement in the standardization process as an engineer, and of having served in a range of capacities from participant to committee chair and secretary in various national and international standards bodies. This involvement began in 1983 when I began participating in the Electronic Industries Association (EIA)[14] CEBus Committee (Consumer Electronic Bus), a project to develop and standardize a low-cost communication network for the home. Initially, my involvement derived from commercial motivation—the need to establish technical standards in order to build a business. Over time, however, the process became an object of personal interest and, ultimately, a focus of academic research.

This interest was due to many factors. One was the rich forum standardization practice provided among technical peers for the exchange of ideas and as a window on advancing technology. Another was the exposure that participation provided to other cultures and to other ways of thinking. Most importantly, however, were the insights that it gave me on the process of interpersonal, inter-organizational, and cross-cultural communication and decision making. A striking factor was a recognition of the importance of standardization as a form of industrial decision-making and public policy formation. But, in spite of its importance, standardization is practiced in relative obscurity in proportion to its influence in shaping technology and to the high stakes involved for businesses, consumers, and society as a whole.

Throughout my experience, I have observed significant changes in standardization practice and have been disappointed in some recent trends that suggest a risk to, or abandonment of, the basic principles of openness and democracy to which the more traditional standardization process aspired and attempted to institutionalize. Although these principles may never have been fully realized in practice, they were at least memorialized and available to serve as a reference model for policy-making and analysis. Recently, I have seen short-term economic priorities and market logic progressively erode financial support for the practice within the United States, while others have been able to turn the practice to their own beneficial purposes on a global scale.

My "insider" perspective has enabled me to access key elements of the discourses of, both verbal and written, and to contextually assess their relative importance in ways that an outside observer could not. I was able to make use of my own knowledge, familiarity and contacts to get past many of the barriers that exist for outsiders conducting such a study, including academic researchers who often opine on this topic. I was privileged to participate in candid discussions that guided my research and analysis for the study in a manner that would not have been available to an outsider.

## CONCLUSION

The title of this study, *Standardization and Digital Enclosure: The Privatization of Standards, Knowledge, and Policy in the Age of Global Information Technology,* seeks to reflect the intersection of one body of discourse with another in the context of the current digital age. But the title is somewhat problematic in itself, and is intended to be provocative. What is privatization? What is meant by standards? This study is an attempt to better define these terms. Accordingly, it assumes a discursive position and then attempts to identify consequences and implications.

*Enclosure* is a reference to the oppositional discourse about a commons or public domain that is at risk of being enclosed. The term *enclosure* is preferred to *privatization* because, as will be shown, the word *private* is loaded with its own set of ideological assumptions. *Digital* is a reference to the *digital age*—the age of ICT, wherein the term *digital economy* refers to an economy based more on knowledge and information than on physical assets. Schiller uses the term in this meaning in the title of his book, *Digital Capitalism* (Schiller, 2000), and, for different ends, Negroponte in *Being Digital* (1995) employs a similar usage. *Digital* is also a reference to the 2nd enclosure movement as a metaphor for the earlier land or spatial enclosure—to contrast the present *digital* (knowledge or idea) enclosure with the 1st *spatial* (land) enclosure[15].

In other words, the 2nd enclosure is the Digital Enclosure—enclosure in the digital age. It is a reconfiguration of the spatial logic of the 1st enclosure movement into the digital or information age in which ideas, knowledge, and policy become privatized, compelling entry into a particular social relation and preempting alternative spaces (*i.e.,* the commons). Within the resuting regime, capitalism establishes proprietary access to information that can be accumulated.

Discourse reflects cultural values. Culture shapes technology. Discourse shapes policy decisions. This project seeks to explore and understand a specific current issue, that of standardization privatized by consortia. It undertakes this exploration by looking at the discourse and language—the words and their culturally embedded meanings—and the conditions of the discourse—the institutional setting in which the discourse proceeds—in order to establish how meaning in this particular debate is rhetorically constructed. It also seeks to cast this debate within a broader social and economic context—to bring greater clarity to standards and standardization within society and to related policy choices.

## REFERENCES

Benkler, Yochai (2001). Coase's Penguin, or, Linux and the Nature of the Firm. *29th Telecom Policy Research Conference.* Alexandria, VA. September 24, 2001. Availiable at <http://www.arxiv.org/abs/cs.CY/0109077>

Benkler, Yochai (2002). Intellectual Property and the Organization of Information Production. *International Review of Law & Economics, 22*(1), 81-107.

Boyle, James (1996). *Shamans, Software and Spleens* (p. 226). Cambridge: Harvard University Press.

Boyle, James (2001). The Second Enclosure Movement and the Construction of the Public Domain. *Conference on the Public Domain.* Duke Law School, Durham, NC, November 9-11.

Cargill, Carl (2001). Consortia Standards: Towards a Re-Definition of a Voluntary Standards Organization. Testimony before the House of Representatives, Sub-Committee on Technology, Environment and Standards on behalf of Sun Microsystems. June 28, 2001.

Cargill, Carl. (2001a). *The Role of Consortia Standards in Federal Procurements in the Information Technology Sector: Towards a Re-Definition of a Voluntary Standards Organization.* Submitted to the House of Representatives, Sub-Committee On Technology, Environment and Standards. Palo Alto: Sun Microsystems. June 28, 2001. 30 pp. Availiable at <http://www.house.gov/science/ets/Jun28/Cargill.pdf/>

David, Paul A. (2006). Using IPR to expand the research common for Science: New moves in 'legal jujitsu'. *Intellectual Property Rights for Business and Society* conference sponsored by DIME-EU Network of Excellence, held in London, September 14-15.

DIN (2002). *Strategie für die Standardisierung der Informations und Kommunikationstechnik (ICT) - Deutsche Positionen,* Version 1.0. Berlin: DIN Deutsches Institut für Normung, eV. Availiable at <http://www2.din.de/sixcms/detail.php?id=3871>

Egyedi, Tineke (2001, October). *Beyond Consortia, Beyond Standardization? New Case Material for the European Commission. Final Report to the European Commission* (p. 69). Delft: Faculty of Technology, Policy and Management, Delft University of Technology.

EPO (2007). European Patent Office. *EPO Scenarios for the Future.* (p. 128). EPO: Munch.

Hardin, Garrett (1968). The Tragedy of the Commons. *Science, 162*(1243).

Hunter, Dan (2002). Cyberspace as Place. *TPRC 2002: The 30th Annual Research Conference on Communication, Information and Internet Policy.* September 28-30, 2002, Alexandria, VA.

Iversen, Eric J., Bekkers, Rudi & Blind, Knut (2006). Emerging coordination mechanisms for multi-party IPR holders: linking research with standardization. *Intellectual Property Rights for Business and Society* conference sponsored by DIME-EU Network of Excellence, held in London, September 14-15.

Lemley, Mark A. (2002). Intellectual Property Rights and Standard-Setting Organizations. *California Law Review, 90*(6), 1889-1979.

Lemley, Mark A. (2002a). Place and Cyberspace. *TPRC 2002: The 30th Annual Research Conference on Communication, Information and Internet Policy.* September 28-30, Alexandria, VA.

Lessig, Lawrence (1999). *Code and Other Laws of Cyberspace* (p. 297). New York: Basic Books.

Lessig, Lawrence (2001). *The Future of Ideas: The Fate of the Commons in a Connected World.* (p. 352). New York: Random House.

Litman, Jessica (1990, fall). The Public Domain. *Emory Law Journal 39*, 965. Availiable at <http://www.law.wayne.edu/litman>

Majoras, Deborah Platt (2005). Recognizing the Procompetitive Portential of Royalty Discussions in Standard Setting. *Standardization and the Law: Developing the Golden Mean for Global Trade* conference at Stanford University, Stanford, CA, 23 September. In S. Bolin (Ed.), *The Standards Edge: The Golden Mean* ( pp. 101-107). Chelsea, MI: Sheridan Books.

MIC (2004). *IT839 Strategy: The Road to $20,000 GDP/Capita.* Ministry of Information and Communication, Republic of Korea. Availiable at <http://www.mic.go.kr>

Negroponte, Nicholas (1995). *Being Digital.* New York: Alfred Knopf.

Schoechle, Timothy (2001, fall). Re-examining Intellectual Property Rights in the Context of Standardization, Innovation and the Public Sphere. *Knowledge, Technology & Policy, 14*(3), 109-126.

Schoechle, Timothy (2003). Digital Enclosure: The Privatization of Standards and Standardization. *SIIT 2003 Proceedings: 3rd IEEE Conference on Standardization and Innovation in Information Technology* (pp. 229-240). Delft University of Technology. 22 October.

Stuttmeier, Richard P., Xiangkui, Yao, & Alex, Zixiang Tan (2006). *Standards of Power? Technologies, Institutions and Politics in the Development of China's National Standards Strategy* (p. 52). Seattle: The National Bureau of Asian Research.

Schiller, Dan (2000). *Digital Capitalism: Networking the Global Market System.* Cambridge: MIT Press.

Sherif, Mostafa (2001, April). A Framework for Standardization in Telecommunications and Information Technology. *IEEE Communications Magazine, 39*(4), 94-100.

Skitol, Robert A. (2005). Concerted Buying Power: It's Potential for Addressing the Patent Holdup Problem in Standard Setting. *Antitrust Law Journal, 72*, 727-744.

Van Houweiling, Molly S.(2002). Cultivating Open Information Platforms: A Land Trust Model. *The Journal on Telecommunications Law and High Technology.* Vol. 1. Also presented at the Silicon Flatirons Conference. "Regulation of Information Platforms." University of Colorado, Boulder, CO. January 28-29.

Yamada, Hajime (2006, April). Patent Exploitation in the Information and Communications Sector—Using Licenses to Lead the Market. *Science and Technology Trends Quarterly Review, 19*, 11-21. [original Japanese version published October 2005]

## ENDNOTES

[1] Although this study focuses mainly on standardization within the field of ICT, the principles and issues explored here are applicable generally to all other fields. The ICT field, being fast moving and contentious, is where some of the stresses on the standards system have been the greatest.

[2] The term *volunteer* used in this context refers to the participants who are either individuals, consultants acting as individuals or on behalf of voluntary clients, or as employees sent by organizations acting on a voluntary basis *(i.e.,* the sponsoring standards body in not paying the participants to make standards).

[3] *Voluntary consensus* means 1) compliance with the standard is voluntary, and 2) a consensus process was used to establish the standard. In the context of the United States standardization system, the term *voluntary* also implies a non-governmental institutional basis.

[4] "Formal" institutions and practices here refers to established non-governmental or quasi-governmental national, regional and international standards developing organizations (SDOs) and SDO accrediting bodies; these formal institutions include such organizations as ANSI (American National Standards Institute), ISO (International Standards Organization), ITU (International Telecommunications Union), IEC (International Electrotechnical Commission), and many others.

[5] The term *ideas* is borrowed from Lessig (2001) and is used in a broad sense. It encompasses any product or manifestation of human intellect including texts, writings, musical compositions, art, artifacts, inventions, and technical standards and specifications.

[6] The analogy to land enclosure is drawn by Boyle (2001).

[7] The discourse is *oppositional* because it challenges the prevailing dominant discourse about ideas, ownership, and the relationship between innovation, property and market based production; and about the nature of the commons, freedom, openness, the free market and intellectual property rights.

[8] The term *code* is borrowed from Lessig (1999).

[9] The term *intellectual property* is deeply problematic, as will be expanded further later in this study.

[10] For example, the Rambus case, an EEE (Embrace, Extend and Enclose) strategy, *etc*. The EEE strategy has been described by Van Houweiling (2002) and will be further discussed later in the book. It describes how standards may become a takeover target by private corporations (private appropriation); standardization can also be a takeover target—suggesting the question: what assurance is there of "openness" if there is no accountability or no legitimation?

[11] These particular documents were selected because, based on the experience of the author, they represent influential and respected sources, and they provide articulate expressions of the basic rhetorical arguments.

[12] The *ex ante* debate seems to resemble the medieval ecclesiastical debates over how many levels there are in hell or how many angels can dance on the head of a pen—a discourse that is intellectually challenging but totally removed from reality.

[13] The *Tragedy of the Commons* is biologist Garrett Hardin's classic allegory about the tendency to exploit a common resource in situations of unregulated self interest (Hardin, 1968).

[14] Subsequently renamed the Electronic Industries Alliance.

[15] The use of spatial metaphors in the digital information economy can be complex, challenging and problematic. For an interesting discourse on this topic, see the related Telecommunications Policy Research Conference (TPRC) papers by Hunter (2002) and Lemley (2002a).

# Chapter II
# Standards and Standardization Practice

*In tracing the theoretic history of weights and measures to their original elements in the nature and the necessities of man, we have found linear measure with individual existence; superficial, capricious, itinerary measure, and decimal arithmetic, with domestic society; weights and common standards with civil society; money, coins and all the elements of uniform metrology, with civil government and law; arising in succession and parallel progression together.*

—John Quincy Adams, 1821

## INTRODUCTION

Technical standards have always played a vital role in the development of industrial society. Historically, standards can be traced to origins in the invention of currency and in early human activities such as warfare, trade and printing—in societies as diverse as ancient China and Rome. For example, Venetian war galleys were mass produced: the size, fittings, ropes and even oars were all standardized and interchangeable (Cargill, 1997, p. 18). Throughout history, standardization evolved as a social practice and it tends to reflect the particular political and economic cultures involved.

*The key to the process was that there was a substantial economic advantage to be gained from creating and using a standard. One would like to believe that somewhere there was someone who sat down and thought this out, but it probably occurred over several generations.* (Cargill, 1997, p. 16)

Although a detailed history of standardization practice[1] is beyond of the scope of this treatment, this chapter will attempt to briefly acquaint the reader with the historical context, taxonomy, terminology and present landscape of the global system of technical standardization.

Beginning from a broad sense, the term *standard* may be defined as:

*...something established by authority, custom, or general consent as a model or example, ...something set up and established by authority as a rule for the measure of quantity, weights, extent, value or quality.* (Webster, 1977, p. 1133)

Upon further examination, however, numerous other specific definitions may be found, depending on one's disciplinary perspective or purpose. Of particular relevance here is a more strict sense of the term determined by the nature of the *authority* mentioned previously and the *legitimacy* thus conferred. Formalists will assert that *standards* are distinct from *specifications* (which have less *legitimacy*). For example, a formal *international standard* definition of "standard" is a:

*...document, established by consensus and approved by a recognized body, that provides, for common and repeated use, rules, guidelines or characteristics for activities for their results, aimed at the achievement of the optimum degree of order in a given context.* (ISO/IEC, 1996, sec. 3.2)

Variability in the meaning and precision of the term *standard* is an important factor in how the entire topic is discursively constructed and how various interests are served by its use—an issue recognized in an OTA (Office of Technology Assessment) study.

*The choice of definitions has major policy implications. How the term "standards" is used in this study, for example, determines the terms of the debate and the range of government options developed for dealing with problems in the standard setting process.* (Congress, 1992, p. 5)

In any case, it is important to recognize that technical standards are not value-free—an idea that technologists may be often remiss or reluctant to acknowledge. It has been noted that, "...we do not discover a problem 'out there,' we make a choice about how we want to formulate a problem. That choice reflects certain values and in turn constrains the realm of possible solutions" (Cheit, 1990, p. 150).

## OVERVIEW OF THE STANDARDIZATION SYSTEM

Today, the world of technical standardization is a global but arcane and diverse societies of experts and various interests engaged in setting standards for virtually every aspect of our modern industrial/information society. Standardization engages tens of thousands of individuals that travel widely and confer on a vast range of issues. These issues are mostly technical, but more and more often bridge into the policy realm as society moves into an increasingly technology-dependent social and commercial environment (*e.g.*, issues such as privacy, security, quality, environmental management, consumer policy, *etc.*).

Standardization, however, is wholly a social practice that has gradually developed through pragmatic commercial and social necessity, not through any theoretical impetus. It found its modern origins in the demands of trade, commerce and manufacturing brought on by the industrial revolution. Metrology standards (*e.g.*, weights, measures, *etc.*) were a necessity of early exchange, transportation and trade, and the introduction of mass production depended completely on industrial standardization of components, measurements, and materials.

Standards have also been used for political and economic purposes of exclusion, domination, or protection. Eastern European railroad gauges were deliberately chosen to be incompatible across certain borders for the strategic purpose of impeding potential military invasion. In modern times, television broadcasting standards were deliberately chosen to stake out trade blocks and avoid domination by outside manufacturers or transmitters. An even better illustration may be the choice of telephone and electrical plugs that were designed to protect national manufacturing industries. The telephone signaling and the electrical distribution systems were designed to be compatible, but the plugs and connectors were gratuitously incompatible (Farance, 2002)[2].

Today, the globalization of economic and industrial production and the dramatic acceleration of the pace of technological development have severely stressed the conventional institutions and formal practices[3] upon which standardization practice was previously based. These institutions, traditionally dependent largely on income from publishing standards, have found their economic basis undermined by the Internet and the ascendancy of electronic document distribution. They have also found themselves somewhat eclipsed, particularly in certain industries such as information technology (*i.e.*, computing), by private consortia and forums that have been able to move more rapidly and more responsively to the demands of accelerated competitive markets, mobile capital, and shortened product lifetimes. Another stress exacerbated by globalization has been certain regional, cultural, and institutional differences in views and traditions about the basic role of standardization in a market economy. As a European Commission official expressed this conflict, "In

Europe Standards are not just documents thrown into the market to compete with each other...as if they were merely a 'product'" (Vardakis, 1998).

## TAXONOMY OF STANDARDS

An overview of a taxonomy of standards is shown in Figure 2.1. Standards can be divided into *metrology* standards and *technical* standards.[4] Metrology is the science of measurement. Such standards were initially developed to support systems of weights and measures for purposes of trade and commerce. Metrology standards often involve basic science[5] (*e.g.*, physics, chemistry), which can be quite expensive to establish, and thus were typically developed by scientific institutes funded by national governments. Responsibility for the government to "fix the standard for weights and measures" is embedded in the *U.S. Constitution*.[6] In the United States, the National Bureau of Standards (presently the National Institute for Standards and Technology, commonly referred to as NIST) was established in 1901, as was the British Standards Association (predecessor to the present British Standards Institution — BSI). Although commerce benefited tremendously from the work of such organizations, the primary impetus for their creation came from science and engineering professionals and from government—not business.[7] In Germany the *Normalienausschus für den Maschinenbau* (predecessor to DIN—*Deutsche Institute für Normung*) was established in 1917. The French may trace their standardization tradition back to Napoleon, but the *Association Française de Normalisation* (AFNOR) was formally established in 1918.[8] In more recent times, the role of these national institutions[9] has changed and they have become more involved in *technical* standardization, the primary focus of this study.

*Technical* standardization, in contrast to *metrology*, has typically developed in a closer relationship with industry, including both manufacturers and users. Stan-

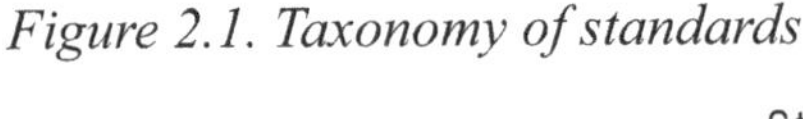

*Figure 2.1. Taxonomy of standards*

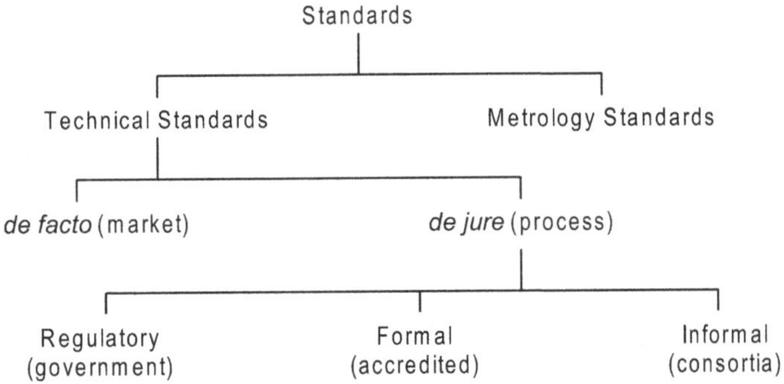

dardization tends to involve standards for product conformity[10], interoperability, compatibility, and adaptability, rather than for basic measurement. Once basic measurements were established, industry then needed to solve a set of problems more closely related to markets and product interoperability. In the United States, such a role fell primarily to *private* trade associations.[11] In this country, the evolution of standards practice became heavily influenced by cultural and historical factors associated with the world's most radically commercial and capitalistic political economy.

*The current U.S. standards process was adopted at the turn of the century, as the Nation entered the industrial age. Its form reflects American political culture and the manner in which industrialization took place in the United States. In contrast to many other countries, where unified national standards bodies were established in conjunction with the State, standards development organizations in the United States first emerged in* ***the private sector****, in response* ***to specific needs and concerns.*** (Congress, 1992, p. 39) [emphasis added]

The emphasis on the *private sector*, and on its specific needs and concern, has significant consequences, as will be seen in subsequent analysis, for the present discursive construction of standardization practice in the United States, and on its tensions with the rest of the world. American political culture firmly embeds the belief that the best way to serve public needs is by fostering private commercial interests. This approach is manifest in policy traditions such as "public interest, convenience, and necessity," found in U.S. regulatory practice from its inception (Rowland, 1997).

*Technical* standards can be further divided into *regulatory, de facto*, and *consensus* categories. *Regulatory* standards are those set directly by government and usually have to do with health, safety, or other areas in which legislators believe regulation is necessary—usually areas the *public good*.[12] There is an increasing trend for government to rely, where possible, on standards set by standardization bodies, and then codified by regulatory agencies into administrative law. *De facto* standards are those set simply by usage or success in the marketplace. Usually they are first established in the form of some successful product or system by a dominant firm or organization in the market place, rather than by a formal process. Examples include Microsoft Windows™ or the VHS™ video cassette recording format standards. The term *de jure* refers to those standards established by some formal process or procedure. *Consensus* standards are developed—agreed upon by *negotiation*—in what is generally referred to as the *voluntary consensus process*[13] under the auspices of formal standards bodies, either based in industry, trade, or professional associations, or in traditional international institutions. *Consensus* is defined as *substantial* agreement, not majority rule or unanimous agreement. What constitutes *consensus* is a topic that will receive further consideration later.

*Consensus* standardization can be further divided into *formal* and *informal*, depending on the legitimacy of the standardizing body. The *formal* bodies are often referred to as Standards Developing Organizations (SDOs) and *informal* bodies are referred to as Standards Related Organizations (SROs). This distinction is made largely by those in SDOs in order to claim superior legitimacy versus non-SDOs. What confers such status of legitimacy is usually "accreditation" by a national body whose legitimacy is, in turn, usually conferred by a national government. In the case of certain international institutions, which may be considered a class of Non-Governmental Organization (NGO), legitimacy rests upon tradition, acceptance, and longevity. The issue of legitimacy and its conferral will be of central importance in the analysis of the discourse later in this study.

## THE INTERNATIONAL LANDSCAPE

While the national institutions were being established, an international framework of institutions was also developing. The oldest international standards institution is the International Telecommunication Union (ITU), which was established in 1886 in Geneva, Switzerland, and is now part of the United Nations. The ITU, perhaps the first international intergovernmental organization, was founded as a treaty organization. Its initial purpose was primarily related to interconnection of telephone networks and radio communications—new technologies at the time. Until the 1980s, the telephone industry was dominated by nationally owned (or regulated, in the case of the United States) monopolies, and technical standardization served a quasi-governmental function, with participants acting as national government representatives. This situation has dramatically changed with the recent trend toward the privatization and deregulation of the telecommunications industry, the development of new technologies, and the convergence of telecommunications (*i.e.*, telephony) with Information Technology (IT) (*i.e.*, computers). Such changes have dramatically affected participation in, and the processes and subject matter of, ITU standardization. But, nevertheless, although ostensibly more open, the new processes still primarily engaged the former participants (*e.g.*, governments and manufacturers) due to the substantial cost of participation and the need for specialized expertise.

Two other institutions are important features on the international landscape of standardization: the International Electrotechnical Commission (IEC) founded in 1906, and the International Organization for Standardization (ISO) formed in 1947 as the post-World War II successor to the International Association for Standardization

(ISA), which was founded in 1926. The IEC standardizes electrical and electronic-related topics and the ISO standardizes everything else. According to the ISO Web site (1999), the organization "...deals with the full spectrum of human activity and know-how, ranging from the specifications of dimensional characteristics of screw threads to the essential features of environmental management systems in companies." The ISO estimates that 30,000 individual technical experts participate in its work in 2,850 various technical committees, subcommittees, and working groups, and that "...there are, on average, a dozen ISO meetings taking place somewhere in the world every working day of the year."(ISO, 1999) The ISO and IEC are essentially sister organizations that share the same building in Geneva, located across the street from the ITU building.

With the development of computing and the need to standardize an increasing variety of interfaces and software functions, the ISO and IEC formed a joint organization in 1986 known as Joint Technical Committee 1 (JTC1) for information technology standards. JTC1 took over management of all computer related ISO (*e.g.,* software) and IEC (*e.g.,* hardware) standards work. Now, with the convergence of telecommunication and computers, the complexity of coordinating work among the ISO, IEC, JTC1, and ITU has become additionally challenging. All four organizations operate under different organizational structures and different governing directives.

Additional forums on the international landscape are the regional institutions, particularly in Europe. The *Comité Européen de Normalisation* (CEN) and the *Comité Européen de Normalisation Electrotéchnique* (CENELEC) were formed in 1965 with the intention to mirror ISO and IEC work,[14] but are limited to European participants. The CEN encompasses not only official national bodies, but also certain trade associations such as European Computer Manufacturers Association (ECMA). In 1988 the European Commission (EC) established the European Telecommunications Standards Institute (ETSI) to focus on telecommunications and IT standardization. ETSI differs significantly from other such regional standards development bodies in that its membership is more heterogeneous, providing for multiple classes of membership, including manufacturers, administrations, phone network operators, users, and researchers. ETSI operates in some ways like a consortium, funding and staffing its own technical work rather than relying entirely on volunteers.

Some other regional groups (which tend to be more loosely organized than in Europe) are the Pacific Area Standards Congress (PASC) and the Pan American Standards Commission (COPANT). The ITU, ISO, and IEC will be examined in more detail in a later chapter. The European national and regional systems, as well as the international systems, stand in contrast to the system that evolved in the United States.

## DIVERGENT SYSTEMS AND CULTURES

Standardization practice developed in a much different direction in the United States than did those in Europe and the rest of the world. This difference has been attributed to a variety of cultural, political, and economic factors. On the lower levels of standardization activity (*i.e.,* the working group or technical committee level), both Europe and the United States adopted a consensus approach, with standards being negotiated in a relatively open process among various experts and interested parties, usually in face-to-face meetings. At the higher management and/or organizational levels, however, U.S. and European practices diverged dramatically. One reason was economics. In the period of rapid industrial development, the United States enjoyed a domestic market that was large enough to sustain mass production, while Europeans were dependent on inter-country trade to support large scale production. "From the beginning, therefore, European standards organizations were viewed as part of the industrial infrastructure...and geared their operations toward trade promotion. Moreover, European governments generally pursued active industrial policies" (Congress, 1992, p. 61).

On the one hand, America embraced the free market paradigm, often attributed to Adam Smith (1776), which is rooted in two principles: that less government is best, and that policy makers should rely on the "invisible hand" of the market. On the other hand, Germany, Japan, Korea, and more recently, China (Keidel, 2008, p. 2), were influenced more by the work of Friedrich List (1841), a lesser-known classical economist who asserted that the state must be an integral player in development with selective protectionism and the nurturing of "infant industries" (Freeman & Soete, 1997, p. 35; 295–299).

In considering these divergent conceptions of development, government roles, and of what consortia are about, it is interesting to look at the work of Gibson and Rogers (1994) on the largest U.S. attempts at cooperative R&D, the Microelectronics and Computer Technology Corporation (MCC), formed in 1982, and Sematech formed in 1987. These ventures were encouraged in part by a government policy initiative, the *National Cooperative Research Act of 1984* (relieving the threat of anti-trust litigation), and driven by anxiety over the phenomenal success of the Japanese semiconductor and electronics industries. This historical account demonstrates the intractable difficulties in translating development and innovation models that had been successful in Asia or Europe[15] into American business and commercial culture. Part of the problem may have to do with the top-down approach of U.S. business (and political) culture. Collaborative innovation and standardization tend to be inherently bottom-up processes, a notion that the Japanese in particular have successfully applied to industrial processes, at least in catching up with the Americans and Europeans.

Thus, the economic merges with the cultural. The aversion in the United States to anything that smacks of industrial policy has it roots deep in American political culture. Katzenstein (1985, p. 19) notes:

*...America's national debate on industrial policy betrays the strength of a liberal ideology. We conceive of the political alternatives that confront us as polar opposites: market or plan. The biases of our ideology are reinforced by a veritable national obsession with Japan, a country that American businessmen in particular view as a statist antidote to America's ideological celebration of market competition.*

*Our political debate typically pits the proponents of government against the advocates of market competition. Fundamentally the debate concerns the character of state involvement in the economy.*

In American standards discourse, this indicates how economic policies are reflected in cultural beliefs. In regard to industrial policy, it not unusual to hear references to the commonplace that government should not "pick winners and losers," presumably because this ought to be the sole domain of the "market." Such rhetoric puts standardization on the defensive, insofar as it lies somewhere between the government (or *public*) and the *market.*[16] Setting standards might be seen as picking winners and losers.

In contrast, Katzenstein characterizes German standardization as a "corporatist" approach (1985, p. 19):

*Democratic corporatism is distinguished with three traits: an ideology of social partnership expressed at the national level; a relatively centralized and concentrated system of interest groups; and voluntary and informal coordination of conflicting objectives through continuous political bargaining between interest groups, state bureaucracies and political parties. These traits make for low-voltage politics.*

Another point of higher level divergence is the tendency for European standards to take on a quasi-legal character. Once established, they come to acquire the force of law, as if they were regulatory standards, and thus may become *mandatory* rather than *voluntary.* Hence, one finds much in American standards discourse about the *voluntary consensus process,* in juxtaposition with the *European process*, although Europeans will often protest, insisting that their standards are also *voluntary.*

The American aversion to industrial planning is not merely drawn from a belief in competition and markets, but also from a deep-seated mistrust of government. In his seminal study on American culture, *American Exceptionalism*, Lipset finds America, "...much more individualistic, meritocratic-oriented, and anti-statist than peoples elsewhere. Hence the values which form the context for public policy are quite different from those in other developed countries" (1996, p. 22).

American politicians generally profess some level of mistrust of government. Lipset traces this recurring theme to America's roots as a society born in revolution, and, citing Max Weber and Antonio Gramsci, to an ability "...to avoid the remnants of mercantilism, statist regulations, church establishment, aristocracy and the emphasis on social class that postfeudal societies inherited" (p. 54).

Lipset's analysis also helps to explain American aversion to socialism and a reluctance to support social welfare programs or a strong labor movement (Steinmo, 1994). In American policy discourse, a sharp dichotomy is drawn between the *public sector* and the *private sector*, where *public sector* means *government*, and *private sector* is everything else. As will be examined later, this prevalent rhetorical dichotomy is superficial, limiting, confusing, and problematic, yet mythically pervasive.[17] In popular political discourse, it is often said that the most important thing that the public sector can do is to "stay out of the way" (of business, commerce, the economy, *etc.*). It is ironic to note that in spite of such mythic rhetoric, the government has actually played an enormous role in American economic life and has done so through *de facto* industrial policy in almost every major infrastructure, such as the development of the railroads, the oil industry, aviation, telephony, highways, and most recently, the Internet and its initial technical standards.

Another point that may be taken from *American Exceptionalism* is the American propensity toward a winner-take-all approach in the structuring of the country's political and economic system. Standardization is basically a cooperative enterprise directed at the creation of a public good or a commons, and may likely be displaced in any perceived contest with market competition, individualism, or freedom.

## THE U.S. STANDARDIZATION SYSTEM

The aforementioned cultural factors have contributed to the creation of a standardization system in the United States that is based on private industry associations and professional societies, with only a minimal role for government. The formal system is structured around the American National Standards Institute (ANSI), the self-appointed "private sector" membership association (now referring to itself as a "federation") that was founded in 1918. ANSI accredits all other standards bodies, publishes standards, and coordinates U. S. representation internationally in the ISO and IEC (although not in the ITU where this function still formally resides with the Department of State). This celebrated commitment to a private *vs.* a government leadership role sometimes obscures the fact that ANSI was initially formed by a combination of government agencies and engineering societies. Government-established policies, meanwhile, played a major role in shaping ANSI and the *voluntary consensus process*. Although today ANSI has no formal relationship with the U.S. government, it does in a sense act on its behalf and does receive some federal government funding.[18] ANSI, in this quasi-public role, administers and accredits a system of international TAGs (Technical Advisory Groups) that meet and develop "U.S. Positions" to be represented at international meetings of the technical committees and working groups for ISO, IEC and JTC1 (ANSI, 2001a)[19].

An important factor shaping the structure of the U.S. voluntary consensus system and validating the legitimacy of the ANSI accreditation role was the National Technology Transfer and Advancement Act (NTTAA) signed into law in 1996 (NIST, 1998). This legislation enabled and encouraged government agency participation in accredited standards committees and encouraged the use of such standards as an alternative to regulatory rule-making and also in government procurement processes (ANSI, 2001, p. 10). This factor has become a focus of rhetoric in the discourse on further enclosure of standardization that will be considered later in this study.

The role of accreditation is particularly important. ANSI does not create standards—it accredits other standards bodies, *i.e.,* SDOs, by maintaining a set of principles of conduct and detailed procedures to support them. These principles are contained in a recent ANSI annual report:

**Open:** Any materially affected and interested party has the ability to participate

**Balanced:** The standards development process should have a balance of interests and participants from diverse interest categories shall be sought

**Due Process:** All objections shall have an attempt made towards their resolution. Interests who believe they have been treated unfairly shall have a right to appeal.

**Consensus:** More than a majority, but not necessarily unanimity (ANSI, 2001, p. 4).

## Principles

The aforementioned principles are codified by ANSI and by its accredited bodies in documents generally called "procedures" (ANSI, 2002a) and there is a great deal of similarity among them, although there are many variations in procedure. The procedures are also very similar to those used by the ISO, IEC, and JTC1, where they are codified in operational "directives." Generally speaking, the ANSI principles mentioned previously are operationalized in the following way:

**Openness:** Meetings are "open door" and any "materially interested," or "materially affected" party may attend (if they pay their own travel expenses) whether or not they are members of the sponsoring trade association or group. Some meetings ask attendees to pay nominal fees to help cover meeting expenses (duplication, catering, hotel meeting rooms, *etc.*). Meeting notices, agendas and any important working documents must be publicly distributed and/or posted well ahead of the meeting date. Minutes of meetings must be distributed in a timely manner. Committee documents are posted on Web sites and may or may not be password protected. This practice varies widely, but if passwords are required, they are often provided to anyone (materially interested) that asks.[20]

**Balance:** A variety of interests should be involved in the activity, particularly competing interests, and the committee must make an effort to engage such broad participation. For instance, participation should include not only manufacturers, but also users, consumer representatives, relevant government agencies, and academic or scientific institutions, depending on the subject matter involved.

**Due Process:** All views are considered. When committee documents are distributed for balloting, all negative votes must state a reason for opposition and also state what is needed to reverse a negative vote. Subsequently, the committee must consider and answer all negative comments, even if sufficient supporting ballots (always significantly more than a simple majority) are already available to consider the document approved. Any party that believes it has been adversely affected by the final outcome must be provided with a right and procedure to appeal.

**Consensus:** Decisions are reached by group consensus or general agreement, sometimes called "substantial agreement," a process described in the *ISO/IEC Guide 2* as:

*...characterized by the absence of sustained opposition to substantial issues by any important part of the concerned interests and by a process that involves seeking to take into account the views of all parties concerned and to reconcile any conflicting arguments.* (ISO/IEC, 1996, sec. 1.7)

## Accreditation

A benefit of the legitimacy conferred upon standards bodies by ANSI accreditation provides substantial protection from potential anti-trust challenges to participants in ANSI-sanctioned standardization activities, and their cooperation with competitors within the context of committee activities. Accredited standards committees are one of the few forums where competitors can legally collaborate and exchange information, as long as certain topics are not discussed—notably pricing. This opportunity suggests another benefit of participation in such committees. It can provide participating individuals with a window on new technological developments and perhaps some indications of competitors' future plans and directions (sometimes called technical or business intelligence). Conversely, non-accredited activities may run greater risk of anti-trust challenge.

## The Sectoral Approach

The U.S. system is organized on a vertical model around specific industry *sectoral* trade associations, rather than on a horizontal national or regional body model that is found almost everywhere else. In other words, ANSI accredits hundreds of trade associations that actually sponsor the standards work through thousands of committees organized by the institute to deal with defined problems in specific industry *sectors*.[21] This structure stands in contrast to that existing outside the United States, where the organization of standardization activity crosses industry sectors and attempts to standardize on the basis of general principles. For example, in the United States, noise standards might be set separately by the associations of various machine manufacturers or users (*e.g.,* chain saws, milling machines, motorcycles, and fire alarms), whereas in Europe, such standards might be set by a general environmental noise level committee or a safety committee. In another hypothetical example, meat processing standards in the U.S. might be established independently by poultry producers, fishing equipment manufacturers, or beef growers, whereas in Europe, such standards might be more centralized in a single committee or distributed among a small number of committees that deal with meat processing as a whole.

Simply put, it is generally understood that the U.S. approach prefers to deal with the specific needs and problems of a given industry sector, whereas the European system seeks to address standardization needs on the basis of overriding principles. This difference may also be characterized as *application vs. principle*, or *results vs. system*. In cultural terms, it may be viewed from a *commercial vs. societal* perspective. In any case, this difference has become a central issue of conflict between the United States and other, particularly European, countries. The arguments are based on deeply embedded cultural beliefs and practices. This issue and the *sectoral approach* discourse will be examined more closely later. In general, it is interesting to note there is a recurring American tendency to focus on targeted solutions rather than on overriding principles. This tendency can be found in the realms of standardization, regulation, and legislation (Schoechle, 1998). The sectoral discourse is brought forward prominently in a broad policy document, ANSI's *National Standards Strategy for the United States* (ANSI, 2000).

## National Standards Strategy

Insofar as U.S. industry has any formal standards-setting policy, it is embodied in a very carefully crafted fourteen-page paper known as the *National Standards Strategy for the United States.* The document was developed by a consensus of the ANSI membership and adopted by the ANSI Board of Directors in August 2000. A detailed analysis of the discourse that constitutes will be made later, but it is interesting to note that the document reveals on its first page the confusing usage of two different meanings for the word *sector*: public sector *vs.* private sector, meaning government versus non-government, on the one hand, and industry sector versus market sector, on the other hand. In such context the document includes the statement, "Most standards are related to specific sectors (*e.g.,* information technology, automotive) and are not applicable to other sectors." Although this is a problematic claim at best, it is a key to understanding some of the current divergent discourse on standardization in the U.S.

The *National Standards Strategy* sets forth a fairly coherent set of policies and ideals to address the future of the U.S. standards system, both domestically and internationally. Along with its 2001 annual report, (ANSI, 2002) ANSI issued an *Implementation Report* on the *National Standards Strategy.* Furthermore, at its annual conference, the institute positioned this *Strategy* document as one of its most significant recent accomplishments[22]. The original *Strategy* and its subsequent *Report* provide a basic discursive re-construction of the U.S. standards system and offer important clues to the meaning of many of the words, phrases, and policies that have become problematic or confusing.

In a sense, ANSI is caught in a quandary. Its constituents do not believe in industrial policy, but are reluctant to provide adequate market-based support for their own common good. ANSI membership is declining, as are its publishing revenues. This latter consequence threatens the institute's basic mission which urges the broadest possible distribution and use of technical standards documents. At the same time, ANSI is pursuing an increasingly crucial international mission on behalf of U.S. industry in the global economy as it faces strong, coherent, standards-based European and Asian industrial policies. The U.S. government, it should be understood, provides minimal financial support to the institute because of an historical commitment to a free market ideology. As a result, the *Strategy* might be seen as an effort to reconcile these conflicting factors (*i.e.,* balance both a public and private agenda) and to gently coax its members into the beginnings of supporting an industrial policy. For this reason alone, the two ANSI documents are interesting pieces of rhetoric and objects of analysis.

A driving force behind ANSI initiatives and an incentive among American interests to adopt (or adapt to) the global standardization system, is the process of globalization, the integration of economic and cultural practices that is a result in large part of the commercial and organizational application of information and communication technologies, including the global Internet. One of the most influential manifestations of this force has been the World Trade Organization (WTO) and its policies regarding Technical Barriers to Trade (TBT). The interdependence of economics, trade, and financial markets has made it virtually impossible for any country or government to act unilaterally, and admission to the global trading club requires conformance with WTO rules.[23] TBT policies require the dismantling of often long established practices of sustaining non-tariff trade barriers in the form of national or regional technical standards that limit foreign competition. Discourse about TBT played a prominent role at the recent ANSI annual conference, in part because it introduced a relatively new dimension of standardization—that of Conformity Assessment (CA) and certification processes—to demonstrate and facilitate TBT policy compliance. TBT policy also plays a visible role in the discourse because various standards bodies, both national and regional, have reacted in differing ways.

One interesting example is the Peoples Republic of China. The Chinese have moved with remarkable speed to embrace WTO policies and to bring themselves into the realm of global standards practice. It has been noted recently that in Beijing, local bookstores stock literature on ISO and IEC standards, and taxicabs commonly display ISO 9000[24] compliance advertising on their seat-backs.[25]

## IT vs. Telecom Standardization and the Rise of Consortia

The final phase of the structuring of standardization practice (and here the most germane phase) in the United States and globally was the emergence of a new institutional paradigm, the *consortia*, sometimes referred to as Market-Driven Consortia (MDCs). To understand the cultural roots of consortia, it is useful to briefly examine the history of the Telecom and IT industries.

Telecommunication standardization is one of the oldest institutionalized industrial practices. It developed in a culture of highly regulated monopolies within a relatively slow-moving and non-competitive utility industry dedicated to the provision of reliable voice telephony service and primarily concerned with issues of network stability. This involved negotiated pricing, quality of service, and the interoperability of national networks. In the United States until the early 1980s, technical standards were set by AT&T (*i.e.,* Bell Laboratories) and then negotiated internationally by the ITU. With the divestiture of the Regional Bell Operating Companies (RBOCs) from AT&T in 1984, Telecom standardization was opened to the voluntary consensus process. Concomitantly, the Committee T1[26] was established for network standards and the TIA undertook primary responsibility for setting Customer Premises Equipment (CPE) standards. Both of these were ANSI accredited bodies.

In contrast to Telecom, IT standardization grew up in the computer and semiconductor industries, a fast moving and highly competitive (traditionally unregulated) industrial culture. Although the industry was dominated initially by IBM Corporation, and *de facto* standardization prevailed, technological advances and entrepreneurs overtook IBM's dominance by the early 1970s through introduction of the minicomputer and then the microcomputer. One might contrast the cultural values of the Telecom and the IT industries as those of cooperation *vs.* competition, monopoly *vs.* market, stability *vs.* change[27], and interoperability *vs.* differentiation.

As this new culture sought to break away from IBM's *de facto* standards and pursue its own "open systems"[28] standardization (Cargill, 1997), the new culture was never very comfortable in the bureaucratic and politically defined Telecom standards régime. Nor was it very comfortable in the relaxed culture of other industry's traditional standardization practices wherein markets were stable and the pace of technological change was relatively slow. As a result, new organizations began to form outside of, or on the doorstep of, the formal structure. They became known as "private *forums* or *consortia*."[29] The cultural motivation for this change is colorfully portrayed by Cargill (1997, p. 69-70), who was an active participant.

*The current standardization model, especially that used by participants in the SDOs, is fundamentally broken. The current model is based upon a myth; [like the one] which is part of French national folklore, ...captured in the Chanson de Roland, written in the eleventh century...which became a powerful agent for coalescing [the] movement [that established] the French King. ...Roland served as an ideal for concepts of loyalty to a king, as well as the dues and obligations of both warriors and the king.*

*In a similar manner, the concept of "Open Systems" has become a convenient icon to express all that is good about computing and the promise that computing can hold. It, too, has undergone significant shifts in its meaning,...[now, yet another] new iconic meaning is emerging.*

Such new meaning ultimately took the form of a divergence into a broad array of groups and institutional practices that are now variously referred to as SROs, MDCs, forums, and/or consortia. Some examples of such groups are the Internet Engineering Task Force (IETF), World Wide Web Consortium (W3C), ATM Forum (Asynchronous Transfer Mode), UPnP Forum (Universal Plug 'n Play), Object Management Group (OMG), and the Open Group. The rules, procedures and practices under which these groups operate (including their *publicness*) vary widely. The only commonality they share is that each lies outside the formalist domain described earlier, where they strive to claim and enclose their own, more private, realms. They purport to answer the problems of obscurity, complexity, implementation difficulty, and, above all, slowness of the traditional system.

Two organizations that lie outside the formal system, yet are not considered consortia in the usual sense, are of special note, the European Telecommunications Standardisation Institute (ETSI) and the IETF. ETSI is a hybrid private/governmental non-profit R&D body that was established by the EC in 1988 and situated in a substantial facility at Sophia Antipolis, France. ETSI functions as a membership organization, with its own professional staff, and develops its own standards in telecommunication, broadcasting, and certain aspects of information technology within Europe that are often then moved into CENELEC, ITU-T, and ISO, IEC, or JTC1. ETSI hosts numerous conferences and maintains active liaison with other consortia as well as many formal committees.[30]

The IETF is a group that has been characterized by its founders as an organization that "does not legally exist"—but as "an 'open, international, voluntary standards organization' operating as an activity of the Internet Society, but which is not separately incorporated" (Bradner, 2001, p. 23).[31] The group first met in 1986, and gathers annually, but does almost all of its work online. It's "parent" organization, the Internet Society (ISOC) was not formed until 1991.[32] IETF issues specifications called Requests for Comment (RFC) that are contributed by a vast worldwide army of individual volunteers, operating in over a hundred working groups, and wherein the only membership requirement is participation. IETF governance might be characterized as following the "philosopher-king" model, by which final decisions are made by a core group of the organization's founders and technical experts, now called the Internet Architecture Board (IAB). RFCs are non-proprietary and IETF now collaborates actively with the ITU-T.[33]

## CONCLUSION

It is within the dynamic and rapidly changing global historical, cultural, and institutional framework of standardization practice that this book proceeds. In particular, this inquiry seeks to explore and understand the rhetorical construction of legitimacy by the diverse set of new institutions and organizations known as consortia, and their discursive interaction with their predecessors. In order to do so, it is useful to survey in the next chapter the literature and discourse on enclosure and on standardization, and on the historical and economic context within which these discourses are conducted.

## REFERENCES

ANSI (2000). *National Standards Strategy for the United States*. American National Standards Institute. New York: ANSI. August 31, 2000, 14 pp.

ANSI (2001). *American National Standards Institute: 2000 Annual Report*. New York: ANSI.

ANSI (2001a). *ANSI Procedures for U.S. Participation in the International Standards Activities of ISO*. New York: ANSI. January 2001.

ANSI (2002). *American National Standards Institute: Annual Report Two Thousand and One*. New York: ANSI.

ANSI (2002a). *ANSI Procedures for the Development and Coordination of American National Standards*. New York: ANSI. Availiable at <www.ansi.org/rooms/room_16/public/gov_proc.html>

Bradner, Scott (2001, spring/summer). "The Internet Engineering Task Force," *OnTheInternet, 7*(1), Reston, VA: Internet Society, pp. 22-26.

Cargill, Carl (1997). *Open Systems Standardization: A Business Approach*. Upper Saddle River, NJ: Prentice-Hall. 327 pp.

Cheit, Ross E. (1990). *Setting Safety Standards: Regulation in the Public and Private Sectors*. Berkeley: University of California Press.

Congress of the United States, Office of Technology Assessment (1992). *Global Standards: Building Blocks for the Future*. TCT-512. Washington, DC: U.S. Government Printing Office. March 1992. 114 pp.

Farance, Frank (2002). Interview and email correspondence with Frank Farance. October 30, 2002.

Freeman, Chris, and Luc Soete (1997). *The Economics of Industrial Innovation* (3rd Ed.), Cambridge, MA: MIT Press, 470 pp.

Gibson, David, and Everett M. Rogers (1994). *R&D Collaboration on Trial: The Microelectronics and Computer Technology Corporation*. Boston, MA: Harvard Business School Press, 605 pp.

Isenberg, David S. (1997, August). "The Rise of the Stupid Network." *Computer Telephony*.

ISO (1999). [Web site for] International Organization for Standardization. January 8, 1999. Availiable at <http://www.iso.ch>

ISO/IEC (1996). *Guide 2: Standardization and Related Activities—General Vocabulary.* Seventh Edition, Geneva: ISO/IEC.

Katzenstein, Peter (1985). *Small States in World Markets: Industrial Policy in Europe.* Ithaca: Cornell University Press.

Keidel, Albert (2008, July). "China's Economic Rise—Fact and Fiction," *Policy Brief 61.* Washington, DC: Carnegie Endowment for International Peace.

Lipset, Seymour Martin (1996). *American Exceptionalism: A Double Edged Sword.* New York: W.W. Norton & Co. 348 pp.

List, Fredrich (1841). *(1904). The National System of Political Economy.* London: Longman.

NIST (1998). *National Technology Transfer and Advancement Act homepage.* Retrieved October 19, 2002 from <http://www.ts.nist.gov/ts/htdocs/210/nttaa/nttsaa.htm>

Random House (1967). *The Random House Dictionary of the English Language.* New York: Random House, Inc., 1960 pp.

Rowland, Willard D., Jr. (1997, autumn). "The Meaning of "The Public Interest" in Communications Policy—Part I: Its Origins in State and Federal Regulation." *Communication Law and Policy, 2*(4), pp. 309-328.

Schoechle, Timothy (1998). *A Public Policy Debate: Standardsmaking Practice and the Discourse on International Privacy Standards.* Standards Policy Research Paper ICSR98-211. Boulder: International Center for Standards Research. Availiable at <http://www.standardsresearch.org>

Smith, Adam *(*1776) *(1976). An Inquiry into the Nature and Causes of the Wealth of Nations.* Campbell, R.H., *et al (*Eds.). Oxford & New York: Oxford University Press.

Steinmo, Sven (1994). "American Exceptionalism Reconsidered: Culture or Institutions?" In Larry Dodd and Calvin Jillson (eds.). *The Dynamics of American Politics: Approaches and Interpretations.* Boulder, CO: Westview Press.

Vardakas, Evangelos (1998). Director, EC Directorate for Industry (DG III), speech before the *Standardization Summit* sponsored by ANSI and NIST, Washington DC, September 23, 1998.

Webster (1977), *Webser's New World Dictionary, College Edition.* New York: The World Publishing Company, p. 1133.

## ENDNOTES

[1] Interesting historical treatments on standards may be found in Congress (1992) and Cargill (1997).

[2] In an operational sense, within standards development, *harmonization* is the removal of gratuitous incompatibilities. (Farance, 2002)

[3] *Formal* institutions and practices here refers to established non-governmental national, regional and international standards developing organizations (SDOs) and SDO accrediting bodies; these formal institutions include organizations such as ANSI (American National Standards Institute), ISO (International Standards Organization), ITU (International Telecommunications Union), IEC (International Electrotechnical Commission), and many others.

[4] This is actually the distinction between *measurement standards* and *normative documents.* Measurement standards are also technical. English uses the same word, *standard*, for both. In French, *étalon* is a measuerment standard and *norme* is a normative document. A similar situation exists in Russian. (Farance, 2002). In common English usage, however, the term *technical standard* has become established in this regard. *Technical Standard* is defined in the basic law covering Federal Procurement with respect to standardization: Public Law 104-113, the *National Technology Transfer and Advancement Act of 1995.* The applicable section of PL 104-113 is *Section 12 (d) Utilization of Consensus Technical Standards by Federal Agencies; Reports*, The fourth subsection, 12 (d) (4) provides a definition of standards as: "the term 'technical standards' means performance-based or design-specific technical specifications and related management systems."

[5] Technical standards (normative documents) also can involve basic science (*ISO Guide 31—Quantities and Units*). A more basic distinction between the two is that metrology standards involve laboratory practice, while *ISO Guide 31* involves knowledge of industry and scientific practice (Farance, 2002).

[6] Section 8, Clause 5.

[7] Engineering professionals (employed in part by business) were the people that faced the technical problems (particularly in the rapidly growing electrical industries of the day); and governments saw standards as a "public good."

[8] Other national institutions include Denmark's DS (Dansk Standardisenngsrad), The Netherlands NNI (Nederlands *Normalisatie Instituut*), Italy's UNI (*Ente Nazionale Italiano di unificazione*), Greece's ELOT (Hellenic Organization for Standardization), Belgium's IBN (*Institut Belge de Normalisation*) and Norway's NEK (*Norsk Elektroteknisk Komite*).

[9] The United States is an exception. While these national institutions evolved to undertake sponsorship of domestic standards activity, legitimation of

standards bodies and representation of national interests internationally, the United States drew a sharp distinction between government and private activity. Standardization leadership responsibility fell to an industry association, ANSI (American National Standards Institute). While NIST remains more focused on science and metrology, it is actively engaged as a participant in technical standardization activities. See <http://www.ansi.org/> and <http://www.nist.gov/>.

[10] *Conformity* is closely associated with the term *conformity assessment*, which often also involves measurement. (Farance, 2002)

[11] As will be discussed later, the term *private* and *public* are highly problematic in the discourse. It is interesting to ask: are trade associations *private* or *public*?

[12] Such *regulatory* standards might include pharmaceuticals, auto safety regulations, food classification standards, agricultural inspection standards, electromagnetic emission limits and spectrum allocations. The term *regulatory* actually refers a distinction between *voluntary* and *involuntary* compliance requirements. (Farance, 2002)

[13] *Voluntary consensus* means 1) compliance with the standard is voluntary, and 2) a consensus process was used to establish the standard.

[14] Some complain that this flow actually works the opposite way, with ISO and IEC work often mirroring CEN and CENELEC work. According to Tony Flood of the Canadian National Committee of the IEC, CENELEC has a program to aggressively promote its work worldwide and frequently the "...CENELEC mantra overpowers everything else." (Flood, 2002)

[15] ESPRIT, the *European Strategic Programme for Research and Development in Information Technology*, was a diversified initiative to promote collaboration between European government-owned and private companies during the 1980s with a view toward internationally accepted standards. Partially funded by the Commission of the European Communities, it acted as an umbrella for a number of unrelated projects, each bringing together industrial, governmental, and academic members to work toward a narrowly-defined goal. It yielded a number of commercially successful technologies and standards, including the MPEG and JPEG standards.

[16] A rhetorical irony: in ancient Greece, the marketplace was the *agora*, which was the essence of the *public* place. But in America, the market is a *private* place and the government is called *public*.

[17] Part of the problem is the way that contemporary discourse confuses the terms *private* and *public*. We speak of *public* schools, meaning government funded and operated, but we also speak of *public* broadcasting, meaning *privately*

operated but funded by *public* contributions (with or without government help) or *public* houses ("pubs") which are *privately* owned and operated, but open for patronage to the general *public* (in contrast to *private* clubs).

18 According to the 2001 ANSI Annual Report, out of $15 million in revenues, it received approximately $500,000 in Government (NIST) grants (primarily for international activity), and it paid over $1.5 million to ISO and IEC in annual dues.

19 ANSI doesn't actually act *formally* on behalf of the U.S. Government. ITU is a treaty-based organization, so its members are countries, as represented by *governments* (although this membership structure is changing, as will be discussed later). ISO and IEC are non-treaty-based organizations, so its members are represented by *nations*. For many countries, there is no distinction between the *government* position and the *national* position. For the United States, the State Department represents the government position in international fora, and ANSI represents the national position (*i.e.,* industry) within international fora (Farance, 2002).

20 Access to working documents and Web sites, and the attitudes of participants and sponsors is an interesting topic in respect to what defines "open." This will be explored in more detail later in this research.

21 Some such associations are the TIA (Telecommunications Industry Association), ASHRAE (American Society of Heating, Refrigerating and Air-Conditioning Engineers), ITI Council (Information Technology Industries), ATIS (Alliance for Telecommunications Industry Solutions), SCTE (Society of Cable Telecommunications Engineers), ASTM (American Society for Testing and Materials), ASME (American Society of Mechanical Engineers), IEEE (Institute of Electrical and Electronics Engineers), and NFPA (National Fire Protection Association).

22 The ANSI annual conference was held October 15-16, 2002 in Washington, DC.

23 The role of WTO and TBT should not be underestimated. A relevant reference is as follows:

- The following principles and procedures should be observed, when international standards, guides and recommendations (as mentioned under Articles 2, 5 and Annex 3 of the TBT Agreement for the preparation of mandatory technical regulations, conformity assessment procedures and voluntary standards) are elaborated, to ensure transparency, openness, impartiality and consensus, effectiveness and relevance, coherence, and to address the concerns of developing countries.

- The same principles should also be observed when technical work or a part of the international standard development is delegated under agreements or contracts by international standardizing bodies to other relevant organizations, including regional bodies. Source: "ANNEX 4, Decision of the Committee on Principles for the Development of International Standards, Guides and Recommendations, *WTO TBT Triennial Review*, WTO Committee on TBT, G/TBT/9; 11 November 2000, p. 24.

[24] ISO 9000 is a family of standards for manufacturing quality management systems. It is a member of a relatively new class of standards known as *technology management system* standards, in contrast to technical standards for similarity, compatibility, interoperability, or adaptability. Another example of management system standards is the ISO 14000 family (environmental management systems).

[25] Comments made at the ANSI annual conference (October 15, 2002) and personal observations of the author.

[26] Committee T1 is now administered by ATIS (Alliance for Telecommunications Solutions).

[27] Standards may be regarded historically as either agents of stability or as agents of change (Cargill, 1997).

[28] Thus the term *open* became prominent in standards discourse, establishing a meaning that could be paraphrased as, "liberated from IBM's proprietary *de facto* standards practice," but still vague in other respects. Farance (2002) disagrees, commenting, "...*open* (even in the days of 1970's IBM mainframes and non-IBM disk drive vendors) means *well-documented, i.e.*, a normative document (a standard or specification) that describes interfaces, interoperability, behavior, compatibility, *etc*."

[29] It is interesting to note that in ancient Rome, a *forum* was the marketplace or *public* square of a city, the center of judicial and business affairs and a place of assembly for the people; or it was a *public* court or tribunal or other assembly for the discussion matters of *public* interest (Random House, 1967, p. 559).

[30] <http://www.etsi.org>.

[31] <http://www.ietf.org>.

[32] <http://www.isoc.org>.

[33] There is some irony in the IETF/ITU relationship in that the two organizations were philosophical opposites, both in terms of their formation/organization/governance and of their technical approach. The ITU represented the conventional government/treaty controlled, centralized, telephony-oriented communication architecture known as the Public Switched Telephone Network (PSTN), while the internet emerged later as the essence of a volunteer, largely

ungoverned, decentralized, data-oriented architecture that has been characterized as the "stupid network" (Isenberg, 1997). Subsequently, the Internet's minimalist technical philosophy has challenged and is gradually subsuming the conventional technology, and the ITU has embraced the Internet.

# Chapter III
# The Global Context of Standardization

*It's the architecture, stupid!*

—Lawrence Lessig, 2002

## INTRODUCTION

The purpose of this chapter is to situate this study in a global economic and social context, and to review the literature and discourses that inform this study and identify its objects of analysis. The discourse on enclosure, including its key concepts, is examined in some detail. The study is couched in an immediate discursive context and then in a greater economic, social, and historical context. The discourse on standards and standardization is briefly surveyed here, but a detailed analysis is left for later discussion in Chapter VII. The key relevant discourses are examined. Related and useful discourses on social construction of technology and on institutionalization are also examined.

## IMMEDIATE CONTEXT OF THE STUDY

One of the defining characteristics of the modern age is the uneasy and ambiguous dichotomy between the *public* and the *private*. Although contemporary policy discourse relies on these terms, it invariably fails to define them in any consistent or meaningful way. Habermas (1962) and Arendt (1958) in particular, have dealt with these terms and traced their historical usage from ancient Greece. A thorough treatment of this topic is beyond the scope of this study, but it is important to visit the general topic and notice the conflicted and conflated use of these terms and the underlying concepts, particularly how how imprecise and confused use of these terms shapes and constrains thought and discourse.

One of the common ways of thinking about *public* and *private*, as noted earlier, is by spatial metaphor. The Dutch architect Rem Koolhaas[1] grappled with this notion in the field of urban design within the context of globalization, commercialization, and commoditization.

*The city used to be something that you get for free. It's been a* ***public*** *space, and it enables the citizens to assemble in a kind of collective sense, but basically through the process, [sic] effects of the market economy and through the withdrawal of the* ***public sector*** *and the kind of complimentary invasion of the* ***private sector****, which is expressed through shopping, the nature of the city has changed from something that is fundamentally free, to something that you have to pay for, so that even in educational establishments, even in religious establishments and certainly in cultural establishments there is always this kind of commercial presence...[the cathedral, the museum] the economies of these institutions [are] dependent on shopping.* [emphasis added] (Koolhaas, 2002)

Koolhaas also notes the difficulty of reconciling public space within a "relentlessly commercial environment" and the cultural differences between American and European levels of public "surrender and resistance" to such forces in design. It is thus an irony of our modern age that the marketplace or *agora*—the classic form of *public* space—has become reconceptualized as a privatized and enclosed space.

It is interesting to note that when Koolhaas refers to the *public sector*, it seems he is not talking about *government*, but about *public space*—something more akin to Habermas' *public sphere*, a non-governmental physical space where private parties may gather. Koolhaas appears to lack the words adequate to describe or define the concept he has in mind and is confined by the *public/private sector* terminology. A similar problem seems to afflict the discourse on enclosure, where the *commons* is the central issue.

In its immediate context, this study engages and applies what earlier has been called the discourse on enclosure. As also mentioned earlier, this discourse can be viewed as an oppositional discourse that is a composite of several related elements: the relationship between intellectual property rights and the intellectual "commons" (Lessig, 1999); the derivative nature of ideas, the intellectual commons, and technical innovation and authorship (Lessig, 2001; Litman, 1990); the enclosure of the intellectual commons (Boyle, 2001); the classic debate on wealth generation and alternative modes of production (Benkler, 2001, 2002; Lessig, 2001a); and forms of such enclosure involving standards and standardization (Van Houweiling, 2002).

## GREATER CONTEXT OF THE STUDY

This enclosure discourse will be examined more closely, but first it is important to briefly step back from the minutiae of case law and theory and begin to examine from a broader interdisciplinary and historical perspective certain driving forces behind the recent trends in law and standards practice. One such force is the notion of intellectual property and the nature and importance of the commons. Another force is the process of globalization, the ascendancy of the ideology of *free market* capitalism, and the processes of *privatization* and *deregulation.*

### Intellectual Property

As noted earlier, Intellectual property rights (IPR), and particularly patents, have in recent years increasingly come into conflict with standards and standardization (Iversen, Bekkers & Blind, 2006). Although a thorough treatment of this aspect of enclosure is beyond the scope of this book, the topic bears some discussion because it is interwoven with the other issues surrounding the enclosure of standardization. In particular, patents are one of the motivations for the creation of consortia, patent pools, general public licenses (i.e., "open source"), and other phenomena. Some brief background is provided later in this chapter on the topic of *property talk.*

Intellectual property is information that derives its intrinsic value from creative ideas. Such ideas may find their expression in many forms including manuscripts, poetry, art, film, music, computer programs, and other works. Western legal institutions have a long tradition of granting the creators of such ideas, inventions, and creative expressions certain rights, thus imbuing their creations with the status of "property"[2] in the form of patents, copyrights, trademarks, and trade secrets. It is easier to understand the present situation by looking at the historical and cultural roots of these concepts and institutions.

The roots of modern conceptions of IPR are embedded in the Enlightenment concepts of liberalism, individualism, and the ownership of property, along with the related concepts of freedom and human rights. These ideas formed the intellectual foundations of modern society. One of the most influential political philosophers of the Enlightenment was John Locke.

In his *Essay on Human Understanding*, Locke built an epistemology that placed the individual at its center and foreshadowed semiotics, a theory of language, signs, and symbols, wherein the creation and communication of ideas became the essential task of Science. In his *Treatises on Government*, Locke described the foundation of private property as the "self owning itself and its labor," and he sought to legitimize both private property and a common realm of "civil society." Locke's focus was on the processes of the conveyance of ideas from the private to the common realm where they could be shared with others. In doing so, he was dealing with the fundamental distinction between the *private* and the *common* (or *public*), a distinction that remains problematic today. This distinction presented a difficulty for Locke, as described by Peters (1989):

*Since Locke has made the privatizing labor of the individual so fundamental, he is left with the task of explaining how different people can have understandings in common, that is, how they may* ***communicate*** *(a term which, of course, etymologically means to* ***make common****). Locke's defense of the individual as a private property holder simply leaves him without good ways to talk about the sources of public order or community: whence the magical power of* ***communication*** *as panacea. The concept results from the failure to theorize the common world.*

This failure to adequately theorize the *public* has returned to haunt us today (Schoechle, 2001: 2003). Locke's problem was a political issue of his time—how to move from the common to the private, or how to establish the individual in an age in which property and rights were the unquestioned domain of the king or church. The problem of IPR in the age of digital communication networks returns us to the problem of completing Locke's account—how to get back from the private to the common. The enclosure discourse attempts to begin such a journey.

## Globalization and the Free Market Ideology

The process of globalization and the attendant ascendancy of the ideology of free market capitalism have been driving forces behind the trends that have inspired the enclosure discourse and shaped standardization practice. As recounted earlier, there can be little doubt that the dramatic growth and global integration of the economy over the past two decades can be traced in large part to the growth and success of the ICT industry and its technical innovations, including the Internet, particularly in the United States. There can also be little doubt that the privatization and deregulation of telecommunications monopolies may have aided this growth in important ways. However, one major consequence of such success has been the reinforcement (until recently) of an unquestioned ideological belief system that some critics find unsustainable and self destructive—that of the unfettered *free market* (Gray, 1998; Frank, 2000; Soros, 1998, 2002).

According to historian John Gray:

*...globalization—which at bottom is nothing more than the spread of science and technology throughout the world—goes back at least as far as the last third of the nineteenth century, when transatlantic telegraph cables were first laid down. Driven by technologies that abolish or curtail time and distance in many areas of activity, it is a by-product of the growth of scientific knowledge, and therefore a wholly inexorable process....* (Gray, 1998, p. xiv)

But Gray goes further, drawing a sharp distinction between *globalization* as an historical phenomenon and the *free market* and other deliberate political and economic constructions:

*...but, it [globalization] has no tendency to bring about the triumph of any specific type of capitalism or political system, to force convergence of liberal values, or to promote peace. The global free market constructed in the last decade of the twentieth century is only one of the several régimes under which globalization has advanced for well over a century **(ibid)**.*

He also argues that American economic and technological ascendancy over the past two decades has promoted an economic ideology that is based on the theory that market freedoms are *natural* and political restraints on markets are *artificial*, a notion he claims is the opposite of reality. He asserts that, "encumbered markets are the norm in every society," and "...that free markets are creatures of state power, and persist only so long as the state is able to prevent human needs for security and the control of economic risk from finding political expression." (Gray, 1998, p. 17). Gray stresses that "democracy and free markets are rivals, not allies." Part of the power of free market rhetoric derives from, the word *free*, a word with tremendous rhetorical force, particularly in American culture. The importance of *free* in the enclosure discourse will be considered in the ensuing discussion of Lessig's and his colleagues' works.

Proceeding with Gray's argument, the principal effect of free market economic policies, propelled by initiatives on all governmental levels, international institutions (*e.g.*, the International Monetary Fund (IMF), World Trade Organization (WTO), World Bank, *etc.*), and corporations is the *externalization of costs*. The ultimate result is an undermining of civil society and, specifically, the weakening or destroying of intermediary social institutions (*e.g.*, trade unions, professional associations, churches, cultural organizations, local authorities, *etc.*) on which society depends for cohesion, order, and stability, but the value of which are not factored into the free market calculus. "In this manner, the innermost contradiction of the free market is that it works to weaken the traditional social institutions on which it has depended in the past—the family[3] is a key example." (p. 29). Another example, as will be argued later, is the standardization system, particularly in the United States, which is based in large part on trade and professional and membership associations.

Similar arguments are made by Frank (2000), Soros (1998; 2002), Stiglitz (2003), and others[4]. Soros argued in 1998 that unrestricted global financial markets and free market IMF and World Bank policies, were bringing about a "lack of social cohesion and the absence of government" (p. x). Furthermore, Soros argues that the policies are making the global capitalist system "unsound and unsustainable"(p. xi) and pose a severe threat to what he calls the *open society*, a notion that Soros has championed[5] based on the philosophical writings of Karl Popper (1945). *Open* is another term imbued with rhetorical power, for it plays a major role in the discursive construction of standards and standardization. As noted earlier, the financial challenges faced by ANSI, and its difficulty in raising private membership support and participation from the very elements that could most benefit from its work, are consistent with the self destructive tendency of extreme free market policies as Gray puts forward. Standardization is a form of intermediary social institution that exists apart from government and the sustenance and functioning of which depend on mechanisms that are not well accommodated within the ideology of free market economics.

Standards may be considered a form of "public goods," and it has been argued (Mansfield, 1970; Congress, 1992, p. 9, footnotes 23 and 24) that public goods tend to be underproduced in capitalist market economies. The free market paradigm, often attributed to Adam Smith (1776), is rooted in two principles—that less government is best, and that policy makers should rely on the "invisible hand" of the market. Gray argues that free market ideology became globally ascendant during the Reagan/Thatcher period and has been dominant in shaping public policy discourse since that time. Stiglitz (2003) makes a similar point about the shaping of global economic institutions and their policies.

It is not the purpose of the present work to argue the validity of Gray's view or to offer debate regarding the economic outcomes of various ideological constructs, but only to situate this study within the rhetorical framework of *globalization*, *market* economics, *privatization*, *deregulation*, the *free*, and the *open*. Regardless of their merits, the power of Gray's arguments and attendant discourses have gained new significance and credibility with the recent collapse of the global economic bubble (Samuelson, 2002) and the exposure of pervasive and large-scale corporate malfeasance, particularly in the United States.[6] In writing about recent public protests outside meetings of the World Bank and the IMF in Washington D.C., *Wall Street Journal* columnist Alan Murray refers to the demise of "market religion":

*...[free market critics] have won the argument. Capitalism now has the black eye they tried so hard to give it. Writing in the Washington Post last week, Robert Weissman of Mobilization for Global Justice exulted:*

*"The era of market fundamentalism is over. Marketization, deregulation, and privatization, and the opportunities for market manipulation offered by inadequate regulation...all central elements in the rise and fall of Enron...are now discredited in the United States. And in developing countries, where their effects have been most devastating, they are the object of widespread public opprobrium."*

*A similar tone was struck at a Friday afternoon seminar inside meetings of the World Bank and the International Monetary Fund, which were the targets of the protests. "Market fundamentalism is gone," said Trevor Manuel, finance minister of South Africa. "The ideology is gone, and that's good," agreed Jan Karlsson, development minister of Sweden.*

*And even Horst Koehler, managing director of the IMF, enforcer of market based practices around the world, said, "We are in the process of searching for new policy concepts that make globalization work for all." Mr. Kohler winced each time someone suggested that market fundamentalism once held sway at his institution, arguing that his home country, Germany, always had recognized that the marketplace had to be tempered with an understanding of the broader needs of society. So much for the triumph of American-style capitalism.* (Murray, 2002)

It will be suggested later that some of the "new policy concepts" that the IMF is seeking may be found in the standardization system, as one of the intermediary social institutions to which Gray refers. Standardization lives in that hazy realm somewhere between the *private* and the *public*. It has been increasingly privatized over the past decade by the rise of consortia, but also in many respects by the traditional institutions, as well as by other means. Standardization is an intermediary institution between the people and the market—perhaps a form of regulation, or a substitute for it—or perhaps a form of privatized governance (Mueller, 2002). In any case, it is that hazy realm that this study attempts to explore. It is within this global economic and social context that the literature and discourse on enclosure and standardization is examined.

## DISCOURSE ON ENCLOSURE

For a seminal book, Lessig (1999) chose the rhetorically potent title, *Code and Other Laws of Cyberspace*. The dual meaning of the word *code,* as *computer code* and as *legal code,* conveys the basic argument that technical architectures and their software implementations embed tacit policy decisions, whether acknowledged or not. Although Lessig focuses particularly on the Internet and related network[7] technologies, his argument is a far more general one. And while Lessig does not focus on the standardization practice, apart from an account of Internet Engineering Task Force (IETF), its history and other anecdotal references, this study bases its underlying thesis on an extrapolation of his argument, *i.e.,* that standards and standardization practice are primary processes by which Lessig's *code* is written.[8] This notion of the policymaking force of technical architectures is not new (*e.g.,* Winner, 1985), but Lessig brings it forward anew in an active and broad public discourse.

### The Commons

In a second book, Lessig (2001) highlights the more general problem of technological innovation and focuses on the increasing encroachment of intellectual property rights, law and industrial practice on what he calls *ideas,* while preserving the *intellectual commons* that he claims is vital to such innovation. Lessig attempts to step outside the concept of *property,* which is so ideologically embedded in contemporary western and American culture. Moreover, he defines the *commons* in terms of *access* as a space where *no permission is required*, and not simply in terms of *ownership* or as *public property (*p. 20). He argues that a commons of ideas is fundamentally different than real property in that it is non-rivalrous, *i.e.,* not depleted by use.

*Language is a commons, though its resource is nonrivalrous (my use does not inhibit yours). What has determined the "commons" then is not the simple test of rivalrousness. What has determined the commons then is the character of the resource and how it* ***relates to a community****. ...the issue for nonrivalrous resources is...enough incentive to create (p. 21).*

Furthermore, Lessig challenges the application of the myth of the *tragedy of the commons* [9] which describes the inevitable ruin of the unregulated commons and justifies imperatives of control of a scarce resource—often a rationale for regulation, privatization, or IPR.

*This "tragedy" [of the commons] consumes talk about "the commons." "Ruin" is taken for granted as the destiny of those who believe in the "freedom of the commons." Hardheaded sorts thus scorn the rhetoric of undivided resources. Only the romantic wastes time wondering about anything different from the perfect control of property.*

*...As researchers have shown, in many different contexts,* **norms** *adequately limit the problems of overconsumption. Communities work out how to regulate consumption* (p. 22). [emphasis added]

*My central claim throughout is that there is a benefit to resources held in common and that the Internet is the best evidence of that benefit. ...the Internet forms an "innovation commons." It forms this commons not just through norms, but also through a specific technical architecture. The Net of these norms and this architecture is a space where creativity can flourish* (p. 23). [emphasis added]

Thus, Lessig challenges the dominant discourse and its mythological assumptions about property and control.[10] He asserts the value of an *information commons* (also called a *platform*), and he attributes its formation to *norms* and *architectures* created by *communities*. This study argues that the *norms*, *architectures,* and *communities* to which Lessig refers to all lie essentially in the realm of standardization practice. In fact, the Internet's *norms* and *architecture* (*i.e., code*) was created by a standards body (*i.e.,* a *community* of interest) that later became known as the IETF (a *consortium* in the context of this study).

Lessig also speaks of the value of the publicness of a resource, and asserts that in certain cases, increasing participation and use enhances the value of a standard rather than diminishing it:

*The value, in these cases, comes from the convergence of many upon a common use, or standard, or practice. And in these cases, keeping the resource in the commons is a way to assure that that value is preserved for all. ...These arguments from tradition are thus grounded in both fairness and efficiency, and economists have extended the arguments from efficiency.* [*e.g.,* Schumpeter's writings , network externalities, interoperability, end-to-end architectures, *etc.*] (Lessig, 2002, p. 88)

Lessig is also concerned about the motivation to enclose a common resource, not merely for immediate economic reasons, but for dominant industries to protect themselves from disruptive innovation in their markets. Referring to the Internet as a platform for innovation, he said,

*Every great idea comes from outside—because the architecture guarantees that outsiders have access to the network, not just the dominant application providers. Dominant players want to protect themselves—that is "fix the bugs." ...to corrupt the original architecture to protect their business models.* (Lessig, 2002)[11]

Standardization practice, depending on how it is organized and managed, can be employed to either help or hinder innovation. Lessig worries about consequences, "when few can make decisions about what kind of innovation will be permitted." (2001, p. 173) His concern raises questions concerning the potential for the enclosure of standardization to be a significant threat to technical innovation.

Lessig asserts that the commons is vital to innovation in large part because of the derivative nature of ideas. He juxtaposes *property* and *control vs. commons* and *innovation* (2001; 2002), He also analyzes the ideology of property and control, and the logic of the *market, i.e.,* that "the whole world is best managed when divided among private owners" (p. 13). This notion of divided and privatized resources is the central tenet of free market ideology—that the generation of wealth in society is better served by holding property in private hands rather than in common. Boyle (1996; 2001) and Benkler (2001; 2002) both challenge this notion directly. The derivative nature of ideas is also emphasized by Litman (1990) in the context of authorship and copyright.

## Enclosure

Boyle likens the increasing emphasis on IPR and the privatization of ideas—the increasing perception in society of information and ideas as an owned commodity—to the land enclosure movement of the 15th through the 19th centuries in England. He names it the "2nd enclosure movement" [12] (2001). The land enclosure movement was an exercise of state power to transform what had traditionally been common land into private property. [13] Although this movement has been celebrated as transforming a peasant society into a market economy, some historians regard this notion as problematic (Gray, 1998). [14]

Boyle begins by presenting both views in respect to the land enclosure movement, and then identifies certain parallels and differences between the two enclosure movements, proposing arguments consistent with Lessig. If standardization can be seen as a sort of intellectual *commons*, it is not difficult to move to the following enclosure argument: that the products (*i.e.,* standards) and the process by which they are made, may be at risk of being drawn into a privatized space under the control of, and at the service of, specific dominant interests, justified by economic efficiency.

## Production

Benkler, in his paper, *Coase's Penguin* (2001) extends Lessig's and Boyle's line of reason, by challenging the economic efficiency argument and the myth, or "dominant paradigm we have about productivity" (p. 3). He examines forms of *peer production* (*e.g.,* open source software), an alternative mode of production to conventional *market-based* and *hierarchical* information production. Benkler argues, offering substantial evidence, that peer production may outperform market-based production in some information industries, because:

*...the primary advantage of peer production is in acquiring and processing information about human capital available to information production projects, and that in this it is superior to both market-based or hierarchical managerial processes. In addition to the informational advantage, peer production more efficiently assigns human capital to information inputs because it does not rely on controlling bounded sets of either factor.* (p. 1)

Here Benkler challenges the central rationale for intellectual property law: incentive to produce. Intellectual property rights (IPR) were institutionalized in the U.S. Constitution in the form of copyright and patent rights, which gives Congress the power "to promote the Progress of Science and useful Arts, by securing for a limited Time to Authors and Inventors the exclusive Right to their respective Writings and Discoveries."[15] Thus the Constitution grants not a property right *per se*, but a commercial monopoly to control use for a limited time. Further, it explicitly states the purpose of such a grant—the public good—thus making patent and copyright, "...the only types of property that have an explicitly utilitarian Constitutional basis." (Boyle, 1996, p. 226) The grant was not about preserving *private* rights, but about bringing ideas into the *public* domain (or the commons) and awarding a concession as an incentive for doing so. In general, the historic legal basis of IPR has been a pragmatic one, emerging from social practice and utilitarian motives rather than from notions of justice or rights, or from social or epistemological theory. [16]

Benkler examines in detail various experimental and theoretical works related to organizational forms and to collaborative behavior and motivation, but his argument against the need for strong IPR is primarily based on two factors:

*[1] an important non-proprietary sector that is burdened by, but does not benefit from strong intellectual property rights. [2]...a cost...not previously identified, which is especially salient in a pervasively networked society. This cost is the loss of the information that peer production efforts generate and the loss of productive resources—human capital that could be used productively and is not because*

*market-based and hierarchical information production are poor mechanisms for identifying and pooling resources.* (p. 2)

These same arguments can be applied to standards and standardization. It can justifiably be argued that standardization practice is essentially a form of peer production—a form of collaboration and of the social construction of technology—and on the technical committee or working group level, at least in part, is outside the market-based or hierarchical modes. In addition, the problems posed by the assertion of IPR in the standardization process have become vexing and intractable issues. It might be argued that seeing standardization as peer production is dubious because technologies, and sometimes draft standards, are brought into the committee by the "volunteer" participants for self interested gains. Of course, human societies reflect a variety of motivations, and such self-serving purposes are not unusual. Nevertheless, Benkler's notion of peer production does not specify or limit individual motivation, but merely that such volunteers are acting as individuals in a discursive, collaborative (and even possibly conflicting) manner to arrive at a mutually agreed-upon end product.

The aforementioned enclosure discourse has been challenged. Post (2000) and others contest Lessig's conclusions from a basic libertarian perspective, standing up for the "invisible hand" in the marketplace. Wagner (2002) argues that inherently "information wants to be free," and that concerns of the enclosure theorists are overblown and have become mythological.

Lessig (2001) and *Vaidhyanathan* (2001) further argue that even the term "intellectual property right" is problematic, and that it is largely a recent rhetorical construction.

*It is essential to understand that copyright in the American tradition was not meant to be a "property right" as the public generally understands property. It was originally a narrow federal policy that granted a limited trade monopoly in exchange for universal use and access. Lately, however, American courts, periodicals and public rhetoric seem to have engaged almost exclusively in "property talk" when discussing copyright.* (*Vaidhyanathan*, 2001, p. 11)

*Vaidhyanathan* explores how this *property talk* constructed a myth that tapped into basic American ethical assumptions and cultural habits, including the notions of rewarding hard work, recognizing genius and creativity, ensuring wide and easy access to information, and encouraging experimentation in both art and commerce (p. 4). Such *property talk* discourse often frames the issues as a dichotomy, in terms of the interests of, and justice for, *inventors* and *artists* (rather than for corporations and cartels) or for society and culture, and particularly in terms of *incentive* and *reward vs. theft* and *piracy* (Schoechle, 2001; 2003).

What are the implications of this *property talk* for standardization? As pointed out earlier, standards are increasingly viewed as a form of intellectual property. Standardization practice deals with production of ideas, documents, and similar intellectual "goods." If the dominant discursive mode is within the *property* paradigm, then all goods are seen as someone's property—either *private* or *public* (*i.e.,* government) property. They must be "owned." Then, however the commons is de-legitimized as a concept because it is not "owned" by anyone. It is not clear, then, who owns or should own the standards or the standardization practice, but they are being increasingly claimed (*i.e.,* enclosed) by various interests, even if only for the publishing revenues (*i.e.,* copyright).

## Free and Open

The terms *free* and *open* are among the most frequently used in the discourses on enclosure and standardization, but they are also among the most difficult to define. Lessig proposes that the question of how resources shall be allocated in society is no longer one of public (government) *vs.* private market, but one of *free vs. controlled (*2001, p. 12). He defines a resource as *free* if "(1) one can use it without the permission of someone else; or (2) the permission one needs is granted neutrally." He explores the difference between *free* and *open*, particularly in the context of the free or open software movement and the advocacy of its principal founder, Richard Stallman, who prefers the term *free.*[17]

> *... Stallman believes that people dilute the insights of the free software movement by minimizing its connection to fundamental values...[but still] I am partial to the term* ***open****—as in* ***open society****; I believe it is properly a reference to values as well as to licenses under which code is distributed; and by "open code" I mean to refer to the values across both technical and legal contexts that promote a world where governing structures—code—are fundamentally free.* (p. 270) [emphasis added]

The term *open* has become dominant in both *open standards* and *open source* (which may also be viewed as a form of standardization) discourse. The literature on the meaning of *open* is extensive, but Lessig's definition provides a good place to begin and is consistent with that used by Popper (1945) and Soros. The meaning of *open* will be explored more fully later.

A final element of the enclosure discourse to be considered deals specifically with the enclosure of standards. Van Howeiling (2002), building on the preceding discourse, has described the standards "pollution problem," wherein public domain protocol standards are appropriated and made proprietary—enclosed—by a process she characterizes as Embrace, Extend, Extinguish (EEE). Lemley (1998) has discussed the notion of standards "pollution", while Lemley and McGowan (1998) have referred to "intellectual property ambush," particularly in regard to the practices of Microsoft Corporation. Subsequently, Lemley has done considerable work related to the intersection between standards-setting, intellectual property rights, and antitrust law (1999; 2001; 2002). Burk and Lemley (2003) deal with patent law, technical innovation, and standards-setting bodies. Iversen, Bekkers, & Blind (2006) and David (2006) deal with some of the various proposed solutions to conflicts between standards and IPR.

Weiser (2001) has moved the useful metaphor of information *platform*[18] forward into the enclosure discourse[19] with a meaning similar to Lessig's *commons*—a standardized base upon which innovation can build—which carries an inherent element of public interest. The difference is that Lessig's *commons* is inherently open or free. Weiser's *platform* however, is simply a standardized (formally or otherwise) hardware or software base for interoperable services and could be open, proprietary, or governed by some regulatory régime. A similar notion in the realm of technical standards is that of *infrastructure*. Messerschmitt (1999) draws a distinction between *applications* and *infrastructure*. He argues that standardization should focus on basic infrastructure and avoid trying to standardize applications in order to encourage experimentation, lower barriers, and increase flexibility. Jakobs (2001) examines the standards-setting process and emphasizes the chronic lack of, and the need, for the engagement of users, rather than having standardization activities driven mostly by manufacturers or service providers.

The importance of standards as platforms for innovation is frequently recounted in standards discourse. One commonly cited case is the Sony Betamax™ *vs*. JVC VHS™ battle over *de facto* standard recording formats for video tape recorders. The struggle delayed market growth until it was resolved by Sony's withdrawal (AFU, 1996). Another classic case is the success of the Global System Mobile (GSM)[20] standard for mobile telephony, cited by Libicki, *et al.* as exemplary of the "power of standards over technology." (2002, p. 8) GSM now dominates the global market except in the United States, where multiple standards still compete, consumers are still confused, and U.S. industry is still disadvantaged. This case has been examined extensively in standards literature from a variety of aspects including technical, business, social, political, and economic (Fomin, 2001; Bekkers, *et al.* 2002; Lehenkari & Miettinen, 2001; Haug, 2002; Rice & Galvin, 1999).

## THE STANDARDS DISCOURSE

The literature in the field of standardization is substantial and growing as it becomes a recognized realm of academic research. The purpose here is not to disclose the entire field of literature, but simply to mention specific elements of the discourse that allow exploration of key issues and the language and embedded meanings, including certain mythological and ideological constructs, that drive policy decisions, both public and private.

The current academic discourse on standards focuses largely on technical, curricular, and instrumental dimensions. This work particularly focuses on case studies and tends to be somewhat anecdotal and highly business sector-specific. Hesser and Czaya (1999, p. 2) have described the current academic debate as concentrating "...on standards and their implications for competitive policy, international trade and company strategies." The literature of the disciplines of business management, engineering sciences, economics, and law has generally treated standards in solely instrumental and industrial process-oriented terms.

The literature on standards and standardization relied on for this study falls generally into two categories: (1) law, economic, and social theory applied to standards, and (2) case studies of standardization projects and related controversies. However, the reader should proceed under the caveat that standardization is an inherently cross-disciplinary and cross-cultural undertaking, and that no general theory or definitive framework that can adequately comprehend the entire topic. Rather, there are many theoretical frameworks that can be applied and yield useful results.

The law and economic literature referred to in the present work includes works by writers such as Balto (2000; 2001), Besen (1990; 1991; 1995), Besen and Farrell (1991; 1994), Bekkers, *et al (*2002), Christensen (1997), David and Greenstein (1990), David and Monroe (1994), David and Steinmueller (1994), David (1985; 1987; 1995; 1996; 2006), David and Shurmer (1994), Demsetz (1967), Economides (1989; 1996; 2000), Economides and Himmelberg (1995), Farrell (1990; 1995; 1996), Farrell and Saloner (1985; 1986; 1986a; 1988), Gaynor, Bradner, *et al (*2001), Greenstein (1992), Greenstein and Stango (2005), Katz and Shapiro (1985; 1986; 1992; 1994), Lemley (1998; 1999; 2000; 2001), Libicki (1995), Libicki, *et al (*2002), Shapiro (2001), Shapiro and Varian (1999), Steinmueller (1995), Taschdjian (2001), Weiss (1990), Weiss and Sirbu (1990), and Weiss and Toyofuku (1996).[21] Much of this literature overlaps disciplines such as economics, law, social studies and political science.

The literature taking mainly social theory, political theory, technical, and policy oriented approaches to standards-setting includes: Bijker (1993; 1999), Cargill (1989; 1997; 1999; 2001; 2001a; 2005), Cargill and Bolin (2007), Cowan (1992), Egyedi (1996; 2001; 2001a), Egyedi and Loeffen (2002), Fomin (2001), Jakobs (2000; 2001), Jakobs, Procter, and Williams (1998), Jakobs *et al (*2001), Hawkins (1999), Krechmer (1998; 2000; 2000a; 2000b); Krechmer and Baskin (2001), Kahin and Abbate (1995), Lehr (1992), Mansell (1993; 1995), Mansell and Hawkins (1992), Mansell and Siverstone (1996), Marks and Hebner (2001; 2004), Oksala (2000), O'Reilly (1999), Robinson (1999), Schoechle (1995; 1998; 1999; 2001; 2003), Schoechle *et al (*2002), Sherif (2001; 2002), Spring (1995), Streeter (1986; 1990), Updegrove (1995; 1995a), van Wegberg (2002; 2003; 2004), Vercoulen (1998; 1999), Williams (1997), and Williams and Edge (1996). Case studies are provided by several of the authors mentioned previously, notably Bekkers, *et al (*2002), Egyedi (2001), Fomin (2001), Sherif (2002), Weiss and Cargill (1992), Weiss and Spring (2000), Weiss and Toyofuku (1996), and van Wegberg (2002; 2003; 2004).

## The Private, the Public and the Open

The discourse that provides the principal object of analysis for this study comes more from trade and industry than from academia. Krechmer (1998), founder and editor of *Communications Review*, a periodical exclusively devoted to reporting on numerous standards committee activities in the ICT field, proposes a set of criteria for defining "open standards" and offers a taxonomy for the basic structure of standards practice and its processes. Warshaw and Saunders (1995) provide a perspective on the issue of defining the public and private interest in standardization, both in North America and in the international arena. The OTA report (Congress, 1992) and Cargill (1997) cited in earlier sections, offer additional perspectives on defining *public*, *private* and *open*.

## Sectoral and Consortia Discourse

The principal objects of analysis in this study are specific reports, policy papers, statements, speeches, presentations, personal interviews, and conversations. In respect to the *sectoral* debate, such discourse is found in recent ANSI Annual Reports (ANSI, 2001; 2002) as well as in the *National Standards Strategy* (ANSI, 2000) and in various works by Warshaw and Saunders (1995). The rise of consortia standardization was noted by Cargill (1989), and is further described by Balto (2000), Cargill (1999), Gifford (1997), Hawkins (1999), Krechmer (1998; 2000; 2000a; 2000b), Vardakis (1998; 2002), Sherif (2001; 2002), de Vries (1999), Updegrove (1995; 1995a), Vercoulen and van Wegberg (1998), van Wegberg (2002; 2002a), and Weiss and Cargill (1992). David and Monroe (1994) discuss the causes and effects of protracted standardization efforts and other "discontents" that are likely motivation for consortia formation (pp. 1-2). Similarly, specific discourse advocating consortia standardization and published in the congressional testimony of Cargill (2001; 2001a), and also in a report to the European Commission prepared by Egyedi (2001), are analyzed in detail. Also, discourses opposing consortia standardization are analysed (DIN, 2002; Sherif, 2002). Analyses of hybrid approaches to standardization, which attempt to combine features of both consortia and traditional standardization, are provided by van Wegberg (2002; 2002a), Schoechle, *et al* (2002), Gifford (1997), Krechmer (1998; 2000; 2000a; 2000b), and Lim (2002).

## DISCOURSE ON SOCIAL CONSTRUCTION OF TECHNOLOGY

The social science disciplines of sociology, philosophy, political science, and communication studies have, until very recently, almost entirely overlooked the phenomenon of technical standardization *per se* (Schoechle, 1999). One example is a developing body of discourse from sociology known as STS (variously as Studies of Technology and Society; or earlier, Science, Technology and Society; or, most currently, as Socio-historical Technology Studies) (Bijker, 1993, 1999; Williams & Edge, 1996; Williams, 1997). One focus of STS is on the *social construction of technology*—the notion that although inventions may often be acts of individual creation or inspiration, their implementation, commercialization, or diffusion into society always involves many other individuals and institutions. The lone inventor working in his or her garage is a powerful myth in American culture. The original invention may ultimately be reshaped in ways that the inventor would not recognize. Standardization is one of the principal processes of the social construction of technology, but is only recently being recognized among STS scholars.

Hawkins provides a summary of the various disciplinary paradigms and approaches:

*Thus for political scientists, the standardization* ***problématique*** *is most closely related to systems of law and governance. Their primary interest is focused on situations in which standards assume regulatory functions, or where they have implications for the conduct of international relations. On the other hand, for sociologists/historians/philosophers of science and technology, questions about standards stem largely from concerns about the creations and sustenance of social power structures. From their perspective, the primary questions tend to focus on institutional processes of standardization, and on the nature of inputs into those processes. ...Economists have tended to concentrate primarily upon the effects that standards have upon the behavior of buyers and sellers of technological products in the marketplace...the realm of information economics—the dynamics of market relationships...network effects and issues of technical compatibility.* (Hawkins, Mansell, & Skea, 1995, p. 3-4)

Mansell has studied the institutional nature and role of standardization from a political economic perspective (Mansell, 1993; 1995; Mansell & Silverstone, 1996). She considers that standards-making institutions provide the "institutional glue" that links technical and institutional change (1995, p. 214). Much of Mansell's analysis is consistent with the perspectives of Benkler and Lessig on innovation and on technical choices embedded in standards. Hawkins et al., interestingly, traces the very early recognition that standards are never value-free:

*In the mid 19th century, for example, Joseph Whitworth [writing in 1882], one of the earliest architects of institutionalized standards-making, observed that for all of the objective technical orientation, choices in standards-making were never purely free of subjective criteria—an observation that remains at the heart of most studies of standards of our time.* (Hawkins, Mansell & Skea, 1995; p. 4)

## INSTITUTIONALIZATION

Codding and Rutkowski (1982) have provided a detailed account of the history, structure, and practices of the ITU, which they assert is the oldest international intergovernmental organization. Cerni (1984) has provided an account of the history, structure, and practices of the ITU, IEC, and ISO, from the perspective of the U.S. government as a participant.

Wagner (1999) has provided a theoretical and methodological framework on institutionalization, bringing together the fundamental work of Veblen (1919) and Innis (1972). Their theories are applied in works by Streeter (1990), Mansell (1993; 1995), and Mansell and Silverstone (1996). These writers provide a useful perspective on the role of discursive practice within institutional settings in the shaping of technologies and policies.

## CONCLUSION

This chapter has sought to situate the present study in a global economic and social context, and to review the literature and discourses that informed the study and that provided its objects of analysis. The discourse on enclosure has been examined in some detail to grasp the salient concepts that relate to standardization. The discourse on standards and standardization has been briefly surveyed. Finally, some of the relevant literature on the social construction of technology and on institutionalization has been discussed.

A further step will be to examine the traditional institutionalization of standardization practice as embodied in the oldest bodies the ITU, IEC, and ISO. With a view of the relative openness of past institutional practice, and from the perspective of the enclosure discourse, it is then possible to analyze the standards discourse, to find what might be the relevant meanings and implications for policy-making. But first, before such institutional history or discourse analysis can be undertaken, it is necessary to propose a theoretical and methodological perspective for the entire study and then to examine some of the basic terms of discourse, including what is meant by *open* and *public*.

## REFERENCES

AFU (1996). The Decline and Fall of Betamax. *AFU White Paper*. Available <http://urbanlegends.com/products/beta_vs_vhs.html>

Albert, Michel (1993). *Capitalism vs. Capitalism: How America's Obsession With Individual Achievement and Short-Term Profit Has Led It To The Brink of Collapse*. New York: Four Walls Eight Windows. 259 pp.

ANSI (2000). *National Standards Strategy for the United States*. American National Standards Institute. New York: ANSI. August 31, 2000, 14 pp.

ANSI (2001). *American National Standards Institute: 2000 Annual Report*. New York: ANSI.

ANSI (2002). *American National Standards Institute: Annual Report Two Thousand and One*. New York: ANSI.

Arendt, Hannah (1958). *The Human Condition*. Chicago: Chicago: University of Chicago Press.

Balto, David A. (2000). Standard Setting in a Network Economy. Speech, *Cutting Edge Antitrust Law Seminars International*. New York, NY. February 17, 2000. Available at <http://www.ftc.gov/speeches/other/standardsetting.htm>

Balto, David A. (2001, June). Standard Setting in the 21st Century Network Economy. *The Computer & Internet Lawyer, 18*(6), 1-18.

Bekkers, Rudi, Verspagen, Bart, & Smits, Jan (2002). Intellectual Property Rights and Standardization: the Case of GSM. *Telecommunications Policy, 26*, 171-188.

Benkler, Yochai (2001). Coase's Penguin, or, Linux and the Nature of the Firm. *29th Telecom Policy Research Conference*. Alexandria, VA. September 24, 2001. Available at <http://www.arxiv.org/abs/cs.CY/0109077>

Benkler, Yochai (2002). Intellectual Property and the Organization of Information Production. *International Review of Law & Economics, 22*(1), 81-107.

Bijker, Wiebe E.(1993, winter). Do Not Despair: There Is Life After Constructivism. *Science, Technology & Human Values, 18*(1), 119.

Besen, Stanley (1990, December). European Telecommunications Standards Setting: A Preliminary Analysis of the European Telecommunications Standards Institute. *Telecommunications Policy, 14*(6), 521-530.

Besen, Stanley (1995). The standards process in telecommunication and information technology. In R. Hawkins, R. Mansell, & J. Skea (Eds.), *Standards, Innovation and Competitiveness: The Politics and Economics of Standards in Natural and Technical Environments.* Brookfield, VT: Edward Elgar Publishing Ltd.

Besen, Stanley, & Farrell, Joseph (1991, August). The Role of the ITU in Telecommunications Standards-Setting: Pre-Eminence, Impotence, or Rubber Stamp. *Telecommunications Policy, 15*(4), 311-321.

Besen, Stanley, & Farrell, Joseph (1994, spring). Choosing How to Compete: Strategies and Tactics in Standardization. *Journal of Economic Perspectives, 8*(2), 117-131.

Bijker, Wiebe E. (1999). *Of Bicycles, Bakelites, and Bulbs: Toward a Theory of Sociotechnical Change.* Cambridge, MA: MIT Press.

Boyle, James (1996). *Shamans, Software and Spleens.* Cambridge, MA: Harvard University Press. p. 226.

Boyle, James (2001). The Second Enclosure Movement and the Construction of the Public Domain. *Conference on the Public Domain.* Duke Law School, Durham, NC, November 9-11, 2001.

Brenner, Robert (1998). The Economics of Global Turbulence: A Special Report on the World Economy, 1950-98. *New Left Review.* no. 229, May/June 1998. 264 pp.

Burk, Dan L., & Lemley, Mark A. (2003). Policy Levers in Patent Law. *Virginia Law Review. 89*(7), 1575-1695.

Cargill, Carl F. (1989). *Information Technology Standardization: Theory, Process, and Organizations.* Bedford, MA: Digital Press.

Cargill, Carl F. (1999). Consortia and The Evolution of Information Technology Standardization. *SIIT '99 Proceedings - 1st IEEE Conference on Standardisation and Innovation In Information Technology*, Aachen, Germany, September 15-17, 1999. p. 37-53

Cargill, Carl (2005). Eating Our Seed Corn: A Standards Parable For Our Time. Available at http://islandia.law.yale.edu/isp/GlobalFlow/paper/Cargill.pdf

Cargill, Carl, & Bolin, Sherrie (2007). Standardization: a failing paradigm. In S. Greenstein & V. Stango (Eds.), *Standards and Public Policy*, Cambridge: Cambridge University Press.

Cargill, Carl (1997). *Open Systems Standardization: A Business Approach*. Upper Saddle River, NJ: Prentice-Hall. 327 pp.

Cargill, Carl (1999). Consortia and the Evolution of Information Technology Standardization. *SIIT '99 Proceedings: 1st IEEE Conference on Standardization and Innovation in Information Technology*. University of Aachen. September 15-16, 1999. pp. 37-42.

Cargill, Carl (2001). Consortia Standards: Towards a Re-Definition of a Voluntary Standards Organization. Testimony before the House of Representatives, Sub-Committee on Technology, Environment and Standards on behalf of Sun Microsystems. June 28, 2001.

Cargill, Carl (2001a). The Role of Consortia Standards in Federal Procurements in the Information Technology Sector: Towards a Re-Definition of a Voluntary Standards Organization. Submitted to the House of Representatives, Sub-Committee On Technology, Environment and Standards. Palo Alto: Sun Microsystems. June 28, 2001. 30 pp. Available at <http://www.house.gov/science/ets/Jun28/Cargill.pdf/>

Cerni, Dorothy M. (1984). *Standards in Process: Foundations and Profiles of ISDN and OSI Studies*. NTIA Report 84-170, Washington, DC: U.S. Department of Commerce. December 1984. 247 pp.

Christensen, Clayton M. (1997). *The Innovators Dilemma: When New Technologies Cause Great Firms to Fail*. Cambridge, MA: Harvard Business School Press.

Christensen, Clayton M. (2002, June). The Rules of Innovation. *Technology Review*. Cambridge: MIT, pp. 33-38.

Codding, George A., Jr., & Rutkowski, Anthony M. (1982). *The International Telecommunications Union in a Changing World*. Dedham, MA: Artech House, Inc. pp. 414.

Congress of the United States, Office of Technology Assessment (1992). *Global Standards: Building Blocks for the Future*. TCT-512. Washington, DC: U.S. Government Printing Office. March 1992. 114 pp.

Cowan, Robin (1992). High Technology and the Economics of Standardization. Dierkes, Meinolf, and Ute Hoffmann (Eds.) *New Technology at the Ourset: Social forces in the shaping of technological innovation*. Frankfurt, New York: Campus-Verlag and Westview. Chapter 14, pp. 279-300.

David, P. A. (1985). Clio and the Economics of QWERTY, *American Economic Review, 75*, pp. 332-337

David, Paul A. (1987). Some New Standards for the Economics of Standardization in the Information Age. In P. Dasgupta & Stoneman (Eds.) *Economic Policy and Technological Performance*. Cambridge: Cambridge University Press. pp. 206-234.

David, Paul A. (1995). Standardization policies for network technologies: the flux between freedom and order revisited. In Hawkins, Richard. Robin Mansell & Jim Skea (Eds.) *(*1995) *Standards, innovation and competitiveness*. Brookfield, VT: Edward Elgar Publishing Ltd.

David, Paul A. (1996). Formal standard-setting for global telecommunications and information services. *Telecommunications Policy, 20*, 789-815.

David, Paul A. (2006). Using IPR to expand the research common for Science: New moves in 'legal jujitsu', *Intellectual Property Rights for Business and Society* conference sponsored by DIME-EU Network of Excellence, held in London, September 14-15, 2006.

David, Paul A., & Greenstein, Shane. M. (1990). The Economics of Compatibility Standards: An Introduction to Recent Research, *Economics of Innovation and New Technology, 1*, pp. 3-41.

David, Paul A., and Hunter K. Monroe (1994). *Telecommunications Policy Research Conference*, held 1-3 October 1994 at Solomon's Island, MD.

David, Paul A., & Steinmueller, W. E. (1994). Economics of Compatibility Standards and Competition in Telecommunication Networks, *Information Economics and Policy, 6*(3-4), pp. 217-241.

David, Paul A., & Shurmer, M. (1996). Formal standards-setting for global telecommunication and information services. *Telecommunications policy, 20*(10), 789-815.

Demsetz, Harold (1967). Toward a Theory of Property Rights. *American. Economic Review, 57*, 347-357.

De Vries, Henk J. (1999). Doctoral dissertation: Standards for the Nation, Erasmus University Rotterdam, published as *Standardization—A Business Approach to the Role of National Standardization Organizations*. Boston: Kluwer Academic Publishers.

DIN (2002). *Strategie für die Standardisierung der Informations und Kommunikationstechnik (ICT) - Deutsche Positionen,* Version 1.0. Berlin: DIN Deutsches Institut für Normung, eV. Available at <http://www2.din.de/sixcms/detail.php?id=3871>

Economides, Nicholas (1989, 5 December). Desirability of Compatibility in the Absence of Network Externalities. *American Economic Review, 79*, 1165-1181.

Economides, Nicholas (1996). Network Externalities, Complementarities, and Invitations to Enter. *European Journal of Political Economy, 12*, 211-234.

Economides, Nicholas (2000). The Microsoft Antitrust Case. Stern School of Business Working Paper 2000-09, New York University.

Economides, Nicholas, & Himmelberg, C. (1995). Critical mass and network size with application to the US Fax market. Working Paper Series // Stern School of Business, NYC EC ; 95,11, Available at <http://www.stern.nyc/networks/95-11.pdf>

Egyedi, Tineke M. (1996). *Shaping Standardization: A Study of Standards Processes and Standards Policies in the Field of Telematic Services.* (Doctoral dissertation) Delft University of Technology. 329 pp.

Egyedi, Tineke M. (2001). *Beyond Consortia, Beyond Standardization? New Case Material for the European Commission. Final Report to the European Commission.* Delft: Faculty of Technology, Policy and Management, Delft University of Technology. October 2001. 69 pp.

Egyedi, Tineke M. (2001a). Why Java Was-not-standardized Twice. *Computer Standards & Interfaces, 23*(4), 253–265. Available at <http://www.elsevier.com/locate/csi/>

Egyedi, Tineke, & Arjan, Loeffen (2002). Succession in Standardization: Grafting XML Onto SGML. *Computer Standards & Interfaces, 24* (4), 279–290. Available at <http://www.elsevier.com/locate/csi/>

Farrell, Joseph (1990). The Economics of Standardization: A Guide for Non-Economists. In Berg, John L., and Haral Schumny (Eds.). *An Analysis of the Information Process: Proceeding of the International Symposium on Information Technology Standardization, INSITS.* Amsterdam: North-Holland, pp. 189-198.

Farrell, Joseph (1995). Arguments for weaker IPR protection in network industries. In B. Kahin& J. Abate (Eds.), *Standards Policy for Information Infrastructure.* Cambridge, MA: MIT Press. pp. 368-402.

Farrell, Joseph (1996). *Choosing the Rules for Formal Standardization.* Berkeley: UC Berkeley.

Farrell, Joseph, & Saloner, Garth (1985, spring). Standardization, Compatibility, and Innovation. *Rand Journal of Economics, 16*(1), 70-83;

Farrell, J., & Saloner, Garth (1986). Installed Base and Compatibility: Innovation, Product Preannouncements, and Predation, *American Economic Review, 76*(5), 940-955.

Farrell, J., & Saloner, Garth (1986a). Standardization and Variety. *Economics Letters, 20*, 71-74.

Farrell, Joseph, and Saloner, Garth (1988, summer). Coordination through Committees and Markets. *Rand Journal of Economics, 19*(2), pp. 235-252.

Fomin, Vladislav V. (2001). *The Process of Standard Making: The Case of Cellular Telepony.* Doctoral Dissertation, Univeristy of Jyväskylä, Jyväskylä, Finland.

Frank, Thomas (2000). *One Market Under God: Extreme Capitalism, Market Populism, and the End of Economic Democracy.* New York: Anchor Books, Random House.

Gaynor, Mark, Bradner, Scott, Lansiti, Marco and H.T. King (2001). The Real Options Approach to Standards for Building Network-based Services. in Timothy Schoechle and Carson Wagner (eds.), *SIIT 2001 Proceedings: 2nd IEEE Conference on Standardization and Innovation in Information Technology.* Boulder, Colorado, October 3-6, 2001. pp. 217-228.

Gifford, Jonathan L. (1997). ITS Standardization: Assessing the Value of a Consortium Approach. in *Proceedings of the ITS Standards Review and Interoperability Workshop.* George Mason University Dec. 17-18, 1997. pp 1-7. Available at <http://www.itsdocs.fhwa.dot.gov/jpodocs/proceedn/2lz1!.pdf>

Gowan, Peter (1999). *The Global Gamble: Washington's Faustian Bid for World Dominance.* New York, London: Verso. 320 pp.

Gray, John (1998). *False Dawn: The Delusions of Global Capitalism.* London: Granta Books. 262 pp.

Greenstein, Shane M. (1992, September). Invisible hands and Visible Advisors: An Economic Interpretation of Standardization, *Journal of the American Society for Information Science, 43*(8).

Greenstein, Shane M., and V. Stango (Eds.) (2005). *Standards and Public Policy.* Cambridge: Cambridge Press.

Greider, William (1997). *One World, Ready or Not: The Manic Logic of Global Capitalism.* New York: Simon & Schuster. 528 pp.

Habermas, Jürgen (1962) *(1991). The Structural Transformation of the Public Sphere: An Inquiry into a Category of Bourgeois Society* (Thomas Burger, Trans.). Cambridge, MA: MIT Press.

Hardin, Garrett (1968). The Tragedy of the Commons. *Science.* vol. 162. p. 1243.

Haug, Thomas (2002). A Commentary on Standardization Practices: Lessons from the NMT and GSM Mobile Telephone Standards Histories, *Telecommunications Policy, 26*, pp. 101-107.

Hawkins, Richard (1999). The rise of consortia in the information and communication technology industries: emerging implications for policy. *Telecommunications policy, 23*, pp. 159-173.

Hawkins, Richard, Robin Mansell, and Jim Skea (Eds.) (1995). *Standards, Innovation and Competitiveness: The Politics and Economics of Standards in Natural and Technical Environments.* Brookfield, VT: Edward Elgar Publishing Ltd.

Hesser, Wilfried, and Axel Czaya (1999, March). Standardization as a Subject of Study in Higher Education—A Vision. Unpublished manuscript, Universität der Bundeswer, Hamburg.

Humphries, Jane (1990, March). Enclosures, Common Rights, and Women: The Proletarianization of Families in the Late Eighteenth and Early Nineteenth Centuries, *Journal of Economic History, L*(1), The Economic History Association, pp. 17–42.

Innis, Harold A. (1972). *Empire and Communication.* Toronto: University of Toronto Press.

Iversen, Eric J., Rudi Bekkers, and Knut Blind (2006). Emerging coordination mechanisms for multi-party IPR holders: linking research with standardization, *Intellectual Property Rights for Business and Society* conference sponsored by DIME-EU Network of Excellence, held in London, September 14-15, 2006.

Jakobs, Kai (2000). *Standardisation Processes in IT: Impact, Problems and Benefits of User Participation.* Braunschweig/Wiesbaden: Vieweg & Sohn Verlagsgesellschaft mbH.

Jakobs, Kai (2001). Broader View on Some Forces Shaping Standardization. *SIIT 2001 Proceedings: 2nd IEEE Conference on Standardization and Innovation in Information Technology.* Boulder, CO, October 5, 2001. pp. 133-143.

Jakobs, Kai, Rob Procter, and Robin Williams (1998). Telecommunication Standardisation - Do We Really Need the User? *Proceedings of ICT '98: the 6th International Conference on Telecommunications*, IEEE Press.

Jakobs, Kai, Rob Procter, *et al* (2001, April). The making of standards: looking inside the work groups. *IEEE Communications Magazine*, pp. 102-107.

Katz, Michael L., and Carl Shapiro (1992). Product Introduction with Network Externalities. Journal of Industrial Economics, *40*(1). pp. 55-83.

Katz, Michael L., and Carl Shapiro (1994). Systems Competition and Network Effects. *Journal of Economic Perspectives, 8,* pp. 93-115.

Kahin, Brian, and Janet Abbate (Eds.) (1995). *Standards Policy for Information Infrastructure,* Cambridge: MIT Press. 653 pp.

Katz, Michael L., and Carl Shapiro (1985, June). Network Externalities, Competition and Compatibility. *American Economic Review 75*(3), pp. 424-440.

Katz, Michael L., and Carl Shapiro (1986, August). Technology Adoption in the Presence of Network Externalities. *Journal of Political Economy*, *94*(4), pp. 822-841.

Koolhaas, Rem (2002). *Newshour Interview by Ray Suarez*. June 25, 2002. Available at <http://www.pbs.org/newshour/bb/entertainment/jan-june02/koolhaas_transcript.html/>

Krechmer, Ken (1998, November/December). The Principles of Open Standards. *Standards Engineering*, *50*(6), p. 1. Available at <http://www.crstds.com/openstds.html>

Krechmer, Ken (2000, July/August). Market Driven Standardization: Everyone Can Win. *Standards Engineering, 52*(4). pp. 15-19. Available at <http://www.crstds.com/fundeco.html>

Krechmer, Ken (2000a, June). The Fundamental Nature of Standards: Technical Perspective. *IEEE Communications Magazine, 38*(6), p. 70. Available at <http://www.crstds.com/fora.html>

Krechmer, Ken (2000b). The Fundamental Nature of Standards: Economics Perspective. *Schumpeter 2000: Eighth International Joseph A. Schumpeter Society Conference*, Manchester, UK, June 28-July 1, 2000. p. 70. Available at <http://www.crstds.com/fundeco.html>

Krechmer, Ken, and Elaine Baskin (2001). Standards, Information and Communications: A Conceptual Basis for a Mathematical Understanding of Technical Standards. In Timothy Schoechle and Carson Wagner (eds.). *SIIT 2001 Proceedings: 2nd IEEE Conference on Standardization and Innovation in Information Technology*. Boulder, Colorado, October 3-6, 2001. pp. 106-114. Available at <http://www.crstds.com/siit2001.html>

Lehr, William (1992, September). standardization: Understanding the Process, *Journal of the American Society for Information Science, 43*(8).

Lehenkari, Janne, and Reijo Miettinen (2001). Standardization in the Construction of a Large Technological System—The Case of the Nordic Mobile Telephone System. *Telecommunications Policy, 26*, pp. 109-127.

Lemley, Mark A. (2002). Intellectual Property Rights and Standard-Setting Organizations. *California Law Review, 90*(6), pp. 1889-1979.

Lemley, Mark A. (1998). The Law and Economics of Internet Norms. *Kent Law Review, 73*, pp. 1257-1288.

Lemley, Mark A. (1999). Standardizing Government Standard-Setting Policy for Electronic Commerce. *Berkeley Technology Law Journal, 14*, pp. 745-752.

Lemley, Mark A. (2001). Antitrust, Intellectual Property Rights and Standard Setting Organizations. In Timothy Schoechle and Carson Wagner (eds.), *SIIT 2001 Proceedings—2nd IEEE Conference on Sandardization and Innovation in Information Technology*, Boulder, Colorado, October 3-5, 2001. pp. 157-169.

Lemley, Mark, and David McGowan (1998). Could Java Change Everything? The Competitive Propriety of a Proprietary Standard. *Antitrust Bulletin, 43*, p. 715

Lessig, Lawrence (1999). *Code and Other Laws of Cyberspace.* New York: Basic Books. 297 pp.

Lessig, Lawrence (2001). *The Future of Ideas: The Fate of the Commons in a Connected World.* New York: Random House. 352 pp.

Lessig, Lawrence (2001a, November/December). The Internet Under Siege. *Foreign Policy.*

Lessig, Lawrence (2002). Remarks at the conference, *The Regulation of Information Platforms*, Silicon Flatirons Telecommunication Program. University of Colorado, Boulder, CO. January 27-28, 2002.

Libicki, Martin (1995). *Information Technology Standards: Quest for the Common Byte.* Boston: Digital Press.

Libicki, Martin, James Schneider, Dave R. Frelinger, and Anna Slomovic (2002). *Scaffolding the New Web: Standards and Standards Policy for the Digital Economy.* Santa Monica: RAND Corporation. Available at <http://www.rand.org/publications/MR/MR1215/>

Lim, Andriew S. (2002). Standards Setting Processes in ICT: The Negotiations Approach. Unpublished working paper 02.19. Eindhoven Centre for Innovation

Studies. Technische Universiteit Eindhoven, Faculteit Technologie Management. 21 pp. Availabe at<http://www.tm.tue.nl/ecis/>

Litman, Jessica (1990, fall). The Public Domain. *Emory Law Journal 39*, p. 965. Available <http://www.law.wayne.edu/litman>

Mansell, Robin (1993). *The New Telecommunications: A Political Economy of Network Evolution.* London: Sage Publications Ltd.

Mansell, Robin (1995). Standards, Industrial Policy and Innovation. In Hawkins, Richard, Robin Mansell, and Jim Skea (Eds.). *Standards, Innovation and Competitiveness: The Politics and Economics of Standards in Natural and Technical Environments.* Brookfield, VT: Edward Elgar Publishing Ltd., pp. 213-227.

Mansell, Robin, & Richard Hawkins (1992). Old Roads and New Signposts: Trade Policy Objectives in Telecommunication Standards. In Klaver, F. & P. Slaa (Eds.). *Telecommunication, New Signposts to Old Roads.* Amsterdam: IOS Press, pp. 45-54.

Mansell, Robin, and Roger Silverstone (Eds.) (1996). *Communication by Design: The Politics of Information and Communication Technologies.* Oxford: Oxford University Press.

Mansfield, Edwin (1970). *Microeconomic Theory and Application.* New York: W.W. Norton

Marks, Roger B. and Robert E. Hebner (2001). Government Activity to Increase Benefits from the Global Standards System. *SIIT 2001 Proceedings: 2nd IEEE Conference on Standardization and Innovation in Information Technology.* Boulder, CO, October 5, 2001. pp. 183-190.

Marks, Roger B. and Robert E. Hebner (2004). Government/Industry Interactions in the Global Standards System. In Bolin, Sherrie, (Ed.) *The Standards Edge: Dynamic Tension.* Ann Arbor, MI: Sheridan Books. pp. 103-114.

McClosky, Donald M. (1972, March). The Enclosure of Open Fields: Preface to a Study of Its Impact on the Efficiency of English Agriculture in the Eighteenth Century, *Journal of Economic History, 32*, The Economic History Association, pp. 15-35.

Messerschmitt, David G. (1999). Prospects for Computing—Communications Convergence. Lecture at the University of Colorado, Boulder, CO. October 24, 1999. Available at <http://www.eccs.berkeley.edu/~messer/talks/99/Convergence.pdf>

Metcalfe, Robert M. (1995, 2 October). Metcalfe's Law. *InfoWorld.*

Mueller, Milton (2002). Interest Groups and the Public Interest: Civil Society Action and the Milton Globalization of Communications Policy. *TPRC 2002: The 30th Annual Research Conference on Communication, Information and Internet Policy.* September 28-30, 2002. Alexandria, VA.

Murray, Alan (2002). No Longer Business As Usual for Forces of U.S. Capitalism. *Wall Street Journal* 1 October, p. A4.

Oksala, Steve (2000, 21 August). *The Changing Standards World: Government Did It, Even If They Didn't Mean To.* ANSI/NIST World Standards Day paper.

O'Reilly, Tim (1999). Open Sources: voices from the Open Source Revolution. O'Reilly and Associates. 280 pages. Available at <http://www.oreilly.com/catalog/opensources/book/toc.html>, read August 27, 2003.

Peters, John Durham (1989, November). John Locke, the Individual, and the Origin of Communication. *The Quarterly Journal of Speech, 75*, pp. 387-399.

Phillips, Kevin (1994). *Arrogant Capitalism: Washington, Wall Street and the Frustration of American Politics.* Boston: Little Brown and Co.

Popper, Karl (1945). *The Open Society and Its Enemies.* 2 vols. London: Routledge.

Post, David G. (2000). What Larry Doesn't Get: Code, Law and Liberty in Cyberspace. *Stanford Law Review, 52*, pp. 1439-1451. Available at <http://www.temple.edu/lawschool/dpost/Code.pdf/>

Rice, John, and Peter Galvin (1999). The Development of Standards in the Mobile Telephone Industry and Their Effect on Regional Industry Growth. *SIIT '99 Proceedings: 1st IEEE Conference on Standardization and Innovation in Information Technology.* University of Aachen. September 15-16, 1999. pp. 153-156.

Robinson, Gary S. (1999). There are no Standards for Making Standards. *SIIT '99 Proceedings - 1st IEEE Conference on Standardisation and Innovation in Information Technology*, Aachen, Germany, September 15-17, 1999. pp. 249-250.

Samuelson, Robert J. (2002). Global Economics Under Siege. *Washington Post* 16 October. p. A25.

Schoechle, Timothy (1995, April). The Emerging Role of Standards Bodies in the Formation of Public Policy. *IEEE Standards Bearer, 9*(2), pp. 1, 10. Available at <http://www.acm.org/pubs/articles/proceedings/cas/332186/p255-schoechle/>

Schoechle, Timothy (1998). *A Public Policy Debate: Standardsmaking Practice and the Discourse on International Privacy Standards.* Standards Policy Research Paper ICSR98-211. Boulder: International Center for Standards Research. Available at <http://www.standardsresearch.org>

Schoechle, Timothy (1999). Toward a Theory of Standards. *SIIT '99 Proceedings: 1st IEEE Conference on Standardization and Innovation in Information Technology.* University of Aachen, September 15-16, 1999. pp. 175-181.

Schoechle, Timothy (2001, fall). Re-examining Intellectual Property Rights in the Context of Standardization, Innovation and the Public Sphere, *Knowledge, Technology & Policy, 14*(3), pp. 109-126.

Schoechle, Timothy (2003). Digital Enclosure: The Privatization of Standards and Standardization, *SIIT 2003 Proceedings: 3rd IEEE Conference on Standardization and Innovation in Information Technology.* Delft University of Technology. 22 October, pp. 229-240.

Schoechle, Timothy, Stephen Shapiro, Michael Rinow, and Barnaby Richards (2002). Evolving Approaches to Technical Standardization: Hybrid Standards Setting. *EASST 2002: Conference of the European Association for the Study of Science and Technology*, York, UK, August 1, 2002.

Shapiro, Carl (2001). Navigating the Patent Thicket: Cross Licenses, Patent Pools, and Standard-Setting, In Jaffe, Adam, Joshua Lerner, and Scott Stern, (Eds.), *Innovation Policy and the Economy, I*, Cambridge, MA: MIT Press.

Shapiro, Carl, and Hal R. Varian (1999). *Information Rules: A Strategic Guide to the Network Economy.* Boston: Harvard Business School Press.

Steinmueller, W. E. (1995). The political economy of data communication standards. In Hawkins, Richard, Robin Mansell, and Jim Skea (Eds.) *(*1995) *Standards, innovation and competitiveness.* Aldershot: Edward Elgar.

Stiglitz, Joseph E. (2003). *Globalization and its Discontents.* New York: W.W. Norton, 288 pp.

Smith, Adam *(*1776) *(1976). An Inquiry into the Nature and Causes of the Wealth of Nations.* Campbell, R.H., *et al (*Eds.). Oxford & New York: Oxford University Press.

Smoot, Oliver R. (2001). *Standards-Setting and United States Competitiveness.* Statement of Oliver R. Smoot, Chairman of the Board of Directors, American National Standards Institute before the House Science Committee, Subcommittee on

Technology, Environment and Standards, June 28, 2001. Available at <http://www.house.gov/science/ets/Jun28/Smoot.htm/>

Spring, Michael B. *et al* (1995). Improving the Standardization Process: Working with Bulldogs an Turtles. In Kahin, Brian, and Janet Abate (eds), *Standards Policy for Information Infrastructure*. Cambridge, MA: MIT Press, pp. 220-250.

Sherif, Mostafa (2001, April). A Framework for Standardization in Telecommunications and Information Technology. *IEEE Communications Magazine, 39*(4), pp. 94-100.

Sherif, Mostafa (2002). *When Is Standardization Slow?* Conference of the European Association for the Study of Science and Technology, York, UK, August 1, 2002. Revised version published in *International Journal of IT Standards and Standardization Research, 1*, (2003, spring), pp. 19-32.

Soros, George (1998). *The Crisis of Global Capitalism [Open Society Endangered]*. New York: PublicAffairs, Perseus Books Group.

Soros, George (2002). *George Soros on Globalization*. New York: PublicAffairs, Perseus Books Group.

Streeter, Thomas (1986). *Technocracy and Television: Discourse, Policy, Politics and the Making of Cable Television*. unpublished doctoral dissertation. University of Illinois at Urbana-Champaign.

Streeter, Thomas (1990, spring). Beyond Freedom of Speech and the Public Interest: The Relevance of Critical Legal Studies of Communication Policy. *Journal of Communication*.

Taschdjian, Martin (2001). Standards and the Velocity of Knowledge: Accelerator or Inhibitor? Paper presented at *SIIT 2001 2nd IEEE Conference on Standardization and Innovation in Information Technology*, Boulder, CO, October 5, 2001.

Techapalokul, Soontaraporn, James H. Alleman, and Yongmin Chen (2001). Economics Of Standards: A Survey And Framework, in Timothy Schoechle and Carson Wagner (eds.). *SIIT 2001 Proceedings: 2nd IEEE Conference on Standardization and Innovation in Information Technology*, Boulder, CO, October 3-6, 2001. pp 193-205.

Updegrove, Andrew (1995). Consortia and the Role of the Government. In Kahin, Brian, and Janet Abate (eds), *Standards Policy for Information Infrastructure*. Cambridge, MA: MIT Press, pp. 321-348.

Updegrove, Andrew (1995a, December). Standard Setting and Consortium Structures. *Standard View.*

Van Houweiling, Molly S. (2002). Cultivating Open Information Platforms: A Land Trust Model. *The Journal on Telecommunications Law and High Technology.* Vol. 1. Also presented at the Silicon Flatirons Conference. Regulation of Information Platforms. University of Colorado, Boulder, CO. January 28-29, 2002.

Vardakas, Evangelos (1998). Director, EC Directorate for Industry (DG III), speech before the *Standardization Summit* sponsored by ANSI and NIST, Washington DC, September 23, 1998.

Vardakas, Evangelos (2002). Director, EC Directorate General Enterprise, presentation before the ANSI Annual Conference, Washington DC, October 15, 2002, and subsequent personal interview.

Veblen, Theodore (1919) *(1990).* The Place of Science in Modern Civilization. In Veblen, T. B., *The Place of Science in Modern Civilization and Other Essays.* New Brunswick: Transaction.

Vercoulen, Frank, and Marc van Wegberg (1998). *Standard Selection Modes in Dynamic, Complex Industries: Creating Hybrids between Market Selection and Negotiated Selection of Standards,* Maastricht: Universiteit Maastricht, pp. 1-14.

Vercoulen, Frank and Marc van Wegberg, (1999). Standard selection modes in dynamic, complex industries: creating hybrids between market selection and negotiated selection of standards. *SIIT '99 Proceedings - 1st IEEE Conference on Standardisation and Innovation in Information Technology,* Aachen, Germany, September 15-17, 1999. pp. 1-11

*Vaidhyanathan,* Siva (2001). *Copyrights and Copywrongs: The Rise of Intellectual Property and How It Threatens Creativity.* New York: New York University Press. 241 pp.

Wagner, Douglas K. (1999). *Discourse and Policy in the Philippines: The Construction and Implementation of a Strategic IT Plan.* Unpublished doctoral dissertation, University of Colorado, Boulder, CO.

Wagner, R. Polk (2002). Information Wants To Be Free: Intellectual Property and the Mythologies of Control. *TPRC 2002: The 30th Annual Research Conference on Communication, Information and Internet Policy,* Alexandria, VA, September 28-30, 2002 <http://papers.pennlaw.net/>.

Warshaw, Stanley I., and Mary H. Saunders (1995). International Challenges in Defining the Public and Private Interest in Standards. In Richard Hawkins, Robin

Mansell and Jim Skea (eds.). *Standards, Innovation and Competitiveness: The Politics and Economics of Standards in Natural and Technical Environments.* Brookfield, VT: Edward Elgar Publishing Ltd. pp. 67-74.

Wegberg, Marc van (2002). Positioning Strategies Of Standard Development Organizations in the Internet: The Case Of Internet Telephony. *EASST 2002: Conference of the European Association for the Study of Science and Technology,* York, UK, August 1, 2002.

Wegberg, Marc van (2002a). Interview, August 2, 2002, at the EASST Conference, York, UK.

Wegberg, Marc van (2003). The Grand Coalition versus Competing Coalitions: Trade-Offs in How to Standardize. *SIIT 2003 Proceedings: 3rd IEEE Conference on Standardization and Innovation in Information Technology.* Delft University of Technology. October 22, 2003. pp. 271-284.

Wegberg, Marc van (2004). Standardization Process of Systems Technologies: Creating a Balance between Competition and Cooperation, *Technology Analysis & Strategic Management, 16*(4), pp. 457-478.

Weiser, Philip J. (2001). Internet Governance, Standard-Setting, and Self-Regulation. *Northern Kentucky Law Review, 28*(4), pp. 822-846.

Weiss, Martin B.H. (1990): The Standards Development Process: A View from Political Theory. *Standard View, 1*(2), pp. 35-41.

Weiss, Martin B.H., and Carl Cargill (1992). Consortia and the Standards Development Process. *Journal of the American Society for Information Science, 43*(8), pp. 559-565.

Weiss, Martin B.H., and Marvin Sirbu (1990). Technological choice in Voluntary Standards Committees: an Empirical Analysis. *Economics of Innovation and New Technology, 1*(1), pp. 111-134.

Weiss, Martin H.B., and Michael Spring (2000). Selected Intellectual Property Issues in Standardization, In Jakobs, Kai, (Ed.) *Information Technology Standards and Standardization: A Global Perspective.* Hershey, PA: Idea Group Publishing.

Weiss, Martin B.H., and Ronald T. Toyofuku (1996). Free-Ridership in the Standard Setting Process: The Case of 10baseT. *Standard view, 4*(4), pp. 205-212.

Williams, Robin (1997). The Social Shaping of Information And Communication Technologies. In Herbert Kubicek, William H. Dutton and Robin Williams (eds.). *The Social Shaping of Information Superhighways: European and American Roads to the Information Society*. Frankfurt: Campus and New York: St. Martins Press. Chapter 18, pp. 299-337.

Williams, Robin, and David Edge (1996). The Social Shaping of Technology. *Research Policy, 25*, pp. 865-899.

Winner, Langdon (1985). Do Artifacts Have Politics? In MacKenzie, Donald, and Judy Wajcman (Eds.). *The Social Shaping of Technology*. Milton Keynes: Open University Press.

## ENDNOTES

1. Koolhaus is an urban designer and teacher and a winner of the Pritzker prize. He is noted for innovative architectural designs including buildings, shopping centers, and other urban spaces.
2. Not fully the same as *property*, a problematic notion that will be discussed later.
3. Gray emphasizes the importance of the destruction of the family as a social institution and lays it at the door of the free market. He argues that by undermining the role of work and social cohesion, and by bringing labor into a world of declining health care, childcare, wages, *etc*., along with uncertainty, flex-time, forced mobility, and the loss of other intermediary social institutions, the family has been sacrificed to the market. It is ironic that much free market populist rhetoric comes from the religious right, a group that celebrates the "family" as an important institution while supporting economic policies that undermine it.
4. Similar or related arguments are made by Greider (1997), Phillips (1994), Gowan (1999), Brenner (1998), and Albert (1993).
5. Soros has established and funded, through the Soros Foundation, a global network of philanthropic institutions around the mission of "Building Open Societies," with a total annual budget of over $500 million.
6. According to various news reports, by August 2002, nearly $2 trillion had disappeared from the U.S. economy since March of 2000, along with 500,000 jobs; leaving $1 trillion in debt outstanding. An August 12 *Wall Street Journal* article under the rubric, "Dialing for Dollars," titled, "Before Telecom Industry Sank, Insiders Sold Billions in Stock: As They Cashed Out Shares, Many Executives Touted Sector's Growth Potential." The story goes on to describe Vincent Galluccio, former top executive of Metromedia Fiber Networks, Inc., now retired and sitting on his tractor in his new Chardonnay vineyard. The article quotes him as saying, "My father taught me that when you play a hand,

put half in your pocket and walk away from the table." He walked away with $27 million, leaving Metromedia in Chapter 11 while the SEC investigates the company's accounting practices. Ironically, a pervasive theme of the business discourse of the time was the management mantra of "build shareholder value" as justification for externalizing costs.

[7] The word *network* has dual meanings. As the engineer designs *networks* of communicating computers, the economist speaks of *network effects*, meaning the exponentially increasing economic value of having an increasing number of nodes or users connected to a network. In the same sense, economists also use the term *network externalities* and engineers refer to *Metcalfe's Law* (attributed to Bob Metcalfe, inventor of the Ethernet protocol, standardized as IEEE 802.3) (Metcalfe, 1995).

[8] Lessig acknowledges the importance of standardization practice, but considers the matter too complex and arcane for him, "as a lawyer" he charged this author to delve into it. Personal conversation (Lessig, 2002).

[9] *Tragedy of the Commons* is biologist Garrett Hardin's classic allegory about the tendency to exploit a common resource in situations of unregulated self interest (Hardin, 1968).

[10] Lessig stresses the danger of using metaphorical models like the *Tragedy* as a foundation for policy decisions. He cites the extensive literature on the commons and its comedies, tragedies, fallacies, *etc.* (Lessig, 2001, pp. 271-272).

[11] Similar arguments are made by Christensen (1997) in examining the difficulties that successful or dominant firms have in dealing with technical innovation in general, and technical standards in particular (Christensen, 2002).

[12] Although Boyle's paper (2001) carries this title, he credits usage of the term to others including Ben Kaplan, Pamela Samuleson, Yochai Benkler, David Lange, Keith Aoki and Hanibal Travis.

[13] Gray also mentions the Enclosure Movement, describing it as an appropriation that "...tilted the balance of ownership in England's agrarian economy away from cottagers and yeoman farmers toward the great landowners of the eighteenth and nineteenth centuries." (1998, p. 8). Gray emphasizes the role of political power in this process and questions the generally attributed historic role of economic efficiency.

[14] Some economic research that has studied the economy of the commons during the enclosure movement challenges the notion that privatization improved societal production efficiency (McClosky, 1972; Humphries, 1990).

[15] *U.S. Constitution*, Article I, Section 8, Clause 8.

[16] Lessig also cites Thomas Jefferson, perhaps the principal architect of the U.S. patent and copyright system, and his arguments against the notion of patent protection as a *natural right* (Lessig, 2001, pp. 94-95).

[17] Lessig (2001, p. 279-280) acknowledges the difficulty in keeping the terminology straight regarding *open vs. free*, of *closed vs. proprietary*, and what is intended by *open source* (*e.g.,* license under General Public License (GPL) or other variations).

[18] The term platform is commonly used in the software industry to mean a standardized base (proprietary or open) for related, compatible or interoperable applications, constituting an element of *infrastructure*.

[19] This was accomplished in part by organizing a conference on the topic (*The Regulation of Information Platforms*, University of Colorado, January 27-29, 2002) that included presentations by Lessig (2002) and others mentioned here.

[20] Originally *Groupe Spéciale Mobile*, the standards body that created it.

[21] A comprehensive survey of the economics literature on standards has been provided by Techapalokul, Alleman, and Chen (2001).

# Chapter IV
# Theoretical and Methodological Approaches

*...a spurious syllogism may...be based on the confusion of the absolute with that which is not absolute but particular. As in dialectic, for instance, it may be argued that what-is-not is, on the ground that what-is-not is what-is-not...so also in rhetoric, a spurious enthymeme may be based on the confusion of some particular probability with absolute probability.*

—Aristotle, 336 BC

## INTRODUCTION

The primary theoretical perspective and framework of analysis for this study are public sphere theory and political economy. Although standardization is a *social practice* that springs from tradition and not from any theoretical grounding, it is situated and institutionalized in a public but non-governmental setting that could be seen as a *public sphere* as described by Habermas (1962). Public sphere theory provides a focus on *publicness* and the *discursive process* that are relevant for analyzing and understanding standardization practice. Political economy provides an approach to understanding the various interests being served by the social practices that are relevant to this study, and to the underlying reasons for their establishment.

The primary research method employed in this project is *discourse analysis*.[1] Furthermore, this study takes a view of standardization as a *social practice*. Discourse analysis provides a method of extracting the meanings in any social practice, which may be defined as *the symbolic means by which people negotiate and share their realities*. In the case of standards-making, standardization practice is a recurring ritualized and institutionalized group interaction and its discourse includes a wide array of oral, written, procedural, and symbolic acts. Standardization as a social practice may be operationally defined as a combination of standardization discussion and the settings within which that discussion proceeds. In other words, practice includes the way meetings are conducted (*i.e.,* rules, procedures, order of business, agendas, *etc*.), who participates or does not, what is discussed and what is not, the language used, documents produced, and the terms of discourse and their underlying assumptions; all these elements taken together, constitute standardization as a social practice. In this study, much of the discourse under analysis is meta-discourse, or discourse about the standardization discourse and practice.

The adapted analysis relies on a definition of social practice offered by MacIntyre (1981) as a cooperative social activity pursuing and systematically extending internal and external goods and standards of excellence. The practice of standards-making is such a cooperative social activity seeking to establish community goods (*e.g.*, communication protocols, product interoperability, certain specialized markets, *etc.*) by a voluntary consensus process conducted under well established procedures, protocols, and rituals.[2] In MacIntyre's terms, such goods (*i.e.*, the technical standards) produced as outcomes of the practice would be considered *external*. Things like cooperation, rational deliberation, practices, procedures, standards of excellence, and other elements inherent to the process itself (including discourse about the practice) would be considered *internal*.[3] Taking this approach of standardization as a social practice recognizes that both the standards and the practice by which they are achieved are human constructions formed by shared activities that, in turn, form social habits and are fundamental to social processes. Other social theorists, including Bourdieu (1990) and Craig (1996), have provided further perspectives on social practice along similar lines.

The problem of enclosure described earlier is complex, and the approach taken in this study is only one possible path of inquiry. In this inquiry, other theoretical perspectives could have been used and other methodological approaches could have been taken. No single approach by itself is complete. In this case, the study of rhetoric offers a starting point, providing theory and method together. For example, a common rhetorical device found in discourse is the *enthymeme*, an incomplete syllogism that persuades not by logical force of argument, but by reliance on pre-existing beliefs and assumptions in the mind of the reader/listener/participant. It also relies on preceding discourse and on the context of that discourse. Discourse analysis is a valuable approach because standardization practice is inherently discursively constituted through words and meanings. Essentially, discourse analysis digs out such meanings and assumptions.

An analysis of discourse can reveal prevailing ideological commitments that lead people to interpret data in conflicting ways. Analysis of a discourse offers to identify the variety of meanings or interpretations found in the various symbols, myths, and ideologies that are embedded in the discourse, and could lead to a better understanding of the policy implications or responses. In particular, this study applies the constitutive paradigm of rhetorical analysis that distinguishes between logical and ideological argumentation. These theoretical perspectives provide a framework within which to analyze and understand the meanings found in the practice.

Secondarily, this study applies an historiographic method to look at existing texts (*e.g.*, scholarly journal articles, congressional testimony, government reports, public speeches, published interviews, and other documents). Furthermore, it employs ethnographic field methods to acquire examples of the discourse in various standards committees, conferences, meetings and other gatherings, as well as in personal interviews. These methodological choices are driven by the author's encounter with bodies of discourse considered in this work, by the pressing need for a better understanding of them and of their institutional origins, and because no such analysis has been recorded.

In summary, this study is an analysis of standards discourse. This is achieved, not in the linguistic sense, although linguistic methods are employed, but as a social practice and a discursive interaction within a public sphere of certain established and evolving institutions. It is an effort to tie rhetorical analysis with social practice in a public sphere. To the ancient Greeks, rhetoric was the instrumental art of persuasion. To some, the content or substance of discussion was relatively unimportant, yet to others, it became an integral part of the development of democratic institutions and citizenship. This research has very much to do with the content of contextual arguments. Consequently, the analysis is grounded in actual social practices situated in existing institutions, and framed as a public sphere and in relationship to political and economic power.

## DISCOURSE AS THEORY AND METHOD

It is useful to begin by defining the term *discourse*. The intent is to use the term as both a verb and a noun, as both a practice and a text resulting from that practice. This duality is important because the language constructs found in the texts must be understood not simply as objects of analysis in themselves, but together with the situated process that produced them. As described by Hartley (1994, p. 93), discourse is "...the social process of making and reproducing sense(s)." The usage here of *discourse* comes from traditions of structuralism, post-structuralism, and semiotics, and relies on concepts developed by Saussure, Barthes, Althusser, and Foucault.

This study relies on the belief that "discourses are the product of social, historical and institutional formations; and meanings are produced by their institutionalized forms" (Hartley, 1994, p. 93).

More specifically, this study borrows from the concept of *discursive practices*, described by Streeter as "...concrete instances of discourse with practical outcomes, or as ways of doing things with words" (1986, p. 17-18). Streeter's sense of *discourse* and *practice* apply directly to this study:

> *The word "rhetoric" often carries the undesired connotation of "mere rhetoric," i.e. of contentless verbiage, while "discourse" connotes substance as well as form. In the less derogatory use of "rhetoric" common to the field of that name, the word is typically based on the model of the public speech, which tends to focus attention on the situation of the individual speaker. "Discourse" connotes the more collective model of a language in use, which is more appropriate in this context. The word "language," however, is too static, lacking the necessary sense of words in context. "Practice" is the most adequate for reference to the behavior of telecommunications institutions in their social context.* (p. 17, footnote 17)

Also, as in the case of Streeter's study of the cable industry, the setting for a discourse is the set of institutional, economic, and social relationships distributed among government, industry, trade associations, engineering societies and standards bodies that define the technologies and architectures on which the industrial system depends. Standardization cannot be reduced to either an institution or a discourse, but is a combination of both (p. 207).

## RELEVANT DISCOURSES AND SOCIAL PRACTICES

Two bodies of discourse are objects of this study: (1) a discourse about enclosure of ideas and counter arguments forming a dialectic, and (2) a discourse and social practice about standards and standardization.[4] Through analysis of these discourses and the institutional frameworks or conditions of discourse within which they proceed, this work attempts to show how the enclosure of ideas shapes and is reflected in the standards discourse. It also reveals how assumptions about intellectual property rights and markets are shaping the standardization process. This work attempts to explain how the dialectic is being resolved in the minds of standards participants since such resolution will have significant material and economic consequences.

## Arguments

Reflected in these discourses are two competing theories or overriding perceptions about the creation of wealth in society. Both theories are couched in terms of the common good: (1) intellectual property rights (held in closed [private] space), versus (2) intellectual commons (held in open [public] space). This competition includes not only the ideas generated (*i.e.,* the standards), but the conditions of the discourse that establishes them, (*i.e.,* the conduct of the standardization process as a discursive practice). The competing theories exhibit a fundamental conflict between the perceptual frames of the *public* (*i.e.,* the common) and the *private* and claims about which of these may result in greater economic value for society—freedom of ideas or ownership of ideas (including proprietary rights to ideas and privatized discourse about them).

## Method

The discourses are examined through documents, narrative accounts, and participation in actual gatherings where a given discourse is conducted. Documents, including historical records, seminal papers, speeches, and constitutive documents (*i.e.,* charters, procedures manuals, institutional directives, membership agreements, white papers, position papers, policy statements, *etc.*) help to establish the initial terms of the discourse. Many of these terms are problematic (*e.g.*, free, open, balanced, public, private, *etc.*).[5] Narrative accounts by participants and informed observers, in the form of correspondence or interviews, and direct observations at gatherings help to reveal the individual and group perceptions that serve to form the discourse and shape its outcomes.

It may be argued that the influence of enclosure, as a discursive condition, shapes the character of a discourse so that enclosure becomes less and less contested. The intent here to show how the discourse encourages a mode of thought with specific implications for policy. Part of what is proposed is an analysis of the attitudes that "openness" and "privatizaton" induce, and the acts that such attitudes encourage.

In summary, the method of this study is to:

1. Examine the enclosure discourse and identify the operative terms of enclosure (external definitions) and posit a hypothesis (*i.e.,* that standards and standardization are being enclosed).
2. Examine the standards discourse and apply the aforementioned methods to a rhetorical analysis of central terms of the standardization discourse to establish internal definitions.

3. Examine specific representative cases to find how these terms are being applied in order to test the hypothesis.

Simply put, the aforementioned hypothesis is developed as follows: (1) enclosure asserts that ideas are being enclosed; (2) standards are ideas; and therefore, (3) some standards may be enclosed (or may be undergoing enclosure). The general hypothesis is:

*The public nature of standards as a commons is being challenged. Privatization for the benefit of specific interests may have deleterious consequences for society as a whole, including the impairment of access to information and communication networks, of competition, and of technical innovation.*

## THEORETICAL PERSPECTIVES

The theoretical perspectives that guide this study include primarily those of Habermas and his concepts of the *public sphere*, *communicative action*, and *practical discourse*; of *discourse as social practice*; and of *political economy* and certain closely related theories of institutional cultures. Also contributing both theoretically and methodologically to this study are theories of rhetoric, the concepts of semiotics and signification, and other aspects of communication theory.

### Habermas

Political philosopher Jürgen Habermas seeks to restore the Enlightenment rationalist tradition. In his theory of discourse ethics and of communicative action, he draws a sharp distinction between *strategic action* and *communicative action*. He emphasizes the concept of the *public sphere*, presumably the realm of communicative action and *discourse ethics*. He sees rhetoric as strategic action, a distorted form of communication.

In his model of, discourse ethics, Habermas (1995) finds a dialogically derived middle ground between (1) the universalistic and rationalistic, but monologic, Kantian ethics, and (2) the situational/contextual Aristotelian ethics which are based on social practice and pursuit of desired virtue. The essence of discourse ethics is that universal ethical principles and the relation between the justification and application of norms can be socially constructed from rational discourse based on language. In this model, Habermas extends the Enlightenment project, committing to the proposition that negotiated norms of truth and good can be determined through *consensus* and group will. This is especially so if procedural principles of rational discourse, which include what he calls "ideal speech," can be adhered to. These notions bear an interesting resemblance to the ideals embodied in the guiding principles of some forms of standards practice.

In his theory of communicative action, Habermas (1962) introduced these aforementioned ideas. He proposed to resolve the conflict between private and public autonomy (*i.e.*, individual rights *vs*. collective rights), which he attributes to individual and collective *subjectivity*, by introducing the concept of *intersubjectivity* whereby the conflicting subjective interests are worked out through a discursive formation of public opinion which recognizes the equal validity of public and private rights. Habermas measures democratic legitimacy not in terms of majority decision-making, but by the discursive quality of the process of deliberation that leads to such decisions. In standardization practice, in contrast to lawmaking and usual democratic governance, simple majority decision is not considered adequate. Only a *consensus* can result in a viable standard. In this sense, standardization practice is potentially a significant example of an attempt to apply communicative action within modern society.

Standardization thus provides an interesting case for testing Habermas' models of communicative action and of the public sphere. In embarking on this test, it is recognized that a fundamental assumption underlying democratic theory is that it offers a path to reach "better" decisions. It is acknowledged that this may be a problematic assumption, but this research relies on applying Habermas' arguments and his application of democratic theory to a difficult evaluation of the material and practical outcomes of discursive practice. That which is "better" is not always known or knowable because it depends on some desired end that may not be understood at the time—or ever. The world and history has no shortage of examples where democracy (in some form) may not have reached a "better" result than would have been obtained had some other system or method been deployed.[6]

Habermas's term for the rational discursive process by which normative disputes are resolved to mutual satisfaction is "practical discourse." He proposes three sets of rules that apply to three levels of argumentation on which practical discourse depends: (1) participants must speak the same natural language and adhere to the same general conventions, (2) participants must desire to reach agreement and must defend only what they believe to be true, and (3) participants must offer only arguments that command assent and not merely seek a desired behavior or outcome. In the third rule, he draws the distinction between *strategic* action, concerned only with external outcome (*e.g.*, possibly employing rhetoric, political influence, threats, bribes, coercion, *etc*.), and *communicative* action concerned with a prevailing "force of the better argument." Strategic action introduces distortions into the ideal speech process. Standardization practice has long sought to institutionalize rules of argumentation similar to Habermas's *ideal speech* and *practical discourse*, with varying measures of success. It should be noted that Habermas's reliance on ideal speech and rational argumentation as the primary model for practical discourse is contested. Hauser argues that "Habermas dampens the range of appearance for controversy—an inherent feature of all public spheres—by excluding rhetorical discourse in favor of the rationalistic argumentation of philosophy" (Hauser, 1999, p. 53). This study examines precisely this issue, and seeks to identify the rhetorical dimensions of standards discourse and explain how it may depart from a rationalistic idealized model. Farrell (1976, 1993) endeavors to extend Habermas's project to incorporate or consider elements of rhetorical discourse. Hauser's and Farrell's arguments in this regard will be considered further.

The formal standardization process today is constituted around a set of rules and procedures that essentially mirror Habermas's concept of ideal speech and his rules of argumentation. The operational directives of the international standards bodies and most national or regional bodies embody similar principles of fairness, adherence to procedures of due process, reasonable notice, openness, inclusion of stakeholders, and decision by consensus that are consistent with Habermas's model of ideal speech and practical discourse. In this sense, formal standardization practice has been characterized as potentially a form of revitalized public sphere within the context of modern technological/industrial society (Schoechle, 1995; Froomkin, 2002).

The aforedescribed concepts are further integrated and developed by Habermas in his more recent works including *Between Facts and Norms* (1996) where he examines the structure and processes of judicial, legislative, and political institutions from the philosophical perspective of discourse theory, ethics, the public sphere, and communication. These vital social institutions and processes are concerned with establishing what "is" and what "ought to be", as are the institutions and processes of standardization.

## The Public Sphere

The practical focal point of Habermas's work for purposes of exploring the consequences of the discourse on standards is grounded in his concept of the *public sphere*. The *public sphere*, as Habermas sees it, is that *conceptual public space where private individuals come together to engage in rational discourse on matters of mutual concern*. It is essentially the forum for discourse ethics and depends upon his principles of ideal speech, practical discourse, and rules of rational argumentation. Such principles include freedom of access, equal rights to participation, truthfulness on the part of participants, absence of coercion, *etc.* (Habermas, 1962, p. 31). Essentially, Habermas provides a central framework for modeling and evaluating the standards practice and its discourse, including *openness*. Inherent in the idea of the public sphere is the notion that, given the principles of ideal speech, reasonable individuals will be able to reach a common consensus or substantial agreement on issues of mutual interest. Inherent in this notion is a required willingness to be objective and to step outside of one's own tradition and frame of reference—a problematic assumption. Nevertheless, Habermas contributes a perspective of a *public* and a model of discursive practice that can be applied to standardization practice and is relevant to many of its underlying assumptions.

## Critiques and Extensions of Habermas

Fraser (1990), while recognizing the importance of Habermas' idea of the public sphere, challenges it as having too limited a conception of a *public* in several respects. She calls for a conception of the public sphere as a multiplicity of publics, asserting that Habermas' valorization of the bourgeois public sphere leaves out marginalized groups of society that ought to also be considered as publics. She also finds his use of the term *private* and *public* as somewhat problematic, or at least in need of clarification, as she identifies several possible meanings for both "publicity" and "privacy" (Fraser, 1990, p. 71). The problematic nature of the terms *public* and *private* is a recurring theme in standardization discourse, and one that is extensively explored in this study.

Hauser also raises the issue of multiple publics and public spheres, going farther than Fraser and introducing the term *reticulate* public sphere, a term referring to a networked or dispersed civic conversation occurring within and among multiple diverse social groups (1999, p. 76). This rhetorical model may have particular relevance in the transnational, multi-institutional, and diverse world of standardization practice. Peters (2002) offers a critique of Habermas's focus on rational discourse, asserting that *dialog* is not democratic but elitist because others must be quiet; that formality and procedure can obscure context; that forced discourse may not be productive; and that the act of leading discourse is not necessarily tyranny, but could be of service. Finally, an important critique of Habermas that is particularly useful in this study comes from Farrell (1993), as noted earlier, who proposed to extend Habermas's project to incorporate or accommodate elements of rhetorical discourse.

Farrell begins by noting that "Habermas' project on discourse ethics, together with his practical placement of the public sphere, offers a basis for synthesizing the normative component of practical wisdom and a rhetoric goaded by emancipatory reason" (p. 188). That being said, he then contends that rhetoric's contribution to coherence in discourse practice is not an accidental feature or a deformation of ordinary discursive practice. Rather he suggests that:

*...rhetoric is an inherent potential of any shared discourse practice. In making this suggestion, I am further distancing myself from specifics of the [Habermas] universal pragmatics position—in particular, its decontextualized privileging of cognitive-centered claims to truth in what Habermas calls "practical discourse" (1993, pp. 233-4).*

*This presupposition of modernity is essential to Habermas's entire orientation to communicative action. While a cooperative principle seems necessary for full reflection on validity claims, the fact remains that such reflection is in order because Habermas envisages communication as having to overcome a problematized distance between subjectivities.* (footnote 5, p. 346)

Farrell further notes that because the discourse or discussion is situated in a specific discursive context, utterances, unlike propositions, are contextually embedded episodes, and their privileged claims, if any, are to plausibility rather than to truth. Thus situated, rhetorically, interdependent discourse becomes a mediation among subjectivities, allowing emergent dimensions of conventional meaning that would otherwise be overlooked or subordinated (*ibid*).

Farrell sees that Habermas, in grounding his brand of pragmatics in universalizable presuppositions, has left the door open to an extension of communicative action into the territory of rhetoric.

*As with many of Habermas's labors, there is a stunning insight here...that the very background of our "taken-for-granteds" in the linguistic turn contains a potentially universalizable normative content. This discovery becomes the procedural scaffolding for the Habermas vision of communicative rationality.* (p. 190)

Finally, Farrell offers a model of discourse based, in part, on Habermas's grounding of the normative validity claims associated with three types of performative utterances: locutionary (communicative action), perlocutionary (strategic action), and illocutionary (symbolic action). As Habermas explains,

*It seems to me that strategic action (oriented to the actor's success—in general, modes of action that correspond to the utilitarian model of purposive-rational action) as well as (still insufficiently analyzed) symbolic action (e.g., a concert, a dance—in general, modes of action that are bound to non-propositional systems of symbolic expression) differ from communicative action in that individual validity claims are suspended (in strategic action, truthfulness; in symbolic action, truth).* (Farrell, 1993, p. 193)[7]

For the analytical purposes of this study, the locutionary, illocutionary, and perlocutionary distinction provides ways of singling out identifiable things that speakers do with the same utterance. This model will be developed further and applied to the standards discourse.

### Democratic Legitimacy

One theoretical perspective that merits mention here is one that is of central importance in the discourse on standardization—the notion of *democratic legitimacy.* Weber (1921-2) has proposed various types of *legitimate* authority, all of which rest basically upon *belief—i.e.,* an authority is legitimate if it is *perceived* or *believed* to be legitimate by those it governs. Habermas has developed this idea further, resting it upon practices of rational discourse and democratic deliberation in a free, open, and fully informed environment (1975, 1984). Chambers (1996) argues that it is often easier for people to agree on what action is right than on why it is right, and that the issue is most often resolved by agreeing on a process rather than on moral validity.

*The argument here is that what we mean by moral truth is highly contested but what we mean, generally speaking, by political justice and democratic legitimacy is not. A just system is one that regulates society fairly and in the general interest. The democratic legitimacy of a just system means that citizens freely consent that their institutions indeed do regulate society fairly and in the general interest.* (p. 144)

The global standardization system is particularly dependent on the discursive construction and perception of its legitimacy because often its institutions do not rely on direct government validation, particularly in the United States.

## Discourse as Social Practice

In recent times, the study of communication has increasingly challenged earlier objectivist epistemological assumptions about society and its communication processes. These assumptions and processes viewed discourse as data that could be observed, collected, and theorized independently of the observer or the content. Increasingly, communication is being re-conceptualized as a human social construction situated in and emerging from specific social practices. In this constructivist concept, the study of rhetoric assumes a role in finding a theoretical understanding of how such constructions are developed and how meaning is established. This conception of the constitutive nature of social practice is seen in its epistemological consequences as characterized by ethnographer Clifford Geertz:

*The concept of culture I espouse...is essentially a semiotic one. Believing, with Max Weber, that man is an animal suspended in webs of significance he himself has spun, I take culture to be those webs, and the analysis of it to be therefore not an experimental science in search of law but an interpretive one in search of meaning. It is explication I am after, construing social expressions on their surface enigmatically.* (1973, p. 5)

Another view of discourse as social practice is expressed by Burke (1950). One of his principal themes is the determinative role of language and symbols, or "how we are invented by our symbols." Burke has developed a notion of action and motion that finds form in his theory of dramatism and the *pentad.* His pentad is a method for visualizing humans as narrative actors in a drama, and he decomposes the drama into elements of purpose, act, scene, agent, and agency. Burke emphasizes the role of ambiguity and looks for the strategic spots at which ambiguity arises. It is through interpretation of such ambiguity that meaning is constructed by the players. He defines a "grammar of motives," (Burke, 1950) wherein motives are sets of rules of usage or the joining of words that lead to meaningful utterances and can be conceptualized as situations or *situated action* (Mills, 1940). As noted earlier, Burke proposes that *identification,* as an alternative to *persuasion*, builds on shared properties and creates new ones. He claims that we already have a multiplicity of identities and the task is to locate both identities and divisions. In summary, Burke's primary contribution to rhetoric is to shift from method to symbol, or, in other words, to move rhetoric from practical art to social practice. Farrell (1993) relies on Burke in bringing symbolic action into the realm of discursive practice and in attempting to extend Habermas's notion of communicative action to include forms of rhetoric (p. 202). The result is the creation of an audience-centered approach to establish a method of invention wherein everyone is a potential rhetor. Standardization evolved as a social practice, and Burke's approach is useful in understanding its discourse, particularly on the lower technical committee or working group level in terms of *identification* and *situated action.*

An account of the constitutive possibilities of discourse would not be complete without mentioning Mikhail Bakhtin. Bakhtin sees human experience entirely in social terms, whereby consciousness is based on symbolic meaning derived through individual and group experience specifically situated in time and space. Rejecting psychological or cognitive approaches in favor of a sociological one, he defines consciousness within a Saussurian semiotic context of *signs*, *myths,* and *ideologies.* Ideological signs, including common sense, are seen not only as reflections of reality, but also as part of the reality itself. Bakhtin notes that:

*Every phenomena functioning as an ideological sign has some kind of material embodiment, whether in sound, physical mass, color, movements of the body, or the like. In this sense, the reality of the sign is fully objective and lends itself to a unitary, monistic, objective method of study. A sign is a phenomena of the external world.* (Bizzell and Herzberg, 1990, p. 929)

In this sense, both the real world and the individual consciousness of it are comprehensible as products rather than sources of language and signification. To Bakhtin, the reality of these signs creates a chain of ideological creativity and understanding, moving from sign to sign and then to a new sign. He continues,

*This ideological chain stretches from individual consciousness to individual consciousness, connecting them together. Signs emerge, after all, only in the process of interaction between one individual and another. And the individual consciousness is itself filled with signs. Consciousness becomes consciousness only once it has been filled with ideological (semiotic) content, consequently, only in the process of social interaction.* (p. 929)

Bakhtin faults philosophy and psychology for localizing consciousness and ideology within the individual, asserting that taken apart from society, consciousness itself becomes a fiction and an improper ideological construct. Within the individual, language or "the word" is the semiotic material of inner life—the medium of consciousness—and it is created only through social interaction. In a sense, Bakhtin goes beyond Burke's later project toward a materialist view of rhetoric, continuing the shift from *method* to *symbol* and he places the audience at the center, constituting a reality. Farrell calls on Bakhtin in his project to bring both conversation (dialectic) and rhetoric into the realm of communication practice (Farrell, 1993, p. 233).

Michel Foucault carries the notion of discourse as social practice even further, seeing history, or social knowledge, as a fragmented narrative that he prefers to call archaeology (Foucault, 1973). This narrative constructed by the writer reflects his or her own institutionalized perspective. Foucault sees discourse as *social practice* and *situated statements*, rather than as texts subject to "interpretation," and as such can only be understood within the context of its *institutionally situated* paradigm. He emphasizes the constitutive nature of discourse and contends that every society is the production of discourse and controlled by discourse. To Foucault, the study of social practice and social will is the study of discourse.

In considering discourse as a social practice, it is important to include Aristotle's *emthymeme*, the syllogism the major premise of which is a popular belief and the intermingling of instrumental and constitutive rhetoric. Edwin Black has suggested that enthymematic discourse "...trades upon a body of settled convictions attitudes and beliefs," (Black, 1965, p. 29) and establishes a *frame of reference*—a *reality* or a paradigm—shared by a body of readers or audience. Enthymematic discourse "...does not proceed argumentatively. Rather its traffic is realities. It does not argue, it describes." Such discourse invokes myths and ideologies by using metaphors and *topos* (commonplaces) that constitute discursive "idiomatic tokens of ideology" (Black, 1970). From a critical examination of such discourse and its style, it is possible to discern a real or implied auditor. That "...auditor implied by the discourse what the rhetor would have his real auditor become" (Black, 1970, p. 113). Thus, it is possible to reveal the *moral character* of the discourse, *i.e.,* the moral beliefs of its authors. Such a perspective is essential to fully making sense of much of the discourse around standardization. Unstated conceptions of reality and moral beliefs about openness, balanced participation, the role of government, and about the role of the market play a major role in the standards discourse.

## Theories of Rhetoric

It is useful to briefly consider traditional *instrumental* views of rhetoric because they provide a contrast to the view emphasized in this study, and because they contribute an insight into how instrumental discourse can become *constitutive* when situated in deliberative and dynamic social circumstances such as standards practice. The ancient Greeks considered rhetoric to be instrumental. To the Sophists, rhetoric was a method of persuasion and content was unimportant. They considered the use of language in heuristic terms and took a playful or experimental attitude toward its use.

Plato objected to such abuse, seeing rhetoric in dialectic terms, as a method of searching for truth. This conflicting view is effectively portrayed in Plato's dialog, *Gorgias*. Although, to Plato, rhetoric was about persuasion, his theory of rhetoric draws a distinction between knowledge and belief, and holds the highest purpose to be the pursuit of truth (good). The persuasive power of rhetoric brings the persuader and the persuaded together, either by persuasion-to-belief (bad rhetoric) or by persuasion-to-knowledge (good rhetoric). If focused on knowledge, rhetoric combats convention/conformance rather than exploiting it, educates rather than satisfies base desires, raises the student/reader to the level of the teacher/writer rather than lowering the teacher to the level of the student, and seeks truth rather than simply conformity or influence.

More specifically, the method of Plato's dialog employs dialectic, the pursuit of truth not simply by articulating a truth position but by approximating or approaching a truth position (concept) by responsive and iterative interlocution or conversation (Farrell, 1993, p. 248) between multiple speakers. They are engaged in a "rite of speech" which adheres to certain principles of sincerity of purpose in search of truth and not merely for the purpose of persuasion alone. In a sense, the discourse practiced in traditional standards committees, especially on the lowest (technical) level, could be seen as striving for this ideal. Farrell discusses "conversational coherence" and applies Habermas and Burke in bringing dialectic into the realm of communicative practice (1993, p. 248).

Aristotle took a much different, although instrumental, view of rhetoric. To Aristotle, rhetoric is not the act or purpose of persuading, but the process of "finding the means of persuasion" in a specific situation. To Aristotle, the discourse is always situated in some specific context. Aristotle sought, through analytical empirical method, to learn the science, or *techne,* of such persuasion and to establish general principles by which the process could be understood. Aristotle relied on invention, dialectic, and knowledge (sensing). His dialectic was different than Plato's, being more focused on process rather than on truth, and so can be viewed as a form of critical thinking. As part of his pragmatic toolkit, Aristotle identified a set of 26 *topoi,* or stock arguments, employing rhetorical syllogisms or enthymemes (Kennedy, 1991).

The enthymeme, or truncated syllogism, is a central device in Aristotle's *techne* of rhetoric. This syllogism is truncated or incomplete because rather than explicitly stating its premises, it deliberately omits an important premise, thus allowing or inviting the audience to provide it. In the process of providing the missing premise, the audience participates in the speaker's project and is induced to "buy in" to the rest of the speaker's assumptions or purposes, thus forming a certain communication link between the speaker and audience. Such engagement of the audience's pre-existing belief structure (perhaps part of its identity) begins to build a common consciousness with the speaker. This method begins to resemble Burke's constitutive *consubstantiality* and Althusser's *interpellation.* Analysis of such enthymemes and identifying their unstated assumptions provide an important and useful method of analysis in approaching the discourse on standardization. Farrell develops Aristotle's enthymeme, maintaining that, "[with]...the moderating reflective tendencies within rhetoric...it is the enthymeme, as a middle-range[8] inferential prototype of rhetorical practice, that allows conduct to become obligatory through an intertwinement of situated interests and perspectives" (p. 232).

## Constitutive Rhetoric

Theories of rhetoric provide a rich method of analysis for understanding the discourse on standardization. Some of these have already been mentioned. Rhetoric has a long tradition, beginning with an "instrumental" view of persuasion and evolving to a later "constitutive view" of identification. Very often these views are intertwined and both can be applied at the same time. Of some interest here is the concept of constitutive rhetoric, variously known as *interpellation* or *consubstantiality*.

Traditional instrumental rhetoric sees discourse largely as a listener or audience being persuaded by a speaker, and assumes that such listeners or audiences are extra-rhetorical or passive and outside the rhetorical process (Burke, 1950). The constitutive view, in contrast, regards the audience as an active participant, embedded spontaneously, intuitively, and unconsciously in the rhetorical process. This view describes a means of bringing the audience into an ideological framework through the process of *interpellation*, a term developed by the French political philosopher Louis Althusser (1971, p. 170). Interpellation is a process of bringing something (*e.g.,* an audience, or a cultural or social entity) into being by simply "calling" or "hailing."

Interpellation is accomplished in the case of the standards discourse by projecting a message at the audience that identifies some *common history* and *purpose* with an additional narrative containing other assumptions about their history, identity, and destiny. Those hearing the message find their own identity embedded in part of the message, but cannot clearly separate out the additional assumptions, and thus tend to adopt the entire identity package. Thus they are "interpellated" as political subjects through a process of identification in rhetorical narratives that "always already" presume the constitution of subjects. They are not *persuaded* in the usual sense, because belief is inherent in the subject position of the rhetoric. In Althusser's terms, the process is an ontological one. The subject is not *persuaded* to be a subject; he or she is always already a subject, by simply hearing the message. This process is very similar to that described by the Burke (1950, p. 50). It works by offering a "consubstantiality" or identity between the would-be subjects. Such subjects may include the participants in the standards process, a public constituency, or those seeking to merge it with a specific political or economic agenda. Charland (1987) has applied Burke's move to identification in an ideological context. He comments that "A consequence of this theoretical move is that it permits an understanding within rhetorical theory of ideological discourse, of the discourse as always only pointing to the given, to the natural, the already agreed upon" (p. 133). Examples of such ideological discourse that are a focus of interest in this study will be about the *free market*, the *sectoral approach*, and *compatibility*.

Simply put, rhetoric can be seen as *constitutive* and incorporated into *social practice*, rather than as a *method of argumentation* that addresses an objective and external audience in the traditional or classical sense. It can also be argued that constitutive rhetoric is also instrumental because it is so effective.

## Theory of Signification and Semiotics

An important and basic method of analysis employed in this study is broadly called *semiotics* or theory of *signification*. Beginning with Saussure (1916) and Barthes (1973) and then extended by Fiske and Hartley (1978), a useful hierarchy has been established consisting of three orders of signification that may be applied to the standards discourse: (1) sign, (2) myth, and (3) ideology.

The first order of signification establishes a "sign" that is constructed by associating a symbol or "signifier" with some meaning or "signified." Sometimes called denotation, it establishes the most basic symbols and signs that constitute the terms of discourse, words or basic procedures through which meaning is conveyed. This order of signification includes the least value-laden terms, including, committee, working group, study group, secretariat, convener, and a vast array of acronyms and other standards jargon.[9]

The second order of signification is constructed by combining first order signs, (which then become signifiers) with additional meaning or other signs to establish a new higher level of connotative meaning. This new sign is known as a *myth*[10] in which cultural images or meanings are packaged and conveyed. Myths encapsulate a set of meanings that reflect the value system of the culture of the person employing them. In the case of standardization, this second order sign can include not only words but also the basic discursive practices of the standards process (*e.g.*, how committees are constituted, organized, and convened). Examples of second order signification found in standardization discourse may include terms such as free, open, balanced, public, private, and SDO.[11] Standards groups and practices are mythologized in a variety of ways that include or employ rules, procedures, organizational structures, administrative or leadership hierarchies, expectations, and limitations.

The third order of signification, *ideology*, combines myths as signifiers to form yet another level of coherent pattern of meaning or belief that constitutes underlying, invisible organizing principles. This order of signification can also include the institutional frameworks within which standardization proceeds, including their economic, social and political contexts. Examples of such ideological signifiers in the present context of standardization might include such strongly value-laden terms as openness, free market, democratic, sectoral, privatization, and globalization.[12]

## Political Economy and Related Theory

A number of scholars have applied methods of discourse analysis to communication policy, often with a theoretical focus on a political economic analysis of the outcomes. Calabrese (1991) has pointed out the problematic nature of terms such as *community* and analyzed the way certain *communities of interest* are discursively constructed.[13] Wagner (1999, 1999a) has investigated the discourse surrounding ICT policy in developing Asian economies from a political economic perspective. Rowland (1986) has examined the role of *discourse*, *myth* and *ideological* perspectives in communication policy research. Streeter has studied communication policy (1990) and, earlier, the history and discursive construction of the early cable television industry (1986) and the role of *technocracy*. This term may be applied critically to some standards bodies or consortia, or to those who would apply and/or justify the results of their work:

*...they tend to involve a continual striving, on the part of a limited and relatively privileged group of people, for a Solution, a Truth, a bedrock of certainty that can then be used to justify making decisions that affect a much larger group of people without embarking on the uncertain and troublesome process of involving the larger group in making those decisions.* (Streeter, 1986, p. 207)

Theories and discourse of political economy are relevant to this study because the discourse on standardization is embedded in a material context. This context is that of policy actions that lock it into economic and political institutions that are constituted by the discourse.

Historical institutionalization, although not the central methodology employed in the present study, does provide valuable insight, especially where it overlaps discourse. Wagner (1999) has studied the role of discursive practice in policy formation from the perspectives of institutionalization established by Veblen (1919) and by Innis (1972). He draws a distinction between the "old" institutionalist school, which emphasized the social entities that provided legal and practical bases for economic activity, and the "new" institutionalist school, which concentrates on firms, markets, communication and transportation systems, and government (p. 95). He points out that the former emphasizes the contingent and culturally constructed nature of information, while the latter emphasizes individual optimization. The former concept is more useful to this study because it is the institutionalized codes of conduct, norms of behavior, and conventions that constitute discursive practice and reveal meaning in standardization. This stands in contrast to the utilitarian or "rational choice" economic models applied by the latter mentioned "new" institutionalism. Wagner notes that the "old" institutionalist method and its inherent anti-teleological ontology (or lack of a predetermined end point or *telos*) renders precise predictions problematic, but does allow a level of objectivity and forecasting, however, because of the tendency for discursive practices or habits to "lock-in" and to become institutions (p. 98). Wagner concludes, with particular relevance to the present study, that the "...kinds of confusion about public and private networks, individual and corporate users of technology, and individual versus collective rights and interests, are not accidental constructions but the consequence of historical predominance of powerful classes within the institutional and discursive realms." (p. 103-4)

Other closely related institutional perspectives that have influenced this study include those put forth by Mansell (Mansell, 1993; 1995; Mansell & Silverstone, 1996; Hawkins, Mansell, & Skea, 1995), Hawkins (Hawkins, Mansell and Skea, 1995), and Feenberg (1991) on the political economy of policy and standards; Bijker (1993, 1999) and Williams (1975) on the social construction of technology; and, of course, Carey (1989) and Schiller (1996, 2000) in communication studies.

## RESEARCH QUESTIONS

The enclosure discourse and its previously mentioned authors, Lessig (1999, 2001, 2001a, 2002), Boyle (1996), Litman (1990), Benkler (1998, 1998a, 2001, 2001a, 2002), and others, have been primarily concerned about the *openness* of *outcomes*—the writings, compositions, inventions, and other works. However, the creative processes of standardization by which many such outcomes are determined have not been examined closely in the context of their arguments. That is the question being explored in the present work by looking specifically at the policy discourses in which the technical standards for ICT are decided, including discourses on the institutional structures and their legitimation. The research questions are: *How public has standardization been in the past and how does that compare with the standardization process of today. What is meant by terms such as "public," "private," and "open," and how are their meanings constructed and applied? If the process is now undergoing increasing enclosure, what are the roots of such enclosure? What are the responses to arguments that enclosure is occurring? What are the institutional responses to the perception of enclosure?*

This study seeks to explain the concept of standards in the context of American and global industrial policy (*e.g.,* what are "standards"? What is "open"?). It then centers the study around an examination of the discourse on standards and considers what enclosure might imply for the future of *open standards.* Next, it considers specific case studies for their relevance to the enclosure hypothesis. Finally, it examines some of the institutional responses and proposed remedies to perceptions of enclosure.

## CONCLUSION

These theoretical perspectives of Habermas and his concepts of discourse, communicative action, and the *public sphere*, of standardization as a *social practice*, and of *political economy* combine to provide a framework for analysis of the standards discourse, the object of analysis for this study. These theoretical perspectives will be applied together, each providing complementary elements of a framework for interpretation of the discourse on standards and standardization—the research "data." For instance, in looking at a specific discourse or practice, the public sphere theory provides a grounding or reference model, the social practice perspective provides an understanding of the linguistic and rhetorical elements of the discourse and how they constitute a reality, and the perspective of political economy provides a way to find the underlying motivations. The enclosure discourse and the research questions identify the specific problematic discursive constructions to look for. This analysis applies the method of discourse analysis and historiography to find the various constructions and meanings embedded in the discourse. The overall purpose here is to provide a way of thinking about standards and standardization.

Research and analysis of standards and standardization discourse is important. The abstract metaphor of discourse and the discursive practices treated above are ultimately reduced to reality through the standards committee process. Perceptions influence policy decisions. Such decisions have material consequences for what is or is not developed (*e.g.*, technologies, standards, intellectual property, innovation, *etc.*) and for who gets access to or control of what (*e.g.*, networks, information, resources). In summary, *the standards discourse contains definitions or meanings. These meanings drive and justify policy decisions with material consequences. By examining the discourse, the meanings and their origins can be established. The final result is greater clarity with respect to the issues and the interests being served or not served. Better policy choices may then be possible.*

The standards discourse is conducted within specific cultural, disciplinary, and institutional paradigms (*e.g.*, science, engineering, business, economics, *etc.*) within which various signs, myths and ideologies shape discourse, thought and action. The power of culture in technical organizations has been recognized and described by Thomas Kuhn (1962) as the "scientific paradigm," concept that is particularly relevant. Standardization bodies and the industrial and governmental groups that constitute and convene them form an institutional context or culture within which certain mythological and ideological constructions are constituted, maintained, and embedded in policy decisions. Wagner has found a very similar situation in other areas of technical and industrial policymaking.

> *...it has been widely recognized since Kuhn's seminal work that adherents to particular disciplinary paradigms maintain orthodoxy within their ranks by means of consensus that are as much sociological as scientific. Disciplinary and ideological boundaries are hence [constituted and] maintained by virtue of discursive practices; tacit and unexamined agreement among members of "epistemic communities," concerning what is properly to be considered valid argument and conclusion. Paradigmatic divisions are further constructed and exacerbated by the selection and promotion of research and ideas by powerful institutions that have a high degree of interest in particular policy outcomes, and in promoting the paradigms of knowledge and discourse that tend to support those outcomes.* (Wagner, 1999a, p. 2)

The discourse of standards and standardization can be understood only in the context of the institutions wherein it is practiced. The discourse is entwined with the social, cultural, economic, and political institutions and practices within which real people and real structures are involved.

From the preceding general historical background, and with the aid of the literature and discourse on enclosure and on standardization that has been presented, this study proceeds to apply the foregoing theoretical perspectives and methodological approaches to the specific standards institutions, discourses, and issues with which they deal. The next chapter will examine how the discourse was situated and observed (*i.e.*, the field research method) and will establish the essential terms of the discourse and their meanings. The ensuing chapter will then apply such meanings in examining the traditional institutions and practices that form the foundation of the present global standardization system.

## REFERENCES

Althusser, Louis (1971). Ideology and Ideological State Apparatuses. In *Lenin and Philosophy and Other Essays.* London: New Left. , p. 170)

Barthes, Roland (1972). *Mythologies.* New York: Hill & Wang (c. 1957, reprint 1972).

Benkler, Yochai (1998). Communications Infrastructure Regulation and the Distribution of Control Over Content. *Telecommmunications Policy, 22,* p. 183.

Benkler, Yochai (1998a). Overcoming Agoraphobia: Building the Commons of the Digitally Networked Environment. *Harvard Journal of Law and Technology, 11,* pp. 287-290.

Benkler, Yochai (2001). Coase's Penguin, or, Linux and the Nature of the Firm. *29th Telecom Policy Research Conference*. Alexandria, VA. September 24, 2001. <http://www.arxiv.org/abs/cs.CY/0109077>

Benkler, Yochai (2001a). Through the Looking Glass: Alice and the Constitutional Foundations of the Public Domain. *Conference on the Public Domain*, Duke Law School, Durham, NC, November 9-11, 2001.

Benkler, Yochai (2002). Intellectual Property and the Organization of Information Production. *International Review of Law & Economics, 22*(1), pp. 81-107.

Bijker, Wiebe E. (1993, winter). Do Not Despair: There Is Life After Constructivism. *Science, Technology & Human Values, 18*(1), p. 119.

Bijker, Wiebe E. (1999). *Of Bicycles, Bakelites, and Bulbs: Toward a Theory of Sociotechnical Change*. Cambridge: MIT Press.

Bizzell, Patricia, and Bruce Herzberg (Eds.) (1990). *The Rhetorical Tradition: Readings from Classical Times to Present*. Boston: St. Martin's Press.

Black, Edwin (1965). Frame of Reference in Rhetoric and Fiction. In Donald C. Bryant (ed.). *Rhetoric and Poetic*. Iowa City: University of Iowa Press. pp. 26-35.

Black, Edwin (1970, April). The Second Persona. *The Quarterly Journal of Speech, 56*(2), pp. 109-119.

Bourdieu, Pierre (1990). *The Logic of Practice*. trans. by R. Nice. Stanford CA: Stanford University Press.

Boyle, James (1996). *Shamans, Software and Spleens*. Cambridge: Harvard University Press. p. 226.

Burke, Kenneth (1950) *(1969). Rhetoric of Motives*. Berkeley: University of California Press.

Calabrese, Andrew (1991). The Periphery in the Center: The Information Age and the 'Good Life' in Rural America. *Gazette*. vol. 48. pp. 105-128.

Carey, James W. (1989). *Communication As Culture: Essays on Media and Society*. Boston: Unwin Hyman.

Chambers, Simone (1996). *Reasonable Democracy: Jürgen Habermas and the Politics of Discourse*. Ithaca: Cornell University Press. 250 pp.

Charland, Maurice (1987, May). Constitutive Rhetoric: The Case of the *Peuple Québécois. The Quarterly Journal of Speech 73*,(2), pp. 133-150.

Craig, Robert T. (1996). Practical Theory: A Reply to Sandelands. *Journal for the Theory of Social Behavior 26*, (1), pp. 65-79.

Farrell, Thomas B. (1976, February). Knowledge, Consensus, and Rhetorical Theory. *The Quarterly Journal of Speech, 62*(1), pp. 1-14.

Farrell, Thomas B. (1993). *Norms of Rhetorical Culture*. New Haven: Yale University Press. 373 pp.

Feenberg, Andrew (1991). *Critical Theory of Technology*. Oxford: Oxford University Press.

Fiske, John, and John Hartley (1978). *Reading Television*. London: Methuen.

Foucault, Michel (1973). *The Birth of the Clinic: An Archaeology of Medical Perception,* Trans. by, A.M. Sheridan Smith. New York: Random House.

Fraser, Nancy (1990). Rethinking the Public Sphere: A Contribution to the Critique of Actually Existing Democracy. *Social Text, 25-26*, pp. 56-80.

Froomkin, A. Michael (2002). *Habermas@discourse.net: Towards a Critical Theory of Cyberspace*. Draft paper. University of Miami School of Law.

Geertz, Clifford (1973). *The Interpretation of Cultures: Selected Essays*. Boston: Basic Books.

Habermas, Jürgen (1962) *(1991). The Structural Transformation of the Public Sphere: An Inquiry into a Category of Bourgeois Society* (Thomas Burger, Trans.). Cambridge: MIT Press.

Habermas, Jürgen (1975). *Legitimation Crisis* (Thomas McCarthy, Trans.). Boston: Beacon Press.

Habermas, Jürgen (1984). *The Theory of Communicative Action*. vol. 1, *Reason and the Rationalization of Society*. vol. 2. *Lifeworld and System: A Critique of Functionalist Reason* (Thomas McCarthy, Trans.). Boston: Beacon Press.

Habermas, Jürgen (1995). *Justification and Application: Remarks on Discourse Ethics* (Ciaran Cronin, Trans.). Cambridge, MA: MIT Press. (portions originally published in 1990 and 1991).

Habermas, Jürgen (1996). *Between Facts and Norms: Contributions to a Discourse Theory of Law and Democracy* (William Rehg, Trans.). Cambridge, MA: MIT Press. 630 pp.

Hartley, John (1994). In John Fiske (ed.). *Key Concepts in Communication and Cultural Studies*. 2nd Edition. New York: Routledge. pp. 92-95.

Hauser, Gerard A. (1999). *Vernacular Voices: The Rhetoric of Publics and Public Spheres.* Columbia, SC: University of South Carolina Press.

Hawkins, Richard, Robin Mansell, and Jim Skea (Eds.) (1995). *Standards, Innovation and Competitiveness: The Politics and Economics of Standards in Natural and Technical Environments.* Brookfield, VT: Edward Elgar Publishing Ltd.

Innis, Harold A. (1972). *Empire and Communication.* Toronto: University of Toronto Press.

Khanna, Parag (2008). *The Second World: Empires and Influence in the New Global Order.* New York: Random House.

Kennedy, George (1991). *Aristotle on Rhetoric: A Theory of Civic Discourse.* New York: Oxford University Press. pp. 190-215.

Kuhn, Thomas S. (1962). *The Structure of Scientific Revolutions.* Chicago: University of Chicago Press.

Lessig, Lawrence (1999). *Code and Other Laws of Cyberspace.* New York: Basic Books. pp. 297.

Lessig, Lawrence (2001). *The Future of Ideas: The Fate of the Commons in a Connected World.* New York: Random House. pp. 352

Lessig, Lawrence (2001a, November/December). The Internet Under Siege. *Foreign Policy.*

Lessig, Lawrence (2002). Remarks at the conference, *The Regulation of Information Platforms*, Silicon Flatirons Telecommunication Program. University of Colorado, Boulder, CO. January 27-28, 2002.

Litman, Jessica (1990, fall). "The Public Domain." *Emory Law Journal, 39*, p. 965. Retrieved from <http://www.law.wayne.edu/litman>

MacIntyre, Alasdair (1981). *After Virtue.* Notre Dame: University of Notre Dame Press.

Mansell, Robin (1993). *The New Telecommunications: A Political Economy of Network Evolution.* London: Sage Publications Ltd.

Mansell, Robin (1995). Standards, Industrial Policy and Innovation. In Hawkins, Richard, Robin Mansell, and Jim Skea (Eds.). *Standards, Innovation and Competitiveness: The Politics and Economics of Standards in Natural and Technical Environments.* Brookfield, VT: Edward Elgar Publishing Ltd., pp. 213-227.

Mansell, Robin, and Roger Silverstone (Eds.) (1996). *Communication by Design: The Politics of Information and Communication Technologies.* Oxford: Oxford University Press.

Mills, C. Wright (1940, December). Situated Action and Vocabularies of Motive. *American Sociological Review.* pp. 904-913.

Peters, John Durham (2002). The Problem of Conversation in Media. *Josephine Jones Lecture.* Department of Communication. University of Colorado, Boulder, CO. October 22, 2002.

Rowland, Willard D., Jr. (1986, April). American Telecommunications Policy Research: Its Contradictory Origins and Influences. *Media, Culture & Society.*

Saussure, Ferdinand de (1916) *(1974). Course in General Linguistics.* London: Fontana.

Schiller, Dan (1996). *Theorizing Communication: A History.* Oxford: Oxford University Press.

Schiller, Dan (2000). *Digital Capitalism: Networking the Global Market System.* Cambridge: MIT Press.

Schoechle, Timothy (1995, April). The Emerging Role of Standards Bodies in the Formation of Public Policy. *IEEE Standards Bearer, 9*(2), pp. 1, 10. Retrieved from <http://www.acm.org/pubs/articles/proceedings/cas/332186/p255-schoechle/>

Streeter, Thomas (1986). *Technocracy and Television: Discourse, Policy, Politics and the Making of Cable Television.* unpublished doctoral dissertation. University of Illinois at Urbana-Champaign.

Streeter, Thomas (1990, spring). Beyond Freedom of Speech and the Public Interest: The Relevance of Critical Legal Studies of Communication Policy. *Journal of Communication.*

Veblen, Theodore (1919) *(1990).* The Place of Science in Modern Civilization. In Veblen, T. B., *The Place of Science in Modern Civilization and Other Essays.* New Brunswick: Transaction.

Wagner, Douglas K. (1999). *Discourse and Policy in the Philippines: The Construction and Implementation of a Strategic IT Plan.* Unpublished doctoral dissertation, University of Colorado, Boulder, CO.

Wagner, Douglas K. (1999a). Embracing Incommensurate Paradigms of Technology and Development: Strategic IT Planning for Developing States. *The 49th Annual*

*Conference of the International Communication Association,* San Francisco, May 27-31, 1999.

Weber, Max (1921-2). *Economy and Society: An Outline of Interpretive Sociology.* 2 vols. G. Roth and C. Whittich (eds.), (1978). Berkeley and New York: Bedminister Press.

Williams, Raymond (1975). *Television: Technology and Cultural Form,* New York: Schocken Books.

## ENDNOTES

1. Discourse analysis is operationally defined here as including observing, transcribing, listening for particulars, selecting excerpts, documenting inferences, linking to scholarly controversies, and identifying or building insightful central claims.
2. Standardization practice could also be seen as being nested within another practice, that of international agreement, policymaking, negotiation, trade, diplomatic protocol, and rhetoric.
3. MacIntyre defines practice as any coherent, complex form of socially established cooperative human activity through which goods internal to that form of activity are realized in the course of trying to achieve standards of excellence appropriate to, and partially definitive of, that form of activity, with the result that human powers to achieve excellence, and human conceptions of the ends and goods involved, are systematically extended (1981, pp. 169-209).
4. The standards themselves and their creation comprises yet another discourse worthy of study, but beyond the scope of the present study. The focus here is on the meta-discourse—*i.e.*, the discourse about the discourse of standards.
5. Other such problematic terms to be considered include "sector," private sector/public sector, sectoral approach, free market, capitalism, democracy, standard, legitimate, accredited, strategic, compatibility, and platform.
6. Khanna (2008) points out the limitations of concepts western liberalism and rhetoric about "democracy" and "freedom" from the perspective of the developing world, commenting that "Liberalism insists on its own universality, but its application across space and time shows that it works best where it already exists" (p. 264). Further observations will be made in the Conclusion.
7. Farrell cites from the chapter, "What is Universal Pragmatics?" (Habermas, 1988, p. 41).
8. By "middle-range," Farrell means between extremes of logical argumentation (*i.e.,* Habermas) and strategic action using belief or ideology.

[9] Additional examples of first order signs include, rapporteur, convener, chairman, draft, contribution, recommendation, and ballot.

[10] It is important to note that the term *myth* is not used here in the layman's sense of "false belief," but as a construct of meaning without regard to assessment of validity or merit.

[11] Additional examples of second order signs include, *consensus, voluntary, market, trade, de facto, expert, standard, SRO, MDC, etc.*

[12] Additional examples of third order signs include development, capitalism, regulation, deregulation, open processes, voluntary consensus, consortium, and relevance.

[13] The term *community* has been appropriated by Sun Microsystems in constituting the *Java Community Process,* a form of standardization consortium. Kuhn (1962) emphasizes the notion of *community* in the establishment of a *paradigm*: "A paradigm governs...not a subject matter, but a group [community] of practitioners."

# Chapter V
# Situated Discourse and Essential Terms

*...communication is a symbolic process whereby reality is produced, maintained, repaired, and transformed.*

—James Carey, 1989

## INTRODUCTION

The research for this work is based on the thesis that standardization is moving from public formal bodies to more private consortia. A relevant question is why? Another basic question is what do these key words mean? In any case, the approach taken in this study is analysis of historical and current discourses about standardization from the perspective of a separate discourse about *privatization*, the *enclosure* discourse. The latter discourse was examined in the previous chapters to establish a perspective for analysis. This chapter seeks to identify the essential terms of the discourse of standards and standardization and to establish meanings. First, however, it is useful to briefly note the way in which much of the current standards discourse is situated and how it has been researched.

## FIELD RESEARCH METHOD

Although this is not an ethnographic study, the research method used was inspired, by Geertz's classic account of the *Balinese cockfight* (Geertz, 1973, pp. 412-453) [1] There may be some similarity between standards committee meetings and cockfights, and the cultural dimensions are no less complex and determinative. As an anthropologist, Geertz sought to gain an understanding of cultural meaning by participating in specific social practices, and by doing so by becoming somewhat of an "insider." By taking the same risks as local villagers, rather than remaining detached, he was able to gain acceptance by, and access to, the participants and their practices. In this Geertz was able to discern the cultural meanings involved in various rituals and activities. The Balinese cockfight is a highly ritualized competition that gives insights into the Balinese culture. My approach here is somewhat similar, having been already accepted as a standards "warrior," an insider who works on standards, administers committee activities, and struggles in the same highly ritualized battles as those that I study.

Much of the discourse discussed in this work was acquired by attending academic and industry conferences and actual standards meetings in which many of the principal actors in the discourse participated. This includes seventeen separate gatherings taking place over a three-year period. A list of these gatherings is provided in the Appendix. Since the author is personally known to many of the participants, it proved to be fruitful to rely on informal, unstructured interviews and discussions in hallways over coffee, around dinner or luncheon tables, and other similar situations. Using this method, the artificiality of structured questionnaires or interviews was absent and did not tilt the answers or inhibit voluntary comments. Although it became common knowledge among the participants at these gatherings that the author was engaged in academic research on standardization, it was also known that the author was attending not solely for research purposes. The author often spoke on, or was a member on, panels dealing with the topic of standards research, or actually was engaged in setting standards. This posture created a fruitful and candid atmosphere that minimized the distortion or biasing of the research data because the researcher was also an insider, a *bona fide* participant. What such an insider perspective provided was access to key elements of the discourse, both verbal and written, and the ability to contextually assess their relative importance in ways that an outside observer might find difficult. The author was able to make effective use of his own knowledge, familiarity, and contacts to get past many of the barriers that exist for outsiders attempting such a study. An example might be candid discussions between the author and participant(s) that would not have been undertaken with an outsider.

## ESSENTIAL TERMS OF THE DISCOURSE

Many terms of the discourse on standards and their usage will emerge in the research below, but several deserve special focus and some initial explanation. Of particular importance are: *public* and *private*, *public sector* and *private sector*, *sectoral,* and, most important of all, *open.*[2]

### Public and Private

The problematic nature of the terms *public* and *private* has been discussed in earlier chapters. The word *public* is usually defined as pertaining to the people as a whole on some unspecified scale (*i.e.,* community, city, state, nation, *etc.*) (Random House, 1967, p. 1162). However, in its successive definitions, *public* comes to refer to *open to* the people, *acting for* the people, *representative of* the people, the *affairs of* the people, and then finally *government.* This gradation or spectrum of meaning is depicted in Figure 5.1.

*Private* begins from the opposite end: the *individual* person, then progressing to *a small group* of persons, to *limited groups or access*, to *not representative* or *not public*, and ultimately to *not government.* A similar spectrum of meaning is depicted in Figure 5.2.

Thus, the meaning of these terms becomes highly contextual and also depends on the direction from which one approaches the terms. In standards discourse, it is not uncommon to find these words, *public* and/or *private,* used in different senses within the same document or even within the same paragraph. For instance, when *public* is directly juxtaposed with *private*, or when the word *sector* is used,

*Figure 5.1. Spectrum of meaning for public*

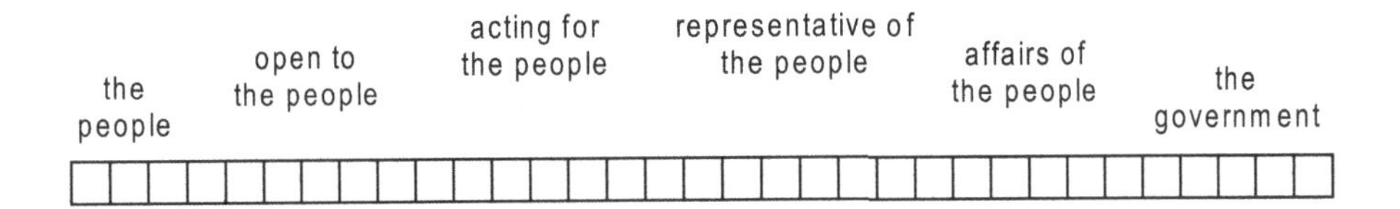


*Figure 5.2. Spectrum of meaning for private*

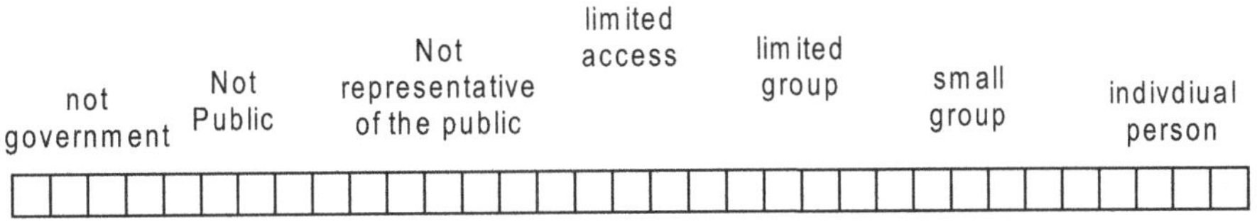

not government
Not Public
Not representative of the public
limited access
limited group
small group
indivdiual person

it signifies *government* and *non-government*, as in the usage *public sector*[3] *versus private sector*. This usage is contrasted with terminology such as *public access, public availability, public forum, public opinion*, or *public exposure*, which does not signify government, but rather multiple private parties considered together as a *public*. This latter case is more the sense of Habermas's *public sphere*.

Habermas traces the evolution of the *bourgeois public sphere* from the Enlightenment period of the 17th and 18th centuries. He sees it as a space that was related to the notion of *civil society,* as distinct from government, "...within which private people assembled to constitute a public and to regulate those aspects of their commerce with each other that were of general concern" (Habermas, 1962, p. 142). Such assembly often took place in London coffee houses and Parisian salons of that period. Habermas's primary concern is the changing rationale for politics. That also was a structural transformation of the public sphere in modern society brought about by a mutual infiltration of public and private spheres, a transfer of public function to private corporate bodies (*i.e.,* privatization), and the extension of public authority over sectors of private realm (refeudalization and/or neomercantilism). He sees this transformation as having lead to a decline of public life and a replacement of the discursive and interactive politics of the earlier *public sphere* by technical, administrative, and commercial discourse and politics that were devoid of genuine public judgement.

This study suggests that the political life of the public sphere may be in a process of restoration to some extent in this highly specialized arena by a necessity of dealing with public policy issues in the *private-yet-public* realm of standardization. Such policy discourse has found its way into standards discourse by the interwoven nature of technical and policy issues, as claimed by Lessig. This public/private ambiguity is relevant because its various perceptions are reflected in the access to participation, and in the interests being served. When Habermas in 1962 recognized the decline of his generalized concept of "the public," he also reflected on the possibility that some of the benefits of the principle of publicness might be realized within more specialized forms of discourse that take place within complex modern intra- and inter-organizational spheres (1962, p. 248). Hence, it can be suggested that the process of standardization approximates this sort of discourse.

The consequence for standards discourse—struggling to deal with notions of what is or ought to be *public* or *private*, and to what degree—is that the situation is ripe for confusion and strategic manipulation. Standardization practice, it is suggested, dwells somewhere between the poles of public and private as in Habermas' *public sphere* or Lessig's *commons*. Since there is no accepted term in the general discourse for such a place—only an unspecified, but nevertheless sharp, *public sector/private sector* dichotomy—the public sphere or commons is thus delegitimized[4] and rendered invisible and, therefore, cannot be considered. This is a case of rhetorical reverse *appellation*, to use Althusser's term. The end result is an inability to stabilize the concept of whether standardization is a *public* or *private* function, or what the role of government, industry, or the public ought to be. Much of the discourse struggles with that question, particularly the tension between the U.S. and European concepts of standardization.

## Sectoral

The term *sector* is likewise afflicted with multiple meanings and also with limiting connotations. This term occupies a prominent place in standards discourse in the United States, particularly in ANSI discourse, when distinguishing between U.S. and European practices. The initial words in the National Standards Strategy (ANSI, 2000, p. 2) set the general tone and acknowledge the dual meaning of *sector*:

*Voluntary consensus standards for products, processes and services are at the foundation of the U.S. economy and society. The United States has a proud tradition of developing and using voluntary standards to support the needs of our citizens and the competitiveness of U.S. industry. The American National Standards Institute (ANSI), the coordinator of the U.S. standards system, has brought together public and private* ***sector*** *interests to make this happen.* [emphasis added]

The aforementieoned passage contains a prominent footnote on the word *sector* that reads,

*The word "sector" is used in two different meanings in this document. The first use divides the world into the "public sector" and the "private sector," distinguishing between the roles of government and non-government. The second use refers to a technology area where customer needs dictate a coherent and consistent approach to standardization. This use is sometimes called "industry sector" or "market sector" but we have chosen to use the simpler term "sector" to include all interested parties and not just commercial interests. Most standards are related to specific sectors (e.g., information technology, automotive) and are not applicable to the needs of other sectors.*

The basic meaning of *sector* is a piece of pie, sliced radially. Its use in geometry, military service, and astronomy imply a rather precise spatial demarcation. In both meanings of the term cited previously, no such precision is possible, yet it is still implied. The use of the term delineated in the ANSI statement fails to acknowledge the ambiguity or overlap between "sectors." The problem with the first usage (government versus non-government) in this respect has already been discussed in this chapter. In regard to the second usage (industry or market sector), one problem is that industry or market *sectors* are not only unclearly defined and overlapping, but often mobile, particularly in high technology areas. For instance, with the convergence of technologies, markets, and business organizations in the ICT field, many older sectoral divisions of telecommunication (*i.e.,* telephony), IT (*i.e.,* computers), software, cable, broadcasting, Internet, *etc.,* have become limiting and conflicting in respect to standardization and to regulation, as noted in Chapter II.[5]

One of the greatest points of conflict between European and U.S. standards discourses has been this *sectoral approach* issue. The ANSI claim that "most standards are related to specific sectors and are not applicable to the needs of other sectors" is problematic and directly in conflict with the European tradition of standardizing and regulating on the basis of overriding principles, wherever possible. This European tradition seeks to base policies and standards on general cases or unifying principles rather than on narrow sets of rules. The European, or more broadly the non-U.S., view is that the sectoral discourse is about the avoidance of establishing broad principles that might later somehow limit the options of players in the market.[6] Such a *sectoral approach* does not provide for "coherence" or allow for the possible restructuring of industries (Vardakis, 2002). In the footnote cited previously, the ANSI claim is that it is necessary to narrow the scope of a standard in order to provide "a coherent and consistent approach" to meeting "customer needs." The implication is that non-sectoral approaches may be of value where "customer needs" are not in question. It is not clear when such a situation would occur. The opposite of ANSI's claim is that the sectoral approach may result in conflicting, overlapping, and competing standards, and that "coherence and consistency" can only be fully achieved by a broader, more comprehensive, approach (Vardakis, 2002, 1998). Thus, both sides claim their path to coherence and consistency is the correct one.

ANSI's commitment to the *sectoral approach* may be seen as accommodating the traditional "turf" divisions of its diverse constituents and the necessities of holding such a voluntary private federation together, absent any overriding mandate (*i.e.,* government regulatory authority).[7] It has been suggested that this structure may also be related to the need to preserve revenues from the sales of standards documents (Cargill, 1997, p. 245). In any case, it may be that the sectoral approach is an inevitable consequence of a standardization régime based on *private* associations rather than one with a greater *public* mandate. The sectoral *vs.* non-sectoral approach has been characterized as a vertical *vs.* horizontal structure. This topic along with the issue of narrow rules *vs.* broad principles will be considered in greater detail in Chapter VII.

## Open

The term *open* is perhaps one of the most important and yet also most imprecise concepts.[8] In the earlier discussion regarding *Free and Open*, it was noted that Lessig interpreted *open* as in *open society.* A good place to start refining the definition is with philosopher Karl Popper's notion of *open society* in his seminal work, *The Open Society and Its Enemies* (Popper, 1945). In Popper's sense, an open society is one where

> *...no ideology or religion is given a monopoly, where there is critical interest in new ideas wherever their source, where political processes are open to public examination and criticism, where there is freedom to travel and where restrictions on trade with other countries are minimal, and where the aim of education is to impart knowledge rather than to indoctrinate.* (Outhwaite & Bottomore, 1993, p. 430)

Much of the aforementioned description seems directly relevant and applicable to standardization as a social practice. Popper contrasts *open* society and its reliance on critical thinking with *closed* groupings where there is a reliance on pre-existing moral authority, tribal loyalty, or corporate unity. He notes that life in an open society can be quite arduous and not necessarily happier than in a closed society because of the burden on individual responsibility and critical thought. Popper's *open society* has no historical determination or destiny. It is only the sum of its members, and their actions serve to fashion and shape it, not the converse. The social consequences of intentional actions are often, and largely, unintentional.

As in the case of other key terms in the lexicon of standardization, the distinction between *open* and *closed* is not a sharp dichotomy, but a gradation of multiple dimensions. Each of the attributes noted by Popper could be applied to standards activities as found in their various institutional or organizational practices from the *very open* to the *not-so-open.*[9] The concept of Popper's *open* applied to standards practice is consistent with both the social constructionist (*i.e.,* Social Construction of Technology) perspective and with Habermasian notions of discursive practice and democratic deliberation (*i.e.,* communicative action) mentioned earlier. With this theoretical perspective, it is possible to more directly define *open* standardization practice.

As discussed in an earlier chapter, a recent ANSI annual report (ANSI, 2001, p. 4) defines *open* as a condition wherein "any materially affected and interested party has the ability to participate." This wording is an abbreviated version of the phrase used in the *ANSI Procedures* (ANSI, 2002a) under the category of "due process," a phrase that derives from the Fifth Amendment to the U.S. Constitution. The full text of the subsection "openness"[10] reads:

*Participation shall be open to all persons who are directly and materially affected by the activity in question. There shall be no undue financial barriers to participation. Voting membership on the consensus body shall not be conditional upon membership in any organization, nor unreasonably restricted on the basis of technical qualifications or other such requirements.*

*Timely and adequate notice of any action to create, revise, reaffirm, or withdraw a standard, and the establishment of a new consensus body shall be provided to all known directly and materially affected interests. Notice should include a clear and meaningful description of the purpose of the proposed activity and shall identify a readily available source for further information. In addition, the affiliation and interest category of each member of the consensus body shall be made available to interested parties upon request.* (ANSI, 2002a, p. 1)

In ascertaining the full meaning of this definition and its historical evolution, it is useful to go back to an earlier discourse, the *National Policy on Standards for the United States*. This document was a precursor to the present *ANSI Procedures*, released by the National Standards Policy Advisory Committee in 1979 (NSPAC, 1979). Much of the wording is identical, but with an interesting difference. The earlier document reads:

*...shall be open to all persons who* ***might reasonably be expected to be, or who indicate that they are****, directly and materially affected by the activity in question. ...Organizations shall give reasonable notice of standards development activities and actions.* (p. 1) [*emphasis* added]

The current language seems to have migrated somewhat over the issue as to who has a legitimate interest on the basis of being "materially affected."[11] The aforementioned citation also has a footnote that reads:

*One member, (George Papritz) expressed the view that the first and last sentences should read as follows: "Participation in national standards activities should be open to all interested persons and groups." "Organizations shall give* ***interested persons and groups*** *reasonable notice of standards development activities and actions.").* [*emphasis* added]

George Papritz at that time was the chief of standards for Consumers Union. It appears that there was at that time a controversy over who should or should not be included in *open* standardization. Since that time the language has migrated in the direction of privatization. It is also interesting to note that the 1979 National

Standards Policy Advisory Committee document used the phrase *government sector*, rather than *public sector*. There are a number of other interesting differences between the two documents.[12]

A useful, if not rigorous, taxonomy of open standards has been proposed by Krechmer (1998) in *The Principles of Open Standards*, a paper[13] that expands on the ANSI principles. Krechmer proposes ten criteria:

*Open Standards is a changing concept, molding itself to the evolving needs of an open, consensus based society. Currently ten concepts are considered, at least by some, to constitute part of the principles of Open Standards. Openness - all stakeholders may participate in the standards development process. Consensus - all interests are discussed and agreement found, no domination. Due Process - balloting and an appeals process may be used to find resolution. Open IPR - holders of Intellectual Property Rights (IPR) must identify themselves during the standards development process. Open World - same standard for the same function, world-wide. Open Access - all may access committee documents, drafts and completed standards. Open Meeting - all may participate in standards development meetings On-going Support - standards are supported until user interest ceases rather than when provider interest declines. Open Interfaces - interfaces allow additional functions, public or proprietary. Open Use - low or no charge for IPR necessary to implement an accredited standard. The first three principles are at the heart of the existing ANSI Open Standards concept. These are required procedures of the American National Standards Institute for all accredited standards organizations. The fourth principle (Open IPR) has been formally added to the standards development process by ANSI, its SDOs and many international standard development organizations. The fifth principle (Open World) is supported by ANSI but not required. The additional five procedures represent Open Standards concepts which are emerging but which are not yet supported by most accredited SDOs. To what extent should Open World and the additional five procedures be considered principles of Open Standards?* (p. 1).

Krechmer then proceeds to attempt to map various activities and organizations according to his proposed criteria, as listed. Although he encounters some difficulty, due to the limitations of the *public/private*, *sector*, and *openness* terminology, he shows how various organizations apply the multidimensional term *open* differently. This line of analysis will be pursued further in respect to certain specific discourses making claims of *openness*.

## CONCLUSION

This chapter has examined how standardization practice and discourse is situated and how it was observed. It also has sought to identify the essential terms of the discourse and establish their meanings. These terms included *public*, *private*, *sector*, *sectoral*, and, most importantly, *open*. The next chapter examines some of the most traditional standardization institutions and practices, with particular attention given to the application of the aforementioned terms and their meanings applied by those institutions.

## REFERENCES

ANSI (2000). *National Standards Strategy for the United States*. American National Standards Institute. New York: ANSI. August 31, 2000, 14 pp.

ANSI (2001). *American National Standards Institute: 2000 Annual Report*. New York: ANSI.

ANSI (2002a). *ANSI Procedures for the Development and Coordination of American National Standards*. New York: ANSI. Available at <www.ansi.org/rooms/room_16/public/gov_proc.html>

Cargill, Carl (1997). *Open Systems Standardization: A Business Approach*. Upper Saddle River, NJ: Prentice-Hall. 327 pp.

Geertz, Clifford (1973). *The Interpretation of Cultures: Selected Essays*. Boston: Basic Books.

Habermas, Jürgen (1962) *(1991). The Structural Transformation of the Public Sphere: An Inquiry into a Category of Bourgeois Society* (Thomas Burger, Trans.). Cambridge, MA: MIT Press.

Krechmer, Ken (1998, November/December). The Principles of Open Standards. *Standards Engineering*, *50*(6), 1. Available at <http://www.crstds.com/openstds.html>

NSPAC (1979). *National Policy on Standards for The United States and a Recommended Implementation Plan*. Washington, DC: National Standards Policy Advisory Committee (reproduced in Cerni (1984), appendix c.2: pp. 223-226).

Outhwaite, William, & Bottomore, Tom (Eds.) (1993). *The Blackwell Dictionary of Twentieth-Century Social Thought*. Oxford: Blackwell Publishers Ltd.

Popper, Karl (1945). *The Open Society and Its Enemies.* 2 vols. London: Routledge.

Random House (1967). *The Random House Dictionary of the English Language.* New York: Random House, Inc., 1960 pp.

Schoechle, Timothy (1995, April). The Emerging Role of Standards Bodies in the Formation of Public Policy. *IEEE Standards Bearer, 9*(2), 1, 10. Available at <http://www.acm.org/pubs/articles/proceedings/cas/332186/p255-schoechle/>

Schoechle, Timothy (1998). *A Public Policy Debate: Standardsmaking Practice and the Discourse on International Privacy Standards.* Standards Policy Research Paper ICSR98-211. Boulder: International Center for Standards Research. Available at <http://www.standardsresearch.org>

Vardakas, Evangelos (1998). Director, EC Directorate for Industry (DG III), speech before the *Standardization Summit* sponsored by ANSI and NIST, Washington DC, September 23, 1998.

Vardakas, Evangelos (2002). Director, EC Directorate General Enterprise, presentation before the ANSI Annual Conference, Washington DC, October 15, 2002, and subsequent personal interview.

Wegberg, Marc van (2002a). Interview, August 2, 2002, at the EASST Conference, York, UK.

## ENDNOTES

1 Chapter 15: "Deep Play: Notes on the Balinese Cockfight," p. 412-453, in Geertz (1973).

2 Other terms include balanced, free market, capitalism, democracy, standard, legitimate, regulation, co-regulation, harmonize, specification, essential requirements, coherent, accredited, strategic, compatibility, and platform.

3 The use of *sector* implies a sharp distinction that is left undefined—a slice of the pie, but of unknown dimension—and fails to recognize or acknowledge the ambiguity or overlap between *sectors.*

4 This term was suggested by van Wegberg (2002a).

5 Examples of conflicting regulatory and standardization paradigms might include telephony and VoIP, cable and DSL regulation, broadcasting and Internet radio. The Telecommunications Act of 1996 is very conflicted in this respect and the structural organization of the Federal Communications Commission (FCC) is under re-consideration.

[6] The issue of data privacy is a good example. The EC adopted broad privacy directives and an effort was made to bring these into the ISO standardization arena—an effort that was successfully fended off by ANSI. A detailed account of this discourse on privacy standards has been provided by Schoechle (1998, 1995).

[7] This is the author's interpretation as suggested by the introductory remarks of Oliver Smoot (ANSI Board Chairman) and of Mark Hurwitz (ANSI President and CEO), other participants and audience comments at the *Breaking Down Borders: business, standards and trade: ANSI Annual Conference*, Washington, DC, October 15-16, 2002. Smoot's remarks emphasized the "sectoral view." Hurwitz' remarks emphasized the "diverse nature of U.S. delegations to ISO meetings" that were "representative of the unique character of the ANSI Federation," the need to "get high level corporate executives to support standardization," and the important "issue of membership" and funding.

[8] It is interesting to note that in one particular dictionary, the definition of *open* is one of its longest, offering 82 separate meanings (Random House, 1967, p. 1008).

[9] Virtually all standards bodies claim some measure of *openness* and virtually none admits to being *closed*.

[10] Section 1 (Due process and criteria for approval and withdrawal of American National Standards), subsection 1.2 (Due process requirements), paragraph 1.2.1 (Openness).

[11] The use of the phrase, "any materially affected and interested party," is interesting. What does *materially* mean—holding a financial stake? Does *and* mean that the party must be both materially *affected* and materially *interested*? What does *interested* mean? The passage following indicates that some proposed to simplify this wording to only *interested*, but were unsuccessful in doing so. What was the purpose in a restricted definition? The term *stakeholder* is also often used in the standards discourse with similar ambiguity.

[12] For instance, the paragraph V. (General Principles), 2. (National Standards Writing Activities), e. (Consumer/User Views) in the older document mandating pro-active soliciting of consumer participation was dropped, as well as V. 6 (Consumer /Small Business Funding) urging financial support of smaller participants.

[13] This paper was awarded 2nd place in the annual *World Standards Day* paper competition for 1998, sponsored by ANSI and NIST.

# Chapter VI
# Traditional Institutional Structures and Practices

*The noisiest of those competitive battles [between suppliers] will be about standards. The eyes of most sane people tend to glaze over at the very mention of technical standards. But in the computer industry, new standards can be the source of enormous wealth, or the death of corporate empires. With so much at stake, standards can arouse violent passion.*

—The Economist, 23 February 1993

## INTRODUCTION

In order to understand how traditional processes and discursive spaces may be undergoing enclosure, it is useful to examine the baseline from which recent changes have taken place. In other words, it is important to ask, How open were these standards-setting institutions and their practices in the past? This is not an easy question to answer. The diversity among various institutions is considerable. Even the question, What institutions should we be talking about? is not easy to answer. A detailed historical account, analysis, and comparison of standardization bodies is beyond the scope of this study, but the three oldest and most established international standards-setting organizations, the International Telecommunication Union (ITU), International Electrotechnical Commission (IEC), and International Organization for Standardization (ISO), have been selected here because they are stable, they are respected, and their authority and legitimacy are well established. These organizations were mentioned briefly in an earlier chapter, as part of an historical overview of the global standardization system. They will now be examined in more detail.

In the previous chapter, certain key terms of discourse were identified and considered. These terms included *public*, *private,* and *open*. The present chapter will examine the history, structure, and practices of the three abovementioned standard-setting organizations in an attempt to form a profile for each. It will then draw comparisons between them on the basis of these terms. The assumption here is that history can help reveal the meaning embedded in this social practice. Only sufficient history will be visited here to find the contextual meaning of the terms. This examination will proceed chronologically, beginning with the oldest of the three organizations—the ITU, then the IEC, the ISO, and finally JTC1, a collaboration between IEC and ISO. The ITU differs significantly from the others because it is a governmental treaty organization, now part of the United Nations, while the others are voluntary non-treaty and non-governmental organizations (NGOs). Nevertheless, the standards[1] that come out of these three organizations are all voluntary in their application. They may later become "national" standards if adopted by an appropriate national body and may subsequently acquire force of law or regulation on a national or regional basis.

## ITU—INTERNATIONAL TELECOMMUNCIATION UNION

### ITU History

The ITU dates from the formation of the International Telegraph Union at a conference called by Napoleon III and held in Paris in 1865. The Paris conference produced a treaty known as the International Telegraph Convention, which established an organization and a set of regulations. The "electric telegraph"[2] had previously been introduced in a number of nations and was immediately employed to improve the operation of their railway systems, but international operations were encumbered by a lack of uniformity and interoperability. One of the earliest tasks of the ITU was the establishment of technical standards, in particular for deciding on a code to be used—the international Morse code, The ITU also addressed matters such as the interconnection of important cities, hours of reception for telegrams, the obligation for foreign message delivery, and the commitment of transmission capacity (Codding & Rutkowski, 1982, p. 4). Another outcome of the 1865 Paris conference was the concept of periodic meetings, known as conferences, of delegates from national telegraph administrations to keep the standards, rules, and regulations up to date. This hierarchical conference-based structure became the basic organizational model of the ITU.

The early application of the telegraph was seen not simply as a means of operating railroads, but more generally, according to the ITU, as "...an efficient means of establishing central control and government" (p. 4). Except in the United States, telegraph systems were generally institutionalized as monopolies, owned and operated by national governments. In the U.S., the telegraph was dominated by Western Union, a private corporation that established an early monopoly (DuBoff, 1983, Horwitz, 1989). News services and other early commercial users were welcomed because their revenues were used to cross-subsidize government and railway users, who often enjoyed the use of the telegraph without charge. In many countries, the telegraph system was combined with the postal system and administered by organizations that later became known as PTTs (offices of Post, Telephone, and Telegraph). With such a system of combined post and telegraph operation and cross-subsidization, it became virtually impossible to identify the true cost of telegraphy. Ordinary citizens had little access to the telegraph due to high charges and low priority. Thus the basic practices of the ITU and its close relationship with member nations and with commercial entities became institutionalized at an early stage. Inherent limitations on openness and public interest representation were cast in institutional structures and practices from the outset, while ordinary citizens or their direct interests were not really involved.

The ITU grew rapidly, from its initial twenty members to twenty-four by its London Conference in 1879 and to forty-eight by 1914. The United States refused to join the ITU because it had never nationalized its telegraph and telephone networks and it could not participate in an organization that regulated their use (p. 11). Nevertheless, the U.S. did send observers to ITU meetings and participated informally in the developing global network, and the U.S. role should not be underestimated for lack of formality. U.S. industrial policy in telecommunications and its role in shaping the international system were reflected in many ways, including the Communication Act of 1934, in ownership policies, and in the way that notions of "public" interest and service militated against the public agency approach of PTT nations.

As the ITU grew, bringing in many developing countries and the interests of their national PTTs (a growing source of revenue for national governments), another basic organizing principle of the ITU took form—the importance of telecommunications in national development and the influence of developing countries as a political force within the ITU.

Except for the development of the telephone, very few technical innovations took place in the telegraph world for nearly four decades. The telephone escaped much of the telegraph's regulatory regime because of its expense and unreliability, the predominance of the telegraph for formal communication (having a written record), and a lack of interest by telegraph monopolies preoccupied with protecting their investment in telegraph systems. It was not until the 1925 Paris Telegraph Conference that, with the perfection of the telephone and of radiotelegraphy, the need was recognized for a separate telephone consultative committee to deal with technical standards for telephony.

Radio also developed along a much different pattern than telegraphy. The new medium found its early application in communications between ships at sea and land and played an important role in the world's merchant fleet for both safety and commercial purposes. The Marconi Wireless Telegraph Company gained an early dominance due to its technological and commercial adeptness (p. 12). When the company, challenged by U.S. and German firms, overplayed its hand by refusing interoperation with the equipment of other manufacturers, it evoked pressure for international agreement that eventually resulted in the 1906 International Radiotelegraph Convention. The 1906 agreement was ultimately incorporated into the ITU structure, although not fully until 1938. The underlying struggle was basically between corporations (*i.e.*, RCA, Siemens, and Marconi), but regardless of the winner, the principle of the primacy of private interests was firmly established.

With the gradual development of telephony and radio, additional technical standardization bodies were established in various forms and eventually incorporated into the ITU structure. The CCITT (*Comité Consultatif International Telegraphique et Telephonique*) became a formal permanent study committee of the ITU in 1956. Its mission was to ensure interoperability of national networks.[3] The CCITT became the principal standardization arm of the ITU for telecommunications. The CCIR (*Comité Consultatif International de Radiocommunication*) was formed as a comparable technical standards body for radio.

An important characteristic that evolved in the ITU was that of the relationship between private interests, public interests, and government interests. The early domination by governments of telegraphy, and later of telephony and to a certain extent of radio, was justified by the importance of these technologies and technical systems to their respective national public interests. Soon, however, as aforedescribed, the interests of government diverged from those of the public. Monopolistic rate structures, impairment of innovation, and lack of competition served governmental institutional interests but worked against much of the general public. Private, largely corporate interests were often served by the development of symbiotic alliances with government institutions. In the United States, arguments were typically advanced against government regulation and ownership, even as monopoly corporate interests were justified by government regulation in the "public interest."[4] In such cases, a general pattern of the "capture" of regulators by those corporate interests they regulated, often produced essentially the same result as the national PTT monopolies (*e.g.*, AT&T) (Horwitz, 1989).

Here, as seen earlier in this study, the limited terminology of the public/private dichotomy proves inadequate in describing the reality of the situation—masking the important distinction between the *public* and the *government*. This concern was reflected in a report of the American observer to the St. Petersburg Telegraph Conference in 1875:

*The interests of the public who use the telegraph seemed to be entirely subordinated to the interests of the state and to the administrations: that is, to a fear lest any improvement (in the rate structure) might produce less revenue than is got at present, and lest it might throw more work on the telegraph bureau.* (Codding & Rutkowski, 1982, p. 8)

The history of the ITU is one of an institution, initiated by governments, but proceeding with simultaneous tension and accommodation relative to private interests. Standardization in the ITU reflects a classic debate over the relationship between standardization and innovation.

*Standards are, in the abstract, both a help and a hindrance to the development of telecommunications. By obtaining common agreement on a standard, business and governments can be induced to implement new facilities and systems. On the other hand, the adoption of such a standard is usually based on existing technological assumptions which may well change...In the real world of competition among providers of telecommunication equipment and services, the adoption of standards also represents a way of creating market opportunities.* (Codding & Rutkowski, 1982, p. 226)

How the market opportunities are reflected in standardization activity can play out in various ways. The standards wars over television standards represented an interplay between standards bodies and national commercial interests that resulted in competing regional standards, much like the history of railroad gauges or of electrical power plugs. For example, a controversy arose over the formation of the radio committee at the 1927 Washington Radiotelegraph Conference: "...the U.S. argued that such a body might hinder the evolution of radio by adopting premature standards. The French were afraid that [through such a committee] some private companies might obtain unfair advantage over others by gaining approval of their technology" (p. 17). Codding reflects that "The two surviving international consultative committees [*i.e.,* CCITT and CCIR] are unique in the ITU experience, and in the experience of most other international organizations, because of their participatory blend of government and private enterprise" (p. 105). For instance, the 13th International Telegraph Conference and the 4th International Radiotelegraph Conference meetings in Madrid in 1932 included eighty countries and sixty-two private companies and organizations (p. 18). In reality, government positions in the ITU are usually driven by private industry research and development priorities.

The result is that industrial interests end up defining the "public interest," despite complaints, particularly heard in the United States, about government bureaucratic interference and national industrial policy.

## ITU Structure

The present structure of the ITU was solidified at Plenipotentiary Conferences held in Geneva in 1992 and in Kyoto in 1994 and is shown in figure 6.1. The governance of the ITU is a top-down structure, with ultimate authority resting with its Plenipotentiary conferences. The CCITT and the CCIR were renamed the ITU-T and the ITU-R sectors respectively and are governed by the World Telecommunication Standardization Conference (WTSC)[5] and the World Radio Conference (WRC), with each meeting held on a four-year cycle. The ITU's international development role was institutionalized through the formation of the ITU-D sector as a basic part of the structure. All standards (known as Recommendations) developed by ITU-T committees (called Study Groups) must ultimately be approved by a WTSC Confer-

*Figure 6.1. ITU and ITU-T organizational structure*

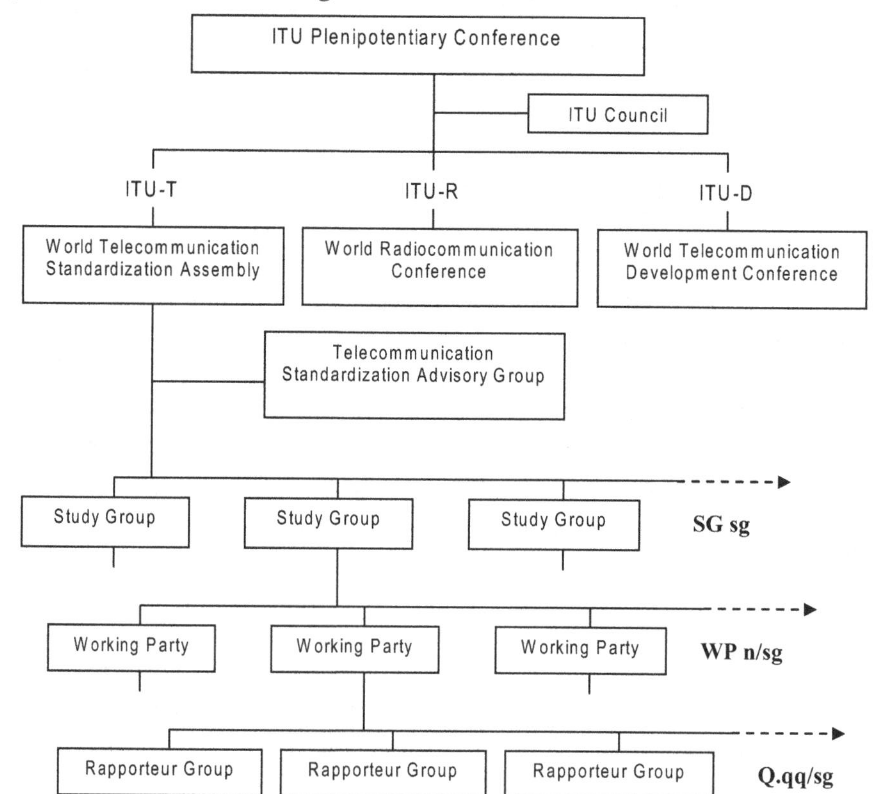

ence that meets every four years before they are considered final. All new work items or projects (known as Questions) must also be so approved every four years.

The ITU-T organizational structure is also shown in figure 6.1 under the ITU-T block. Standardization activity takes place within the eighteen Study Groups (SGs) that have been established to date (see table 6.1). They are administered by the Telecommunication Standardization Bureau (TSB) acting as a central secretariat in Geneva. Each SG is divided into Working Parties (WPs) that are assigned general topics, including groups of specific Questions. Each Question is then assigned to a Rapporteur Group, the lowest level, where, as in most standards bodies, the actual work is done. As in other standards bodies, the actual study work is carried on largely through contributions (*i.e.*, documents) that are submitted by group participants, distributed by the TSB, and then discussed either by correspondence or at scheduled international meetings. This work is often the result of collaboration with other organizations. Such contributions often require substantial resources to produce, and represent an avenue by which private corporate interests are represented in ITU activities at the lowest working levels. In most cases, the WTSC approves the Recommendations as presented or makes only minor changes. (Cerni, 1984, pp. 140-141). In practice, the work generally proceeds without waiting for formal approvals. A useful summary of the working procedures and practices in ITU-T committees is provided in ITU-T/ISO/IEC (1996).

Participation in SGs is formally established by national bodies as members. In the U.S., this role is under the Office of International Communications of the Department of State and is administered through a U.S. National Committee structure

*Table 6.1. ITU-T study groups*

| SG | Title |
|---|---|
| 1 | (discontinued) Definition and Operational Aspects of Telegraph and Telematic Services |
| 2 | Operational Aspects of Service Provision |
| 3 | Tariff and Accounting Principles |
| 4 | Telecommunication Management |
| 5 | Protection Against Electromagnetic Environmen Effects |
| 6 | Outside Plant and Related Indoor Installations |
| 7 | (discontinued) Data Communication Networks |
| 8 | (discontinued) Terminal Equipment for Telematic Services (Fax, Telex, Videotex, etc.) |
| 9 | Integrated Cable and Television Networks |
| 11 | Signaling Requirements and Protocols |
| 12 | Performance and Quality of Service |
| 13 | Next Generation Networks |
| 14 | (discontinued) |
| 15 | Optical and Other Transport Network Infrastructures |
| 16 | Multimedia Terminals, Systems and Applications |
| 17 | Security, Languages and Telecommunication Software |
| 18 | (discontinued) Digital Networks |
| 19 | Mobile Telecommunication Networks |
| TSAG | Telecommunication Standardization Advisory Group |

that mirrors the ITU's SG structure. U.S. National Committee SG meetings are "open to all interested persons" (Cerni, 1984, p. 142). Contributions (documents) are submitted on a national level and then advanced to the ITU in Geneva, either as a U.S. contribution, representing a consensus of one of the U.S. mirror SGs, (*i.e.*, a formal U.S. position) or as an individual member contribution, representing the position of a private U.S. organization or of individual experts.

Although voting membership in the ITU is formally organized on a national basis, in practice, participation in technical work has expanded over time to include a wide array of organizations and interests. These participants are classified as Recognized Operating Agencies (ROAs), Scientific and Industrial Organizations (SIOs), and others. ROAs are basically telecommunications operators or service providers, nationally or privately held (*e.g.*, Nippon Telephone and Telegraph, AT&T, Telenor, British Telecom, Deutsche Telecom, Telecom Italia, *etc.*). The SIO category includes manufacturers and other corporations (*e.g.*, Lucent, Ericsson, Siemens, Nortel, Nokia, Telcordia, *etc.*), and other standards bodies, including both formal Standards Development Organizations (SDOs) and private consortia (*e.g.*, ETSI, IETF, ECMA, IEC, ISO W3C, IEEE, *etc.*). Also included are many non-governmental organizations (*e.g.*, European Space Agency, International Committee of the Red Cross, International Criminal Police Organization, Inter-American Association of Broadcasters, Intelsat, the Internet Society, *etc.*). A recent dues-paying membership tally showed 189 member states, 180 ROAs, 232 SIOs, 19 Associates, and 42 others (Zhao, 2001, p. 18).

## ITU Practice

In the past, ITU practices have reflected control by governments, participation by industry, and coordination with a few formal SDOs. Today, this practice has evolved into a one that is basically driven by industry, is market oriented, and includes participation by governments, regulators, and many SDOs and consortia (Zhao, 2001). The evolving and growing influence of non-governmental participants was noted by Codding (1982) over twenty years ago:

*One of the aspects of the work of the consultative committees that has intrigued its observers over the years is the participation of individuals from private entities along with individuals from government agencies in important aspects of decision-making. While the logic of including manufacturers and private users of telecommunications equipment in the search for acceptable standards is strong, government entities are rarely willing to share international decision-making responsibilities so openly.* (p. 99-100)

*It seems clear that the participation of private operating agencies and the manufacturers of telecommunications equipment in the establishment of international standards is more than merely symbolic. While national administrations tend to keep a close rein on the non-governmental groups at the time of the final decision-making in plenary assemblies, these groups are given a wide opportunity to affect the work of the study groups where basic research takes place.* (p. 102)

The previous paragraphs illustrate a certain mutual tension-yet-reliance that exists between government and private participants. The inherent tensions that existed between these elements is further illustrated by the following passage:

*Particularly controversial in this international dialogue is the matter of private leased circuits...A new Recommendation has recently been adopted...[regarding] the use of private international telecommunication networks by specialized organizations. This matter is controversial because if private businesses (i.e., "specialized organizations" in CCITT jargon) obtain their own circuits, they could potentially compete with the existing monopolies which own and operate the public networks.* (p. 232)

Another aspect of ITU practice that has evolved is the accessibility of documents. Traditionally, ITU working documents, as well as standards, were not accessible to the public. A bewildering mix of incompatible, inconsistent, duplicative, and expensive CCITT "Yellow Books" and (even more obscure) CCIR "Green Books" published in arcane formats made ITU standards work available only to the most determined observer. This mix of documents and formats reflected the historic evolution and jurisdictional struggles over various subjects within the institution. This situation is changing with the introduction of limited document access via Web sites and the inclusion of other standards bodies, students, and academic researchers in ITU activities. Such changes also include the introduction of alternative "fast track" standardization processes such as the Industry Agreement, the Alternative Approval Process (AAP), the MoU (Memorandum of Understanding), joint management team meetings, common texts, *etc*. The recent changes reflect a struggle for relevancy and adaptation by this oldest standardization institution to the rapid advance of technology and markets and to the resulting competition from "private" forums and consortia.

## ITU Summary

The ITU has faced a dramatic challenge to re-invent itself—as a large institution with its legacy as a clearinghouse for monopoly telecommunications carriers and with representation by national governments under treaty obligations. It has been characterized as a "public" organization by Codding and Rutkowski (1982) and by Cerni (1984), in contrast to other standards organizations, which are seen as "private."

*It is interesting to note that a* ***public*** *international organization like the ITU isn't really necessary for the adoption of international standards. Indeed, several important* ***private*** *international standards-making bodies have existed for many years. Some of the more prominent include the International Organization for Standardization (ISO) and the International Electrotechnical Commission (IEC)...These organizations tend to serve the needs of manufacturers for rather detailed equipment standards.* [emphasis added] (Codding & Rutkowski, 1982, p. 225)

In what sense is the ITU public, while the ISO and the IEC are private? Clearly, the underlying unstated assumption appears to be that the terms *public* and *government* are synonymous. This study seeks to show the limitations of such a concept and the need for more dimensionality of meaning for these terms. While parts of the ITU process are ostensibly open, the complexity and detail of the technical work tend to favor well-heeled interests and place nonprofit public, NGO-like entities at a substantial disadvantage. It is not clear that the new inclusiveness of the ITU will or can remedy this problem. In fact, it may exacerbate the problem by making the process more complex, enabling access for certain parties who have adequate financial resources, but perhaps not others. The Plenipotentiary-based top-down governance structure may not be consistent with a quest for openness either.

## IEC—INTERNATIONAL ELECTROTECHNICAL COMMISSION

### IEC History

The IEC, like the ITU, originated from an international conference, but it took a much different path—one of voluntary, non-treaty and non-monopoly orientation. During the late 1800s and early 1900s, several international congresses were held as a result of growing interest in international cooperation in electrical standards. As early as 1881, meetings were held to agree upon basic units of measurement, establishing the Volt, the Ampere, and the Ohm. In 1904, the International Electrical Congress held in St. Louis resulted in a decision to establish a global body to "facilitate the coordination and unification of national electrotechnical standards" (Cerni, 1984, p. 21). The IEC was actually formed at a subsequent conference in London in 1906, and a central office was established there. The motivation was primarily to "prevent the kind of divergence in national electrical standards that had already resulted in European nations operating electrical systems at 220 Volts while North American nations operated at 115 Volts" (p. 130). Another important early motivation was electrical safety. In addition, gratuitous and unnecessary variation in national standards for electrical plugs, partly perhaps intended to favor and protect certain markets and suppliers, was threatening to impede trade and commerce. The IEC's mission is now considered to be to "promote international co-operation on all questions of standardization and related matters in the field of electrotechnology" (IEC, 2003). Its central office was moved from London to Geneva in 1948.

## IEC Structure

The IEC, organized much like the ITU-T, is governed by national committees and employs a hierarchical structure of committees to carry out its technical work. A major difference between the two organizations, however, is the strong governing role of a centralized IEC executive committee, rather than large plenary assemblies. The IEC organizational structure is shown in Figure 6.2.

IEC membership is based on national committees. Each member country has a body responsible for representing relevant interests within the particular country such

*Figure 6.2. IEC organizational structure*

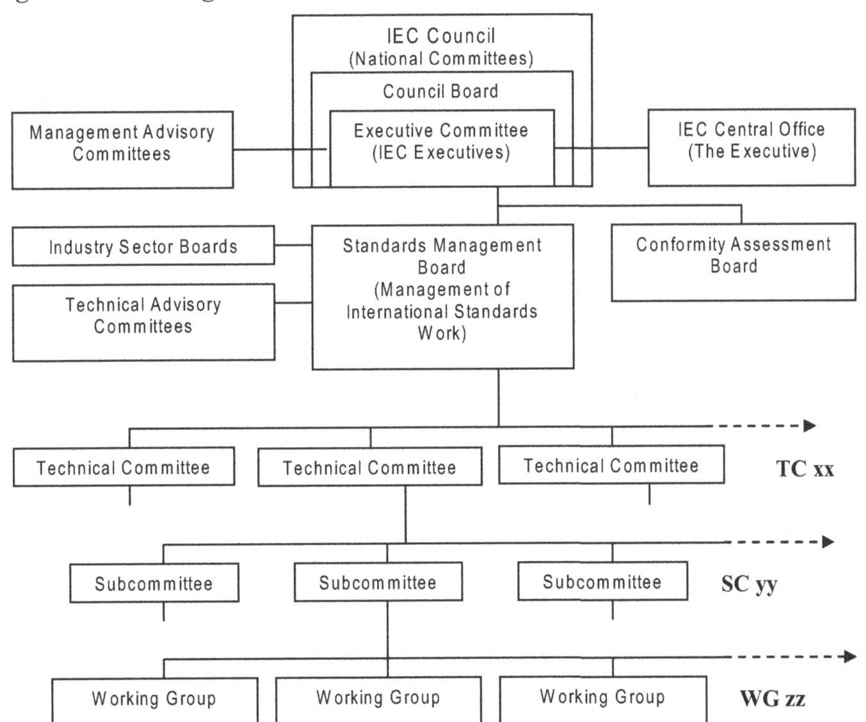

as industrial, commercial, government, trade association, professional society, test laboratory, research laboratory, academic, customer, *etc*. Government participates on a basis generally comparable with other interest groups within a member country. The U.S. National Committee for the IEC is administered as part of the American National Standards Institute (ANSI) and has no connection with the U.S. Department of State. In other countries, that role is generally maintained by traditional national bodies that may or may not have a direct government involvement (*e.g.*, *Deutsche Institute für Normung* (DIN) for Germany, British Standards Institution (BSI) for the United Kingdom, *Association Française de Normalisation* (AFNOR) for France, *etc*.). The national committees coordinate each nation's consensus position and reflect it in consensus standards development and in IEC governance.

A structural feature unique to the IEC is its four special Advisory Committees. These include the Advisory Committee on Safety (ACOS), the Advisory Committee on Electronics and Telecommunications (ACET), the Advisory Committee on Electromagnetic Compatibility (ACEC), and the Advisory Committee on Environmental Aspects (ACEA). These committees monitor and provide advice with regard to certain topics across all standardization activity in other IEC committees (including JTC1 committees). For instance, ACOS looks at the safety considerations of standardization work proceeding in all other IEC and JTC1 committees. A list of IEC TCs and SCs is provided in the Appendix.

IEC membership presently includes fifty countries and eleven associate member countries, for a total membership of sixty-one (IEC, 2003). The IEC is supported primarily by membership dues, but a significant portion of its revenue depends on royalties and sales of publications (such as standards). Membership dues are based on a sliding scale according to each country's economic strength, and associate member dues are greatly reduced in order to attract developing countries.

## IEC Practice

Although national committees are formally responsible for appointment of national delegations to meetings, in practice, meetings are generally open to anyone who knows about them and has the time and resources to attend. This limitation is not a small matter because highly paid technical experts must travel internationally and working meetings can often occupy a week or more. Knowing which meetings to attend is also not trivial because it requires careful monitoring of the committee work over time and maintaining continuity with the ongoing activities. The IEC has implemented a number of mechanisms intended to increase direct industry involvement and influence. These include Industry Sector Boards for setting priorities for standardization and planning, direct industry liaisons for PAS (Publicly Available

Specification) submissions, and a special President's Advisory Committee on future Technologies (PACT) for top-level industry players.

Technical Committee administration is not carried out by a central body, as in the ITU-T, but by various national bodies or committees that undertake the responsibility and expense of serving as "Secretariat." This role is often very influential and seldom relinquished, once established. The same is also true on the Subcommittee and Working Group levels. The national bodies often delegate the secretariat role to trade associations or other entities in the various countries.

In spite of the apparent openness of meetings, working documents and Web sites are required to be password protected and are not easily available to outsiders. This practice seems to be for two reasons: 1) to protect revenues from the sale of final documents, and 2) to shield work-in-process and voting positions from outside scrutiny, perhaps for fear of embarrassment, misinterpretation, or premature adoption by the market. Often, the passwords can be obtained, but the net result is that the openness of the process is questionable and citizens and individuals do not generally have access or influence.

The IEC has sought in recent times to streamline its organization and to introduce a hierarchy of "products" in an effort to successfully compete *vis a vis* the many newer consortia. A recent list of such products includes:

International Consensus Products:

- International Standards (full consensus) (IS)
- Technical Specifications (full consensus not (yet) reached) (TS)
- Technical Reports (Information different from an IS or TS) (TR)
- Publicly Available Specification (IEC-PAS)
- Guides (non-normative publications)

Limited Consensus Products:

- Industry Technical Agreements (ITA)
- Technology Trend Assessment (TTA) (IEC, 2003, p. 25)

The first three, IS, TS, and TR, are the traditional IEC products, developed under the full formal rules of procedure. TS and TR are often the results of standards efforts that are not yet ready for full standardization, but are under consideration or informative to ongoing standardization projects. The PAS, however, is a recent addition. It represents a mechanism for the adoption of a specification that may not have been created by an SDO, but is already in use in the marketplace. This case often represents an effort to "bring the work of consortia into the realm of the IEC" (p. 36). One possible interpretation is that by endorsing the PAS, the IEC seeks to reclaim its "market relevancy" by recognizing market realities, even though the

PAS subject matter was not developed under the formal procedures that traditionally bestow legitimacy on SDO-developed standards.[6] Much the same can be said about the "limited consensus" products on the list. Here, perhaps the IEC is trading some of its institutional legitimacy for some measure of market relevance and the ability to publish and sell additional documents. The ITA process is similarly justified on the basis that it "offers a new and dynamic way of achieving market acceptance of a new technology with the IEC's intrinsic seal of approval" (p. 35).

### IEC Summary

The IEC, like the ITU, is seeking to re-invent itself and to find its place in the present environment of fast moving technology and markets. In many ways, the IEC has an easier task because of its smaller size and its relative freedom from government and politics. Its challenge will be to adapt to the challenges posed by the rise of consortia and other trends without sacrificing its reputation and institutional integrity. A major problem that continues is its dependence on publishing revenues—a factor that works against its basic mission of disseminating its standards widely. The IEC's recent thrust into the area of conformity assessment, testing, and certification may offer alternative sources of revenue. The organization also faces some of the same criticism for slowness and lack of transparency that has afflicted the ITU. In particular, the lack of public access to working documents inhibits or precludes public participation, and has from the beginning.

## ISO—INTERNATIONAL ORGANIZATION FOR STANDARDIZATION

### ISO History

The ISO is, like its sister the IEC, a voluntary, non-treaty international organization headquartered in Geneva. They share a building across the street from the ITU. The ISO was founded in 1946, continuing the work started in 1926 by the International Federation of National Standardization Associations (ISA). The ISA's activities were interrupted in 1942 by the Second World War, then resumed at a conference in London in February 1946, where delegates from 25 nations created a new organization to "facilitate the international coordination and unification of industrial standards" (ISO, 1999). The first ISO standard was published in 1951 with the title, "Standard Reference Temperature for Industrial Length Measurement."

The ISO develops, coordinates, and promulgates international standards that facilitate world trade, contribute to the safety and health of the public and help protect the environment. Its standards cover all fields except electrical and electronic engineering, the domain of the IEC (Cerni, 1984, p. 121). The IEC and the ISO together, according to ISO/IEC Guide 2,

> *...form the specialized system for worldwide standardization. National bodies that are members of ISO or IEC participate in the development of International Standards through technical committees established by the respective organization to deal with particular fields of technical activity. ISO and IEC technical committees collaborate in fields of mutual interest. Other international organizations, governmental and non-governmental, in liaison with ISO and IEC, also take part in the work.* (ISO/IEC, 1996)

Like the IEC, the ISO sees its mission primarily in terms of facilitating international trade and commerce. It has broadened its mission in recent years to support sustainable development, technology transfer,[7] and social concerns such as consumer welfare and the "digital divide" (Smoot, 2003).

Over the last decade, the ISO has introduced a new class of standards, known as "technology management" standards. Rather than specifying products, systems, technologies, or compatibility, this new class defines procedures for the management of technical systems including procedures, documentation, training and record-keeping requirements). The first of this new family was the ISO 9000 series (for which the ISO name is probably best recognized by the public) for quality management standards, primarily applied in manufacturing processes. Another series is the ISO 14000 series for environmental management, also applied in manufacturing processes. A new family, now in its early stages, is the ISO 18000 series that deals with social responsibility (*e.g.*, labor rules, personnel management procedures, *etc.*). These new management standards have been somewhat controversial, primarily due to cultural differences among ISO members.[8]

An example of ISO's social initiatives is the Consumer Policy Committee (COPOLCO), an Advisory Committee that considers issues such as consumer privacy, safety, and education, across all other ISO technical committees. The COPOLCO has been active in establishing guidelines for standards that affect the disabled and the elderly, and a current focus is e-commerce. According to the current ISO president, "COPOLCO caused the launch of the ISO Corporate Social Responsibility project. It was through the involvement of leading international consumer organizations that we learned that they think ISO excludes consumer input." (Smoot, 2003)

## ISO Structure

The ISO's structure closely resembles that of the IEC, employing a similar decentralized management structure with a small central administration. Also, like the IEC, the ISO distributes the responsibility and expense of managing technical committee work among secretariats in member countries. These secretariats are responsible for the appointment of committee chairs and for providing administrative support and assuring procedural integrity in accordance with ISO rules. This decentralized (TC/SC/WG) structure is intended to assure "that decision making is carried out at the level of highest knowledge of consequences, and that decisions are approved with a minimum of bureaucracy and cost" (ISO, 1999).

The ISO organizational structure is shown in figure 6.3. As with the ITU and IEC, ultimate governing authority and establishment of basic policy rest with the members. However, as with the IEC, the actual administration of ISO is effectively delegated to an executive body and central secretariat with essentially all of the technical work distributed among member bodies acting in the role of secretariat. ISO Technical Committees, as in the IEC, are numbered chronologically, beginning from TC1, "Screw Threads" (established in 1947), to TC225, "Market, Opinion and

*Figure 6.3. ISO organizational structure*

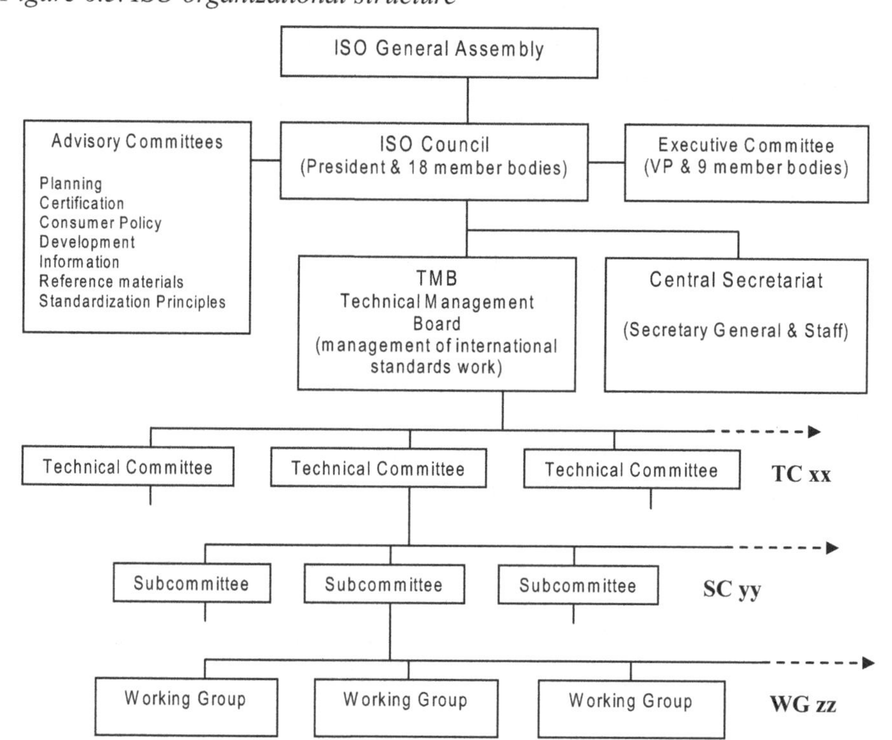

Social Research." Recently, there were 186 Technical Committees, 576 Subcommittees, and 2057 Working Groups in the ISO (ISO, 1999). A list of ISO TCs and SCs is provided in the Appendix.

The ISO sees itself as integral to the new global landscape in which the WTO (World Trade Organization) is a prominent feature. A major purpose of international standards is the elimination of technical barriers to trade (TBT), a significant item on the WTO agenda. Many new WTO members, notably China, also seek to become active players in the standards arena. ISO membership, tracking WTO growth, presently totals 147: 97 full voting members, 35 corresponding members, and 15 subscriber members.

ISO has long played a major role in information and communication technologies (ICT)—particularly in the realm of software, computing, and communication protocols—while the IEC has played a complementary role on the hardware side of ICT, the realm of cables, connectors, signaling methods, *etc.* As ICT matured, these realms became interwoven, "open systems" (discussed here in an earlier chapter) became a vital focus of interest in standardization, and the coordination, competition and overlap between ISO and IEC committees became a more significant problem. In 1986, ISO and IEC took the decision to form a new joint committee known as Joint Technical Committee 1 (JTC1) for "Information Technology" and to fold all relevant ISO and IEC committees into it. Perhaps the largest change was moving ISO TC97, the committee for "Information Processing Systems," and its many subcommittees, into JTC1. The United States, which had held the secretariat for TC97, became the secretariat for JTC1. The most significant effort produced by TC97 during the years prior to the formation of JTC1 had been the creation of what was known as the Open Systems Interconnection (OSI) family of data communication network protocol standards and the closely related family of telephone system standards, developed in cooperation with CCITT, known as Integrated Services Digital Network (ISDN).[9]

## ISO Practice

Within ISO committees, as in IEC committees, participation in meetings is open to any interested party, but voting, usually only done on a TC or SC level, is open only to "P" (Participating) members. Members having "O" (Observing) member status have no obligation to attend meetings and no right to vote. Voting procedures usually require a super-majority that varies, depending on the nature of the ballot. Procedures are defined in "directives." The IEC, ISO, and JTC1 each have their own sets of directives that are similar, yet different, reflecting slightly differing organizational cultures and terminology. In essence, the three organizations operate in the same or very similar manner, with JTC1 operating as a super-TC

*Table 6.3. JTC1 Subcommittees*

| SC | Title |
|---|---|
| SC 2 | Coded Character Sets |
| SC 6 | Telecommunications and Information Exchange Between Systems |
| SC 7 | Software and System Engineering |
| SC 11 | Flexible Magnetic Media for Digital Data Interchange |
| SC 17 | Identification Cards and Related Devices |
| SC 22 | Programming Languages, their Environments and System Software Interfaces |
| SC 23 | Optical Disk Cartridges for Information Interchange |
| SC 24 | Computer Graphics and Image Processing |
| SC 25 | Interconnection of Information Technology Equipment |
| SC 27 | Information Technology Security Techniques |
| SC 28 | Office Equipment |
| SC 29 | Coding of Audio, Picture, and Multimedia and Hypermedia Information |
| SC 31 | Automatic Identification and Data Capture Techniques |
| SC 32 | Data Management and Interchange Services |
| SC 34 | Document Description and Processing Languages |
| SC 35 | User Interfaces |
| SC 36 | Information Technology in Learning, Education, and Training |
| SC 37 | Biometrics |

within both the ISO and the IEC. A list of active JTC1 subcommittees is provided in Table 6.3.

The traditional formal standard development cycle within ISO (and essentially the same within IEC) follows a basic seven-step consensus building process (Cerni, 1984).

1. The new work item (NWI) is included in the program of work (POW) of a TC. The initial document, called a working draft (WD), must be circulated among appropriate members (TC, SC, or WG) with a view to the subsequent presentation of a draft proposal (DP), (sometimes called committee draft ([CD]). The circulation time for the first DP or CD is three months. The DP or CD must have "substantial support" from the P-members of the TC. Members casting negative votes must provide comments and the comments must be answered by the originating committee. A DP or CD may fail and go back around, after modification, for another ballot cycle, until sufficient consensus (substantial support) is achieved.
2. The DP or CD is registered at the central secretariat within 2 months of final approval by the TC.
3. The central secretariat registers the DP as a Draft International Standard (DIS) after checking and editing to ensure conformity with directives.
4. The DIS is approved by the member bodies within six months of distribution by the central secretariat. The DIS must receive a majority approval by the TC members and 75% of all voting members. Two or more negative votes prompt special consideration.

5. The approved DIS and revision are returned within three months to the central secretariat for submission to the Council.
6. The DIS is accepted by Council as an International Standard (IS).
7. The IS is published.

ISO committees have published 13,736 International Standards and other normative documents (including those of JTC1), and 4,437 work items are currently in progress (Smoot, 2003, p. 3). In 2002, 889 of these documents were produced; approximately 40% of them were developed in 3 years or fewer (p. 3).

## ISO Summary

Consideration of the speed at which documents are produced as of recent times reveals a dilemma that faces the ISO, as it does the other institutions described: competitive pressures posed by other standardization bodies and processes compel compromises between "market relevance" and legitimacy. The recent improvement in speed with which documents are adopted reflects the fact that "an increasing number do not go through the traditional full ISO process" (Smoot, 2003, p. 4). Increasingly, alternative "fast-track" or PAS processes are being employed to bring standards in from consortia and other sources in order for ISO to survive economically. "As many entering consortia work have found, it is not the administration or procedures, but the degree of consensus that chiefly determines speed" (p. 4). It will be argued here that in spite of rhetorical arguments from consortia advocates, it is not bureaucratic delays that take up time; consensus building and the open processes that take time. What is the price of speed for the sake of market relevance? This study seeks to examine how rhetoric about speed and market relevancy may work to renew old exclusionary tendencies and further enclose the standardization process in new ways.

In many respects, the ISO shares many of the same challenges as does the IEC, except perhaps more so because of its larger size. The organization's continuing dependency on publishing revenues works against its basic purpose, the widest distribution and use of its standards, and these revenues are continually eroding due to the increasing electronic distribution of documents.

# CONCLUSION

This chapter has examined the history, structure, and practices of the three oldest standards-making organizations, the ITU, IEC, and ISO. It began with the question, how open were these institutions and their practices in the past? As has been discussed in an earlier chapter, openness is a multi-dimensional quality. Most importantly, openness here is not openness in any absolute sense—as an objective measure—but is about the perception of openness. The aforementioned examination has been informed by the ten criteria of openness proposed by Krechmer (1998). In particular, it has looked at such indicators as working document access, access to meetings, balance of participation, voting, governance, and dependence on sale of standards (*i.e.*, access to final standards, technical reports, and other normative documents). A question that emerges is how to operationalize this information in a useful and informative way. The following discussion suggests a method that might provide a path for further research.

Using Krechmer's criteria as a starting point, it would be possible to construct an operational profile of each organization by assigning some numerical index value to each of the ten criteria and plotting it graphically. A suggested format is shown in figure 6.4. The scoring shown is for illustrative purpose only and provides a model of how openness might be evaluated. The limitations in this approach are significant and need to be recognized. It would not be possible to construct a conclusive profile of an organization without further refining each criterion and developing a methodology for quantifying each one. This profile technique is suggested here only as a possible methodology—to show the utility of the criteria and how they might be applied or operationalized. Evaluation and quantification could possibly be based on surveys of some set of observers. It would be important to consider how different evaluators (*e.g.*, governments, regulators, manufacturers, consumers, *etc.*) might judge the criteria. In any case, such a survey would provide subjective measure and would need to be based on some group consensus.

What has been found in the preceding examination of three institutions is that none of these organizations are fully "open" in all respects. They have always had to navigate between the conflicting needs of their members—government, industry,

*Figure 6.4. Suggested openness profile format*

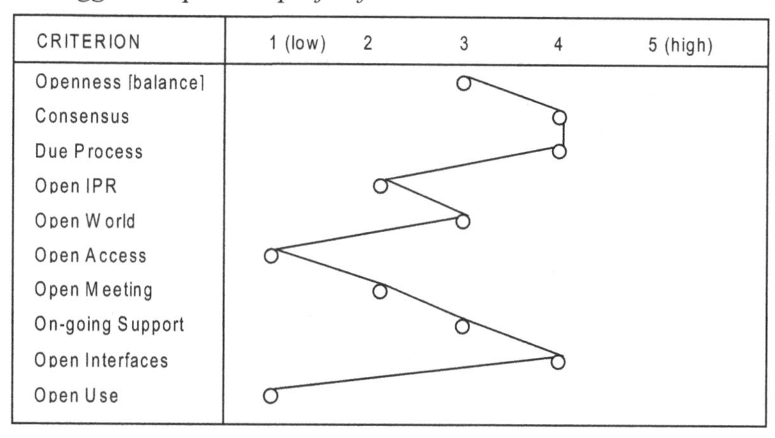

users, and consumers—and their history reflects the compromises they struck. They differ from each other in various ways, and they do not always succeed in fulfilling their own principles. But, they do in many respects earnestly espouse and aspire to those principles and to an open process in general. At the same time they are still supporting themselves by selling documents and, more recently, their legitimacy. The ITU differs most significantly of the three organizations, mainly because of its origins as a creature of government monopolies. All three face similar challenges to their openness and legitimacy presented by the necessities of maintaining or regaining market relevance while not losing their grip on traditional principles. The ISO and the IEC suffer from dependence on publishing revenues and from the limitations that such dependency places on the distribution of their work. These two also aspire to be technical and apolitical, although inevitably they are tossed about by political forces because their work has enormous political and economic consequences. The standards-setting institutions face the same erosion of trust that afflicts the global economy, governments, and industry.. As the president of the ISO noted recently, "The trust crisis affects not only corporations; it affects organizations generally, including ISO. While ISO is a non-governmental scientific and technical organization, it cannot help but be affected by politics" (Smoot, 2003, p. 2). In response, the ITU, IEC, and ISO are searching for new ways to be inclusive, to address the needs of consumers and of developing countries, and to pay increased attention to the transparency of their own procedures, especially where public perception is involved. A remaining dilemma concerns the question of how to be open and engage public participation, given the inherently arcane, technical nature of standardization work and the need for technical expertise backed by substantial resources not generally available to ordinary citizens or public advocacy groups.

Industry and trade have always played a central role in standards institutions and in driving their agendas. It is ironic that industry, which gains so much from the work of standards-setting institutions, often neglects these same institutions, failing to support them adequately, either financially or through earnest, good-faith participation. New technologies depend particularly on standards, but often "...leaders in new technologies frequently are ignorant or disdainful of the benefits of standardization" (p. 2). Thus industry leaves standards-setting organizations to fend for resources as quasi-public institutions having to serve multiple masters. "Public goods" tend to be chronically under-funded. An alternative view is that it may actually be no irony at all that industries under-funds their standards committees. If these organizations were better supported, they could be more open and democratic, and could possibly even undertake technical research. Consortia, on the other hand, are not under-funded.

The next chapter will proceed with an analysis of the actual discourse about standardization. It will identify the principal arguments and other rhetorical elements from the viewpoint of the theoretical perspectives described earlier. It will seek to clarify the significant issues and describe how the process of standardization is being discursively re-constructed.

## REFERENCES


Cerni, Dorothy M. (1984). *Standards in Process: Foundations and Profiles of ISDN and OSI Studies.* NTIA Report 84-170, Washington, DC: U.S. Department of Commerce. December 1984. 247 pp.

Codding, George A., Jr., & Rutkowski, A. M. (1982). *The International Telecommunications Union in a Changing World.* Dedham, MA: Artech House, Inc. (p. 414).

DuBoff, Richard B. (1983). The Telegraph and the Structure of Markets in the United States, 1845-1890. *Research in Economic History, 8*, 253-277.

Horwitz, Robert B. (1989). *The Irony of Regulatory Reform: The Deregulation of American Telecommunications.* New York: Oxford University Press. (p. 414).

IEC (2003). *Introduction to the IEC.* Slide presentation. Geneva: International Electrotechnical Commission.

ISO (1999). [Web site for] International Organization for Standardization. January 8, 1999. Available at <http://www.iso.ch>

ITU-T/ISO/IEC (1996). *Guide for ITU-T and ISO/IEC JTC1 Cooperation.* Annex A to WTSC Recommendation A.23, October 1996; and Annex K to ISO/IEC JTC1 Directives. December 1996.

Krechmer, Ken (1998). The Principles of Open Standards. *Standards Engineering, 50*(6), November/December 1998. p. 1. Available at <http://www.crstds.com/open-stds.html>

Rowland, Willard D., Jr. (1997, autumn). The Meaning of "The Public Interest" in Communications Policy—Part I: Its Origins in State and Federal Regulation. *Communication Law and Policy, 2*(4), 309-328.

Rowland, Willard D., Jr. (1997a, autumn). The Meaning of "The Public Interest" in Communications Policy—Part II: Its Implementation in Early Broadcast Law and Regulation. *Communication Law and Policy, 2*(4), 363-396.

Smoot, Oliver R. (2003). SIIT 2003 Address by Mr. Oliver Smoot, President of ISO. *SIIT 2003 Proceedings: 3rd IEEE Conference on Standardization and Innovation in*

*Information Technology*. Delft University of Technology. October 22, 2003. Visited November 5, 2003. Available at <http://www.siit2003.org/slides/smoot/ISO_Presentation to SIIT.pdf>

Zhao, Houlin (2001). The IT Standardization and ITU. Speech at *SIIT 2001: 2nd IEEE Conference on Standardization and Innovation in Information Technology*. Boulder, CO, October 5.

## ENDNOTES

1 ITU standards are known as "Recommendations" while ISO and IEC call theirs "Standards."

2 Earlier telegraphy systems were visual, based on networks of semaphores (*e.g.*, Siemens).

3 The CCITT, like other ITU-related bodies and conferences, has a convoluted history. Some predecessors to the CCITT were the CCIF (*Comité Consultatif International Telephonique*), 1934-1956; the *Comité Consultatif International des Communications Telephonique a Grande Distance*, 1924-1934; and the CCIT (*Comite Consultatif International Telegraphique*), 1925-1956. Codding and Rutkowski (1982) provides a detailed historical account of these various bodies.

4 The phrase "public interest, convenience and necessity" is firmly embedded in regulatory practice and administrative law in the United States as a justification for government intervention. Rowland (1997, 1997a) argues that such meaning clearly and paradoxically referred to regulatory actions that were primarily intended assure the health of industry and commerce. The rationale in the United States was that the interests of the public and consumers could best be served by protecting the economic well being of private corporations that provided goods and services that benefited the public (Rowland, 1997, p. 320). This approach stands in contrast to the more direct government "public interest" (*i.e.*, public agency) rationale of most other nations and their PTTs.

5 Formerly called the Plenary Assembly of the World Administrative Conference for Telephone and Telegraph.

6 It should be noted that PAS processes and meaning differ between IEC, ISO, and JTC1. The ISO directives define PAS as "a document not fulfilling the requirements of a standard." In contrast, the JTC1 directives define PAS as a path for origination or development of international standards outside of JTC1's subcommittees, and the resulting document can become a international standard.

[7] Standardization itself can be seen as a form of technology transfer (ISO, 1999).

[8] For instance, in the U.S., industry often tends to see these new standards as bureaucratic and interfering with management prerogatives and market forces. ISO 9000 was also initially seen by some as an effort to fortify European industry against foreign competitors by creating a quality documentation hurdle. Subsequently many enterprising firms have turned it into a marketing asset. The area of social responsibility is still particularly thorny.

[9] A detailed history of the OSI and ISDN standardization efforts is provided by Cerni (1984).

# Chapter VII
# Discourse on Standardization:
## Public or Private?

*When what is said does not sound reasonable, affairs will not culminate in success.*
—Confucius, c. 479 BC

## INTRODUCTION

The preceding chapter examined the history, structure, and practices of the principal international standards-setting organizations, the International Telecommunication Union (ITU), the International Electrotechnical Commission (IEC), and the International Organization for Standardization (ISO), with particular attention to their composition, the public/private nature of their organizations, the interests they serve, and their processes. The present chapter will examine current discourse on standardization and the rise of the consortia movement. It will do so first by framing the debate within general structural and openness issues and then by looking at specific rhetorical examples of arguments, claims, and controversies. It will then establish a taxonomy of arguments and rhetorical discourses, focusing on the issue of legitimation of consortia standardization. It will next analyze several important cases, looking at public documents, testimony, and reports. In doing so, it will examine the specific claims to legitimacy made by consortia and traditional bodies. It will seek to clarify how the practice of standardization is being discursively re-constructed. Finally it will consider international, institutional, and industrial responses to these claims of legitimacy and to the political/economic pressures they have brought.

The reader should remember that consortia and the more formal standards-setting organizations come in many forms. This book is really about the rhetorical construction of arguments in light of theoretical areas that can be applied to standards and standardization as a whole—it is not about attempting any general theory of standards or of consortia, nor is it an inventory of their benefits and shortcomings. Rather, it searches for common elements between each of these two groupings and for the most basic principles that distinguish them from each other.

## STRUCTURE AND OPENNESS

This section will examine the discourse on standards and standardization in order to identify the broader issues and the terminology that set the stage for a detailed analysis of the discourse. These issues include the structure of standardization, the openness of its practices and processes, institutional motivation, and forms of enclosure that are interwoven in the standards environment. In particular, this section will focus on the key issue of claims to legitimacy that are made by the various actors in the debate. This focus will provide a framework for establishing a taxonomy of arguments in the following section.

### Discourse on Structure of Standardization

The emergence of standardization outside the traditional formal system raises issues about how to describe it, or even refer to it. The terminology, Standards Development Organization (SDO) *vs.* Standards Related Organization (SRO), as has been noted earlier, comes from the SDO world, whose members generally draw a sharp distinction between *standards* and *specifications*—standards being those documents that the SDOs create and that enjoy the legitimacy of creation by duly accredited bodies. Members of SDOs regard specifications as normative documents that do not meet the rigorous requirements of formal standards. Many members of the non-SDO world also regard themselves as standards developers, but they do not generally draw such a strict distinction.[1] It is difficult to find terminology both worlds accept. One problem is the vagueness of the term *specifications*. It is too general to convey the normative purpose of a document. Additionally, there is no consensus regarding the scope of the term, standards.

Another way of distinguishing these SDO and SRO perspectives, coming from economic discourse about standardization, is *market-based vs. negotiated* selection

(Vercoulen & van Wegberg, 1998) of standards, with consortia or SROs falling roughly into the market-based category. However, this distinction is not precisely the same as the SDO *vs.* SRO distinction. It juxtaposes market-based competition among standards bodies and their standards (much as is the practice in the United States) with the realm of formal standardization where competing standards are avoided and a single standard must be chosen by whatever authority is able (much as is the practice in Europe). The competing-standards model practiced in the U.S. has also been characterized as a "bottom-up approach" (Egyedi, 1996, p. 13; Thomas, 2002). The negotiated approach practiced in Europe has been characterized by critics as "...too academic, too lengthy, and unresponsive to market needs" (p. 13)[2]. Although the market-based *vs.* negotiated dichotomy is perhaps useful, it runs a risk of overlooking commonalities and overlap. Those engaged in negotiating standards often argue that they are serving market needs, and it is likely that those in consortia frequently engage in negotiation. Another problem is that the market-based mode would also include *de facto* standards that may not involve any standardization process, but simply reflect the dominance of a particular successful technology or firm. In any case, the nature of the negotiation process itself has been highlighted and is receiving some needed attention (Lim, 2002), and the new category of *hybrid* selection has been introduced (Vercoulen & van Wegberg, 1998). This hybrid model involves active collaboration between negotiated and market-based selection processes and it has been further resolved into component elements:

*There are two kinds of hybrid standard selection modes. The former consists of attempts to integrate elements of market selection in a negotiated or coordinated mode. The latter consists of including forms of coordination into the market selection mode (such as the development of a market leader).* (p. 18)

The hybrid model has also been examined employing a three-way distinction, SDOs, Market-Driven Consortia (MDCs), and International Standardization (Shapiro, *et. al.* 2001). Of particular importance are the perceptions of the differences in the cultural values between the traditional SDOs and the newer MDCs that are evident in the discourse.

*A hybrid standard setting solution offers a promising way to encompass the positive aspects of SDOs, MDCs, and international institutions, such as the ITU, in the development of technical standards. SDOs tend to most highly value openness, balance, fairness and due process. In contrast, MDCs tend to place more emphasis on speed, efficiency, branding and promotion, and they often have the resources to provide follow through with testing and marketing. A hybrid system may be better able to respond to market needs intrinsic in today's convergent telecommunications market than each of these organizations might accomplish alone. Neither the SDO, the MDCs, nor the traditional international institutions are without significant drawbacks. Such include issues of openness, timeliness and relevance to society and to the marketplace* (Schoechle, et. al. 2002, p. 10-11)

It might be observed that MDO values are far more diverse that those of SDOs, although a great deal of variation can be found in both. These values may be discerned from an analysis of the structure and procedures of each organization based on each of the criteria provided by Krechmer (1998) discussed previously. It is increasingly evident that no one can credibly claim a monopoly on *openness*, and that some consortia may be considered more open than some SDOs, an observation supported by Egyedi and by Krechmer.

*...in certain respects consortia appear more open than the formal bodies. For example, while the latter usually keep access to committee drafts restricted to participants—and, thus, seek consensus within a limited group—consortia more often post their drafts on the web and actively seek comments from outside* (Egyedi, 2001, p. 57)

Krechmer has observed that accredited organizations, by not always appearing open to new directions or new members, have provided an impetus to consortia.

*Currently openness is deemed to be met if all stakeholders can participate in the standards development process. This may present the appearance of a closed committee to those who are not current stakeholders—a subtle but real way that incumbent stakeholders dissuade potential future stakeholders* (1998, p. 3)

The term *stakeholder* is frequently used but never defined. Presumably it means what ANSI refers to as "any materially affected and interested party." Krechmer seems uncertain as he continues:

*Possibly standards development meetings should be open to all (which the IETF [Internet Engineering Task Force] offers)*[3] *as well as provide open access to committee documents. In this way, informed choices may be made about bringing new work to an accredited standards committee. Too often stakeholders in a new technology are reticent to bring their ideas to an accredited standards committee they have no experience with, or access to.*

*The Open Standards concept may not be well served by stakeholder-only committee meetings. Ultimately, as technology use expands, technical standards stakeholders are everyone. Using the Internet, access to committee discussion can be opened to almost all. This review appears to argue for open standards development meetings. However participation in standards meetings is a significant reason why some organizations join SDOs. So offering free meeting participation to all may well have negative economic consequence to some SDOs* (p. 4)

Lessig would probably argue that as standards come to define our modern world, it is increasingly true that there is no one who is not a stakeholder.

Another structural distinction is that between *de jure* and *de facto* standards—*de jure* being those created by SDOs. While the term *de facto* standard is widely used, there is little agreement on its meaning, which is caught up in the rhetoric about openness and legitimacy.[4] On one extreme is a strict constructionist view. De Vries defines *de facto* standardization as "standardization carried out by non-governmental parties other than [SDOs]" (1999, p. xvii). That is, as carried out by "...companies, sectoral or specialized standardization organizations, and consortia" (p. 175).

Another author concerned about the importance of the distinction is Gifford, who defines *de facto* standard as, "a design or protocol that dominates the market through the unorganized actions of suppliers or consumers, without any formal[5] adoption or enforcement by non-market or regulatory bodies...usually associated with a single manufacturer..." (1997, p. 3) Gifford associates *de facto* standards with "market power" and is concerned about its dependence, "...on factors other than technical performance." Gifford places consortia in a separate category from *de facto* standards, SDOs, and regulation. Even though all three of these meanings are different, they all relate to openness, participation, and legitimacy.

Looking to the future, Jakobs predicts the de facto/de jure dichotomy will loose importance:

*This very popular classification should primarily refer to the different ways the respective standards emerge; this includes the characteristics of the respective originating organisations. ...the attribute "de jure" is associated with standards that emerge from an SDO through a voluntary process, and which are indeed only voluntary in nature. Standards that emerge purely through market forces (maybe through the dominant position of one or a group of players) are referred to as being "de facto." With the increasing complexity of the world of standards, and the cross fertilisation between SDOs and consortia, this distinction will eventually become obsolete* (2000, p. 13-14)

One recurring theme in the Europe *vs.* U.S. standards discourse is a fundamental difference in the way the two societies approach standardization and regulation—that is, *principles vs. rules.* This distinction is essentially the same as the discourse about the *sectoral approach* and the *horizontal vs. vertical* structure. This topic has been mentioned previoulsy with respect to privacy standards and to the U.S. preference for sectorally targeted solutions rather than for broader solutions based on overriding or common principles. One other particular case may be informative—that of accounting standards.

Two different sets of accounting standards are used at present: 1) Generally Accepted Accounting Practices (GAAP) in the United States (set by the FASB—Financial Accounting Standards Board), and 2) the International Accounting Standards (IAS) in Europe and other countries (set by the IAS Committee).[6] In commenting on the roots of the recent accounting scandals in the U.S. and particularly the collapse of Enron, the seventh largest U.S. corporation, Goran Tidstrom of the European Federation of Accountants, explains,

*...in the U.S., accounting standards are "rules-based," while in Europe, standards are based on principles. In other words, in the U.S., what is not specifically forbidden is allowed. In Europe, an accounting decision must be justified on the basis of sound business principles...[In] the U.S. [one is] living under the impression that everything that is not regulated in detail is allowed. And as long as you are not breaking what is stated in a specific standard, it is [allowed]. That would definitely not be the case in Europe, if an accounting decision were at the same time against sound interpretations of the business deal* (Baker, 2002)

Tidstrom continues, adding that the enormous debts hidden by Enron in off-the-books partnerships were fully legal under U.S. law and were apparently sanctioned by Enron's auditor, the firm of Arthur Andersen LLP.

This basic difference in approaches has its roots deep in American and European cultures and is reflected throughout the standardization system. It may have to do with the primacy of commerce in American culture (Schoechle, 1998; Lipset, 1996). In any case, the differences will not be resolved easily and will continue to be reflected in standards discourse in such terms as sectoral approaches, government, regulation, bureaucracy, free markets, market-driven, relevancy, consumer interests, coherence, harmonization, *etc.*[7]

## Discourse on Openness: Framing the Debate

The structure and practices of the international standards system are often the subject of discourse about openness and participation, and here the collision between the U.S. system and the European and international systems becomes apparent. A U.S. perspective is provided by Warshaw and Saunders (1995).

*In contrast to the U.S. system, standardization in the European Community (EC) has significant government influence. The EC system is much more integrated at both the national and regional levels. It is also generally a closed system. Participation in formal standards development activities at the regional level is limited to authorized representatives of European national bodies, thus excluding participation by non-European countries, even signatories to the GATT code.* (p. 69)

*Which system for international standards development can best meet the changing needs of global business? One which resembles the European regional system—monolithic, integrated, formalistic and policy-driven; or a system more like that which has evolved in the U.S.—pluralistic, sometimes fragmented,* ***ad hoc*** *and market-driven.*

*...superficially the international system looks like the European model. However, international standardization activities extend well beyond ISO/IEC, and there is real competition among international standards bodies in several areas. As in the U.S. system, competing organizations seek to respond to changing priorities and interests.* (p. 70)

*The U.S. model, which provides for broad participation of all interested parties and which allows competition among parties to determine the best technical approach may be the best one to enable international standards bodies to respond to the growing demands placed on them.* (p. 71)

Although these words were written in 1995, they still very much reflect the discourse at recent American National Standards Institute (ANSI) annual conferences (ANSI, 2002b). The following is an account of a conference, taking the form of a debate, that shows how rational arguments are interwoven with fundamentally different underlying cultural identities and beliefs (*i.e.,* clashing mythological and ideological frameworks). The opening keynote speaker, William Lash III, Assistant Secretary for Market Access and Compliance for the U.S. Department of Commerce (a Bush administration appointee), struck a confrontational posture, celebrating, "American principles of openness, transparency due process and the [proper] role of government," and assailing the European "15 votes to [the United States'] one," "block voting," and penchant for regulatory standards (Lash, 2002). Lash called himself the "assistant secretary for hard messages [to Brussels]," insisting that "we don't want a club," "international standards do not mean European standards," and that "ISO does not mean ISO-lation to European companies." He then described Japanese standards and trade practices in even less salutatory terms.

The rhetorical content of the remainder of the conference, though not as colorful, reflected a collision of systems and cultures (*i.e.,* United States *vs.* the world), and also reflected rapid change and adaptation on all sides. As the conference proceeded, it became clear that the organizers, deeply concerned about U.S. global competitiveness, were attempting to create a provocative multilateral discourse and had invited speakers from all over the world suited to that purpose. Following Lash, an international panel convened on the topic of "Competing in the Global Stage: Knowledge is Power," that provided the general flavor of the international discourse that is relevant today.

Dr. Setsuo Harada , head of Sony's trade relations office, denied the validity of certain comments by Lash characterizing standards as weapons. He portrayed the *privatization* of standardization as beneficial, detaching it from traditional government bureaucracy. Tony Flood of the Canadian National Committee of the IEC (and IEC presidential candidate) spoke about *global relevance* (*i.e.,* reducing the domination of regional [*i.e.,* CENELEC] standards)[8], about *harmonization* (*i.e.,* throw out your standard and accept ours), about the influence of the IEC over developing countries (that lack technical expertise of their own), and about *essential requirements* (*i.e.,* European infrastructure-related standards mandated by the European Commission). Flood concluded by commenting that, "The reason we got into this situation was no participation [by North Americans]; we've got to get everyone involved—[especially] manufacturers." He added later, "Going it alone as a country is not the best way to get things done today...conflicting [competing] standards confuse the market." Veit Ghiladi of Daimler Chrysler (Germany) commented that his firm dedicates many senior level people and $20 million per year to standards.[9] In response to the complaints about ISO one-country-one-vote policies, Neil Reeve of Shell Global Solutions International (The Netherlands) commented that, "the vote is irrelevant, the words that end up in the document are what counts and they get put in by the working group—not by vote, but by consensus."

The influence of World Trade Organization (WTO) policies was evident in the conference discourse. Ghiladi commented about openness, stating that the European Telecommunications Standards Institute (ETSI) accepts corporate members as well as national members, and posts all its documents for free on the Web. Flood responded that the WTO defines open standards in terms of "open to all stakeholders and consensus." He said, "WTO does not give preference to country and regional interests when other needs exist in other countries or regions."

On the issue of structural differences between the United States (based on business) and the European Union (based on government), the question was asked, how can the United States even compete with these government-sponsored people? Flood responded that the problem did not exist at technical levels; the participants all come from companies, not government. Ghiladi added that "About 20,000 come from companies and others from universities—not from government, not experts anyway—some from unions and some from consumers." "Companies," Setsuo agreed. Reeve commented that, "Government participation is small, one of many—[it's] mostly industrial people doing the actual work."

A prominent speaker at the conference was Evangelos Vardakis, Director, Directorate General Enterprise, of the European Commission. He spoke of sweeping changes in the role of regulation and standardization beginning with the *New Approach*, a major initiative launched in 1985 to facilitate economic integration of the European Community. He remarked on European recognition of the WTO Technical Barriers to Trade (TBT) principles and particularly on the importance of *coherence* for standards: "Competing standards are OK, but not conflicts—a difference with the American System." Vardakis bemoaned the "egoistic" hunting of sales by standards bodies (based on copyrights). Looking into the future, he recommended, for the sake of global convergence, readying for change (not merely exporting a national system), recognizing a single international body, agreeing on good regulatory practices, agreeing on levels of safety, interoperability, consumer information practices, and on cooperation not conflict. When questioned about the essential requirements, he commented on the problem of the vertical *vs.* horizontal structures of standards and regulation (a reference to the U.S. emphasis on the vertical, or sectoral, approach). Vardakis emphasized the importance of "allowing new organization of industries."

One puzzle that remains at the end of the day is that—for all the U.S. rhetoric about broad participation, responsiveness to market needs, decentralization, limited government role, competition, and the merits of the sectoral approach, *etc.*—it seems difficult for U.S. industry to step up to fund and participate in its own system, or to go to the international meetings.

## Rise of Consortia: Clash of Systems

The emergence of consortia in standardization practice is a relatively recent phenomenon, probably beginning in the late 1980s in the IT industry.[10] Carl Cargill provides a brief history of the consortia movement in a conference paper on the topic (1999) as well a more extensive history (1997), written from his perspective as an active participant and vocal advocate in the process.[11] The primary driving forces are most often considered to be the need to "keep pace with rapid market change" and the "extra time needed to achieve consensus" in SDOs (Krechmer, 2000, p. 1). A fairly accepting yet circumspect account of consortia is provided by CEN[12], in providing an extensive survey of consortia that lists 225 organizations in its seventh edition (CEN, 2002).

*The role of formal standards organizations (globally, ISO, IEC and ITU) has become somewhat muted in the world of [Information and Communication Technology (ICT)]. This is perhaps principally because the processes of the formal organizations have been deemed too slow to keep up with the furious pace of market change. Thus, for example, in the 1980s much effort was expended in the formal organizations on [Open Systems Interconnection (OSI)] standardization, whilst the market was developing in a rather different direction.*[13]

*The relative decline in the role of formal standardization has been matched by the development of "consortia standardization", in which interested parties club together to produce "standards" without the so-called drawbacks of the formal standards process. These consortia have their own drawbacks, not least in terms of ensuring broad consensus and subsequent visibility, but they are an established "feature of the landscape."*

David and Monroe (1994) discuss the various forces at play, performance problems, protracted duration, and they characterize an "array of discontents with the present standardization régime" (p. 1).

*Excessive delays in the drafting and approval of standards, rising costs of the resources engaged directly and indirectly in support of their deliberations, needless uncertainties and confusions caused by the multiplicity of organizations asserting "jurisdiction" or self-assigning responsibility for standardization in the same or closely related technical areas, alleged biases and lack of due process in the constitution of standards committees, non-responsiveness to the economic interests of under-represented consumers and users of the technologies in question—all have been cited as defects in the process.*

Although the impetus for the ascendancy of consortia is complex, the primary perceived motive seems to be speed. This is followed closely by the ability to focus effort among like-minded participants on a relatively narrow task, to dispense with unnecessary procedures, and often to have the money to follow through with branding, marketing, testing, certification, and maintenance—functions that would not be within the charter of most SDOs.

In Cargill's view, "The market no longer really cares where a standard comes from as long as there is a standard" (1999, p. 1). Cargill argues that the formal bodies were incomplete—that the real task is to standardize, to test, to market, then to promote. He points out that the old SDO system was not set up to do that. He asserts that the consortia among IT firms "are not cabals," that "companies have limited money to spend," and that "standards are competing for bandwidth" and need a more complete process/lifecycle—a more efficient way in a market economy. He adds that the old system was based on monopoly, not on a market.[14]

Cargill traces the early roots of consortia to a crisis that arose in U.S. standardization practice in the late 1960s and early 1970s. During that time, increasing regulatory use of standards by the government and a 1997 antitrust suit alleging anti-competitive behavior in setting plywood standards called into question the general fairness of standardization practice. Particular issues were the lack of due process and lack of consumer protection in existing voluntary standards organization procedures (Cerni, 1984, p. 53). Partly in response to the threat of government regulation and of legislation establishing government management of standardization,[15] ANSI stepped forward with a "private sector" solution, assumed accreditation responsibility and, in 1977, formed an independent 30-member body, the *National Standards Policy Advisory Committee*, that drafted the policy document, the *National Policy on Standards for the United States* (NSPAC, 1979). This document became the basis of the present system of due process that defines the voluntary consensus *process*. Essentially what was happening was a privatization of regulation—vesting in ANSI regulatory responsibility that might elsewhere have been a government, or *public* responsibility. In present day ANSI rhetoric, this is the "basic principle of the U.S. standardization—that standards-setting is a partnership process between government and the private sector" (Smoot, 2001). But, as Cargill summarizes the end result, "Once the chronology is traced, it becomes apparent why the U.S. voluntary system is the way that it is, with all of the "due process" features that make it very cumbersome for the IT industry" (1999, footnote xiv).

Another key formative document that came out of that early crisis period, described in Chapter II, was *OMB Circular A-119* (Office of Management and Budget), drafted over a 6-year period from 1976 to 1982 and last revised on February 10, 1998. Like the NSPAC document, it helped solidify ANSI's role as a quasi-regulatory authority in the accreditation of standards bodies. This document has now become a focal point for the discourse about consortia legitimation because it defines federal policy with respect to regulation and procurement. The document directs agencies to use voluntary consensus standards in lieu of government-unique standards "except where inconsistent with law or otherwise impractical." But in doing so, its wording raises a question for consortia, described by Gerald Ritterbusch, Director of Standards for Caterpillar Inc., in his recent testimony before a Congressional committee:

*Because the OMB Circular No. A-119 is silent on consortia standards, some have interpreted the meaning of the circular to not allow consortia standards to be used by government. The emphasis for the circular was to make inroads in the usage of formal standards development process standards by government because the process already has performed the due diligence that would be required for government promulgation of a standard.* (Ritterbusch, 2001, p. 3)

Ritterbusch, an ANSI board member, wants to preserve ANSI's accreditation role, but at the same time probably does not want to exclude consortia standards from procurement, thus possibly provoking increased pressure for Congressional legitimation of consortia in *OMB Circular A-119*. He continued with conciliatory words:[16]

*The circular lumps all other standards as "other" and delineates the private sector standards as "non-consensus", "industry", "company", or "de facto" standards. As consortia standards are "non-consensus", they would clearly be in this group. ...the purpose of the circular was to create greater interest with government agencies to look to the formal standards development process standards, but not limit them to the use of only these standards. Thus, the use of consortia standards by government is not prohibited in any way.*

Contrary to Cargill, Ritterbusch asserts that the process matters; that timeliness is important, but not at the sacrifice of openness, information availability, balance, consensus, and due process. Rather than relying only on "like-minded interests," Ritterbusch prefers "multiple minds and hands." He characterizes the debate in terms of risk. According to Ritterbusch, the formal voluntary consensus process provides a form of due diligence evaluation that the user would otherwise need to perform, and that although consortia standards may be appropriate in some cases, the burden of risk is then on the user, and the preference in the long run should be toward the consensus standards.

## Over-Inclusiveness Argument

An argument favoring consortia is made by Balto in describing what he calls over-inclusive standard -setting organizations and in defending the tendency by consortia to limit participation. He bases his argument on issues of competition, efficiency, consensus, and innovation:

*Obviously denying access means that some firms will be excluded, and that raises questions whether those firms can use the antitrust laws to claim an illegal group boycott. Some might suggest that the "safe" course is to admit all to avoid antitrust disputes. Such an approach may be unnecessary and counterproductive. It is unnecessary because for the most part, the courts have accepted the argument that the ability to restrict membership to a standard setting body may be efficient and procompetitive.*

*It is counterproductive because a relatively small group may function more effectively than a more inclusive one and the resulting efficiency will increase incentives to establish standards, thus increasing output...Overinclusiveness may pose more significant problems than exclusion.... Standard setting bodies may end up in delay or stalemate. Adoption of a new technology may be delayed as the body tries to achieve consensus...Overinclusive standard setting may actually result in less innovation [through] reduced differentiation, ...dampened incentives to innovate, [and]...entrenchment of an inferior standard.* (Balto, 2001, p. 6-7)

Balto's arguments are reminiscent of arguments that have long been advanced against democratic forms of governance; inefficiency, delay, gridlock, *etc.* Although Balto's arguments sound reasonable and plausible, they do not address the basic concerns of the *enclosure* discourse: the downsides of exclusiveness are unchecked power, lack of accountability, and the depredation of the public commons. The recurring argument is that: because the resulting standard document is open (if not the *process* by which it was made), it is good enough to be called an open standard. This is an enthymematic argument relying on assumptions about what is open. Lessig would likely argue that who gets to *define* the code and the architecture is important, not simply who gets to *use* it, because the former inherently shapes the latter. Balto's arguments are pragmatic and rhetorically persuasive, and what he describes may be reality in practice; but they do not address the fundamental issue of legitimacy.

## Institutional Motivation

It is possible that one of the driving forces for consortia formation is simply money. ICT corporations have money to spend and problems to solve. Consortia do address real problems and do reflect industry needs. Each consortia is a corporate structure, even if non-profit, and may take the form of a highly focused trade association[17] whose activities may complement those of SDOs. Through such institutionalization, organizations often tend to acquire a life of their own, which may persist well after the original mission has been accomplished. Consortia membership fees are often huge, especially for large corporate members, with some paying over $50,000 per year (compared to the rather meager budgets for SDOs). This funding enables consortia to hire a paid cadre of management, consultants, and staff, who may then establish a significant ongoing organizational economic and cultural structure.

## Two Systems, Solutions, and Motives

The ascendancy of consortia, as described previously, has led to a global standardization system that has bifurcated into roughly two kinds of organizations—the traditional formalistic bodies and the newer, diverse consortia, or fora—and into two resulting solutions to the problem of standardization. These two systems and solutions are driven by two contrasting sets of characteristics. These are depicted in Table 7.1. It should be remembered that the consortia category is much more diverse than is the formal category, so the characteristics depicted in the table should be considered as general guidelines, with all of the caveats with which generalizations should be treated.

The principal characteristics described in this table are classified according to organizational motivation, member participation and activities, resources and their sources, and approach to intellectual property. These contrasting characteristics are manifested in the arguments that are advanced by each side about their relative merits and the basis of their claims to legitimacy.

*Table 7.1. Principal characteristics of consortia vs. SDOs*

| Consortia | SDOs |
|---|---|
| *Motivation* | |
| market-driven | process-driven |
| speed | consensus |
| strategic focus | technical focus |
| *Participation and Activities* | |
| like minded participation | balanced participation |
| pay-to-play | open access |
| high membership dues | little or no fees |
| private club, responsible to members | public responsibility/obligation |
| defensive participation | participation only where relevant |
| full service (marketing, promotion, branding, testing, certification, *etc.*) | specialized (to standards-setting only) |
| *Resources* | |
| membership dues, license fees | document sales revenue, trade association sponsorship, some government funding |
| significant budget and staff support | all volunteer/little or no budget |
| *Intellectual Property Rights (IPR)* | |
| embrace IPR issues (*e.g.*, licensing) | avoid IPR issues (*i.e.*, RAND statement: Reasonable and Non-Discriminatory) |
| specifications distributed free | standards documents for sale |

## TAXONOMY OF ARGUMENTS AND DISCOURSES

The basic enclosure process that has been observed here is the shift of a significant amount of standardization activity and industry financial resources, at least in the ICT industries, from the accredited voluntary consensus process to consortia of various types. The reasons are varied, but speed and control are clearly important factors. The more public process is often perceived to be slow, uncontrollable, and, to some, unnecessary. But for now the consortia process still lies outside the fold—lacking in the legitimacy that is conferred by the formal institutional process and by government procurement and trade policies. Some consortia have carved out their own niches and gained enough respect through technical and market performance to be satisfied with the situation. These consortia have hammered out satisfactory working relationships with the formal institutions (*e.g.,* the IETF). Others are clamoring for legitimacy, for example, in federal procurement processes, in the United States, the European Union, and in the WTO policy arena, where legitimacy boils down to money.

### Arguments

The basic arguments regarding standardization are analyzed here for *topoi*[18], or lines of reasoning, and classified into a taxonomy; then the arguments are summarized; and finally, specific cases and texts of characteristic discourses are analyzed in detail for their use of these arguments. The first level of classification found here is between (1) arguments for and against consortia legitimation and practice, and (2) standardization system structural[19] arguments. All of these arguments are also classified according to the principal issues that they address. Many arguments address multiple issues and there is considerable overlap among them. These issues are described in terms of the essential questions that they propose to answer. The basic issues identified here are:

- Openness—How open should the process be?
- Expediency—How should the process be done?
- Administration—Who should do it?
- Purpose—Why should it be done?
- Legitimacy—How legitimate is the process or the outcome?

The consortia and formalist legitimation arguments constitute one distinct set of arguments. The system structural arguments constitute a second distinct and more general set of arguments. In the debate over consortia, the two sets often become interwoven. This is in part because the consortia phenomenon was largely initiated in the U.S. and carries with it significant cultural aspects that carry over between the two sets of arguments.

## Consortia Legitimation Arguments

Each consortia legitimation stock argument, or *topoi*, is given a short mnemonic tag and briefly paraphrased in its preferred reading, followed by its oppositional reading. The preferred reading is the meaning intended by the presenter of the argument. The oppositional reading is the meaning attributed by an adversarial reader, and it is not the same as a rebuttal or counter-argument, although it may be similar. Each argument is also summarized in the first half of Table 7.2.

The *open standard* argument: Because the outcome is open, the process is irrelevant. Oppositional reading: This is simply trying to justify an exclusionary process.

The *compatibility aim* argument: Because compatibility is the real aim of standardization, the process to achieve it is unimportant except where there is a regulatory justification. Oppositional reading: This is simply trying to justify an exclusionary process.

The *over-inclusiveness* argument: Too many "cooks in the kitchen," slow down the process and hurt competition, efficiency, consensus, and innovation. Oppositional reading: This is simply trying to justify an exclusionary process.

The *market-driven standards* argument: The market knows its needs better than government or self-appointed technocrats. Oppositional reading: This is simply trying to justify an exclusionary process serving limited interests, and to avoid common infrastructure needs which are not adequately served.

The *relevance* argument: The market knows what needs are relevant better than government or self-appointed technocrats. Oppositional reading: This is simply trying to justify an exclusionary process serving relevant (favored) interests.

The *speed* argument: SDOs, being removed from the market, are slow and unresponsive to the real needs of the market. Oppositional reading: Slowness is a myth that is not justified and is used as an excuse to justify an exclusionary process and hide other agendas.

The *lesser evil* argument: Consortia may not be perfect, but they are better than living with proprietary standards that are not open at all. Oppositional reading: This is simply trying to justify an exclusionary process.

The *multi-party* argument: Consortia are better than having standards set by a single party monopolist that is not open at all. Oppositional reading: This is simply a variation on the aforementioned *lesser evil* argument, trying to justify an exclusionary process.

The *common good* argument: Consortia standardization is a reality, so we should accept and live with it while we work together for a common good, the avoidance of proprietary or monopoly-owned (single-party) standards. Oppositional reading: This is simply trying to justify an exclusionary process and is perhaps a licensing cartel or cabal.

The *managed innovation* argument: SDOs are not equipped or suited to deal with the difficulties of managing Intellectual Property Rights (IPR), while consortia can aggregate, license, enforce, and limit IPR royalties. Oppositional reading: This is simply trying to justify a licensing cartel.

The *monopoly's child* argument: SDOs have inherited the culture of the old telecom monopoly world and many of its bad habits, such as slowness, insulation from market forces, procedural excess, *etc.*, and they have no special claim to legitimacy. Oppositional reading: This is simply trying to justify an exclusionary process.

The *regulatory relevance* argument: the public's range of legitimate interest in standardization is limited to areas with regulatory relevance (*e.g.*, health, safety, *etc.*); outside of that, let the market work on its own. Oppositional reading: This is simply trying to justify an exclusionary process.

The *democratic failure* argument: SDOs are not really more democratic in actual practice, having many failures and shortcomings in meeting their ideals; often consortia actually are better in this regard. Oppositional reading: That principles of process may not always be met is no reason to throw them out; this is simply trying to justify an exclusionary process.

## Formalist/Traditionalist Legitimation Arguments

The formalist/traditionalist arguments, many of which respond to the aforementioned consortia arguments, are briefly paraphrased as follows and are also summarized in the second half of Table 7.2, along with their oppositional readings.

The *accreditation* argument: Accreditation establishes the openness and legitimacy of the standardization process as well as the standards that are produced, assuring that principles of due process, consensus, and balance were followed in recognition of a public purpose. Oppositional reading: Accreditation, as structured, is unnecessary and is a means of protecting territory from other bodies; it preserves unnecessary bureaucracy and protects traditional revenue streams.

The *standard vs. specification* argument: Only standards established through an accredited process should enjoy the title of standard, a term that conveys an assurance of openness and quality; other normative documents should be called specifications. Oppositional reading: We all can make standards, and SDOs have no special claim to the term, not even to the term SDO—we also are SDOs.

*Table 7.2. Taxonomy of consortia legitimation arguments and counter-arguments*

| Argument | Issue | Preferred reading | Oppositional reading |
|---|---|---|---|
| *Consortia Legitimation Arguments* | | | |
| open standard (outcome) | openness | open outcome, process irrelevant | justify exclusionary process |
| compatibility aim | openness | compatibility matters, not process | justify exclusionary process |
| over-inclusiveness* | expediency | too many "cooks in the kitchen" | justify exclusionary process |
| market-driven standards | administration | market knows needs best | infrastructure needs not served |
| relevance | administration | market knows needs best | justify exclusionary process |
| speed | expediency/legitimacy | unresponsive to market | myth/an excuse |
| lesser evil | legitimacy/administration | consortia better than monopoly | justify exclusionary process |
| multi-party | legitimacy/administration | consortia better than monopoly | justify exclusionary process |
| common good | legitimacy/administration | avoid proprietary/monopoly systems | justify exclusionary process/cartel |
| managed innovation | expediency | manage IPR | licensing cartel |
| monopoly's child | legitimacy | SDO are just old monopolists | justify exclusionary process |
| regulatory relevance | legitimacy/administration | limitation of the public interest | justify exclusionary process |
| democratic failure | legitimacy/administration | SDOs are not democratic in practice | justify exclusionary process |
| *based on competition, efficiency, consensus, innovation | | | |
| *Formalist/Traditionalist Legitimation Arguments* | | | |
| accreditation | legitimacy | open process/public purpose | protecting territory/bureaucracy |
| standard *vs.* specification | legitimacy | open process/quality assurance | protecting territory |
| stakeholder | legitimacy | open process/fairness | protecting territory |
| consensus | legitimacy | open process/technical quality | protecting territory/bureaucracy |
| infrastructure | purpose | public goods, market failure | technocracy and bureaucracy |
| resource sink | expediency/administration | consortia grab all the resources | envy, inefficiency, unresponsiveness |
| poker game | purpose | consortia defensive position | envy, poverty, inefficiency |
| accommodation | expediency/administration | SDO are adapting/accommodating | protecting territory |
| open process | openness/legitimacy | build common interests | cumbersome and unnecessary |

The *stakeholder* argument: An open and fair process must include all stakeholders to be considered balanced and legitimate. Oppositional reading: Stakeholders are those with a direct financial interest. Those not willing to pay-to-play, are not sufficiently interested to contribute. This argument is just a way of protecting territory.

The *consensus* argument: Only through an open and balanced consensus process can all interests be properly represented; superior technical quality comes from striving for broad agreement with many interests represented. Oppositional reading: Excessive consensus takes too much time and hurts innovation; consensus can be reached more quickly with fewer people around the table. This argument is just a way of protecting territory and bureaucracy.

The *infrastructure* argument: Market-driven consortia may be good at applications but they are not good at building basic common infrastructure; public goods are chronically under-produced. Oppositional reading: Moving away from the market is an excuse for bureaucracy and for technocrats who think they can pick winners and losers.

The *resource sink* argument: Consortia are extremely expensive and proliferating, using up all the financial resources and leaving none to support the standards system. Oppositional reading: The SDOs are envious; their business models don't work and they are trying to excuse their poverty, inefficiency, and unresponsiveness.

The *poker game* argument: Consortia participation is often driven by defensive motives and distrust and does not reflect genuine interest in standardization. Oppositional reading: SDOs are envious and trying to protect their territory.

The *accommodation* argument: SDOs are adapting and accommodating the consortia by working together in hybrid arrangements, Publicly Available Specifications (PAS), workshops, and other means. Oppositional reading: The SDOs owe an obligation to consortia which are really doing the work. SDOs are simply protecting their territory.

The *open process* argument: Standardization should be about building common interests, so process is as important as outcome. Oppositional reading: The SDO interpretation of an open process is cumbersome and unnecessary.

## European System Structural Arguments

The European System Structural arguments are briefly paraphrased as follows and are summarized in the first half of Table 7.3, along with their oppositional readings:

The *negotiated vs. market-driven* argument: Standards should be negotiated among many stakeholders, not simply those with market-oriented or financial concerns, such as government agencies, regulators, consumers, users, *etc*. Broader public needs should be addressed, not simply those of manufacturers. Oppositional reading: A focus on negotiation ignores vital economic interests and becomes an excuse for bureaucracy and technocracy.

The *principles vs. rules* argument: Standardization should be guided where possible by general overriding principles that can be applied consistently across industry sectors. Oppositional reading: Principles are subject to interpretation that invites uncertainty, and they may needlessly constrain sectors where they are not needed.

The *horizontal vs. vertical* argument: Applying the same principles horizontally across sectors makes the system of standards coherent. Oppositional reading: This is trying to force solutions from one sector onto another where they may not appropriate and may needlessly constrain sectors where they are not be needed (similar to *principles vs. rules* argument).

The *top-down vs. bottom-up* argument: Top-down management and coordination of standardization provide guidance and assure coherence, avoiding conflicting standards. This is very similar to the *horizontal vs. vertical* argument. Oppositional reading: This imposes bureaucracy and technocratic judgments, removing it from the market.

The *harmonization* argument: It is important to provide coordination to reconcile potentially conflicting or competing regional standards in order to establish a coherent and consistent system that is the same worldwide. Oppositional reading: Harmonization in practice means "throw out yours and use ours," due to imbalances in voting rights.

The *coherence* argument: It is important to provide enough centralized leadership to assure a coherent system of standards. Coherence means possibly competing, but not conflicting. Oppositional reading: This is an excuse for imposing bureaucracy and technocratic judgments, removing standardization from the market.

The *essential requirements* argument: Rather than setting regulatory standards, government should establish essential requirements to guide standards bodies and thus hand off much of the role of regulation. Oppositional reading: This masks resistance to deregulation.

## U.S. System Structural Arguments

The U.S. System Structural arguments are briefly paraphrased as follows and are summarized in the second half of Table 7.3, along with their oppositional reading.

The *rules vs. principles* argument: Standards need to meet real needs in a particular market, and they should establish specific rules that provide certainty to producers and users in the marketplace. Oppositional reading: Narrow rules are simply a way of complying with a requirement while limiting constraints and keeping open the maximum number of options for manufacturers or providers.

*Table 7.3. Taxonomy of standardization system structural arguments and counter-arguments*

| Argument | Issue | Preferred reading | Oppositional reading |
|---|---|---|---|
| *European System Structural Arguments* | | | |
| negotiated *vs.* market-driven | openness/purpose/legitimacy | broad (public) needs addressed | government bureaucracy/technocracy |
| principles *vs.* rules | administration/legitimacy | general case guides the system | uncertain and constraining |
| horizontal *vs.* vertical | expediency/administration | coherent standards | conflicting standards |
| top-down *vs.* bottom-up | expediency/administration | coherence and guidance | government bureaucracy/technocracy |
| harmonization | administration/purpose | coordination and consistency | throw out yours and use ours |
| coherence | administration/purpose | competing but not conflicting | government bureaucracy/technocracy |
| essential requirements | purpose/administration | regulatory hand-off | resisting deregulation |
| *U.S.System Structural Arguments* | | | |
| rules *vs.* principles | administration/legitimacy | market needs certainty | limit the damage, maximize options |
| sectoral approach | administration/expediency | responsive to market needs | limit damage, keep constituents happy |
| bottom-up *vs.* top-down | expediency/administration | responsive to market needs | incoherent, allows conflicting standards |
| global relevance | administration | reducing European domination | excuse lack of participation |
| voluntary | administration/expediency | government meddling | public interest un-represented |
| block voting | administration | unfair representation structure | voting against the U.S. position |
| privatization | openness/legitimacy/admin. | economically efficient | public interest un-represented |
| competition | purpose/legitimacy | consumer choice | insincere, consumer confusion |

The *sectoral approach* argument: Each market sector has different needs, and participants in that sector best know what their needs are and should set their standards accordingly. Oppositional reading: This is a way for specialized trade associations to maintain their territory and keep their constituents happy, limiting constraints, and keeping open the maximum number of options.

The *bottom-up vs. top-down* argument: Standards should be responsive to needs that are identified by those engaged in the market, not by bureaucrats, technocrats, or planners. Oppositional reading: This is just about diverse trade groups maintaining control, while often allowing incoherent or conflicting standards.

The *global relevance* argument: Standards must be relevant to global markets and not allow regional domination. Oppositional reading: Complaints of regional (*i.e.,* European) domination often are an excuse for a lack of participation in the international system.

The *voluntary* argument: Through the voluntary consensus system the market sets standards with government acting only as an ordinary participant. Compliance is voluntary, not mandated, and does not take on the force of law or regulatory policy. Oppositional reading: This allows the public interest to be under-represented; standards very often must also serve a public or common purpose, beyond simply a regulatory function.

The *block-voting* argument: The international system is unfair, with each European country having a vote, and the United States having only one vote. The Europeans often vote as a block. Oppositional reading: This is really fear about any voting against the U.S. position. The actual work is done by consensus at the working group level, not by voting.

The *competition* argument: Standards should be able to compete, leaving it to the market and to consumer choice to determine which one is best. Oppositional reading: This is insincere and is really about each trade association or standards body trying to claim territory; it leads to consumer confusion, inefficiency, and lost opportunity.

## Summary of Principal Arguments

The foregoing taxonomy of arguments provides a set of specific elements, or *topoi*—lines of reasoning out of which the principal arguments are constructed. These elements may then be combined with more general arguments, or elements, to form the principal arguments. Such general arguments might include efficiency (expediency *vs.* delay), consensus (fairness *vs.* gridlock), standardization (innovation *vs.* stalemate), IPR (innovation *vs.* monopoly/cartel), sector (private *vs.* government roles), sectoral standardization (market/industry *vs.* coherence), and openness. Some more general arguments are mythological or ideological (or moral) in nature, relying on sets of cultural beliefs or conceptions of reality that are unstated and may be interpreted differently, depending on the audience. Typical mythological arguments might include *slowness*, *openness*, *publicness*, and *compatibility* as the aims of standardization. Typical ideological arguments might include democracy, the free market, the role of government, the role of the market, and the role of law (regulation/law *vs.* volunteerism), wealth creation (commons *vs.* private), innovation (freedom of ideas *vs.* ownership of ideas), the relationship between IPR and innovation, rules *vs.* principles, the sectoral approach, and elitist *vs.* egalitarian conceptions of governance. Some of the *topoi* listed in the aforementioned taxonomy articulate portions of these mythological and ideological arguments.

Two strawman arguments will be constructed and presented in the following two sections. They will draw on the elements described previously. The first strawman argument will be a principal argument for consortia standardization as a challenger to the traditional formal system. The second strrawman argument will be a response of the traditional formal system. These examples can be considered only as sample, or typical, arguments, because an almost infinite number of variations could conceivably be constructed.

## Consortia Principal Strawman Argument

The rise of consortia serves an unmet need in the standardization system. It provides an alternative process that is needed for a number of reasons. Consortia are more focused on specific market needs and thus faster at addressing them than the general-purpose formalist traditional system of standing bodies. Consortia are full-service organizations that not only establish a technical standard, but then go on to provide conformity testing, certification, branding, and promotion—all necessary elements to bringing a new technology into the marketplace, and not provided by the traditional bodies. Consortia can meet these needs because they are structured to raise the needed financial resources and, at the same time, to focus the group around those with a true financial and market stake.

The aim of standardization is to achieve technical compatibility and market coordination. In cases in which standards touch on public policy issues (*e.g.*, health, safety, *etc.*), the inclusiveness and procedural formalities provided by the traditional system may be appropriate, but generally such a focus results in an over-inclusiveness that actually works against efficiency, and that can impair development of competition, make consensus more difficult to achieve, and hamper technical innovation.

Consortia are fully capable of expeditiously delivering open standards. Open standards define open systems and foster a competitive and open markets. With SDOs, the market may get stuck with proprietary or monopoly standards. Consortia create standards through a multi-party process that, while not being overly inclusive, may actually be more open than many of the SDOs. SDOs are frequently not good at fulfilling their democratic ideals, are subject to political pressures, and are often more restrictive than consortia (*e.g.,* working document access, distribution of standards, *etc.*). Consortia, being adequately supported by membership fees, are not motivated to restrict access to working documents, and are able to more freely distribute their standards. SDOs are chronically under-funded and presently have no generally viable business model. They still are often dependent on traditional publishing revenues, a factor that works against their basic mission, the broadest possible distribution and use of standards.

Another problem that afflicts the formalist traditional system is the difficulty it has dealing with IPR issues. With the unfortunate trend of participants claiming IPR on standards, SDOs are adverse and ill-equipped to enforce their own RAND (Reasonable and Non-Discriminatory licensing) policies. Courts are also finding such policies inadequate. In contrast, consortia are in a good position to deal with and manage this increasing problem by forming patent pools and/or licensing authorities, capping royalty obligations on standards, negotiating with IPR holders from a position of strength, and enforcing IPR disclosure policies among their members.

Consortia standardization is a reality and needs to be recognized. Its standards are already and increasingly in the marketplace, and formal purchasing, procurement, and trade systems must and do utilize them, albeit informally at present. Some criteria must be found to recognize and incorporate consortia standardization into the global system, perhaps through some form of "differentiated" standards policy that recognizes the difference between regulatory governance and market coordination in meeting the separate legitimate interests of both the public and private markets.

## Formalist/Traditional Principal Strawman Argument

The formal standardization system is being challenged by the rise of consortia, but it is adapting and improving itself accordingly. Consortia have often represented themselves as a panacea and have cloaked themselves in mythology. They claim to be faster and more focused on their goals than SDOs. An examination of the record shows examples that contradict this myth and demonstrate that consensus-building takes time, whether it be in a consortium or an SDO. Certainly, time can be saved by limiting the consensus to a small number like-minded people, but what is the cost in terms of the consequences for the market and society? Such focus could also be characterized as "group-think" or "tunnel vision."

Openness and balanced representation of interests are critical elements of standardization, and the process by which standards are made can be as important as the standards themselves. It is a fallacy to claim that the only legitimate stakeholders are only those with a direct commercial interest. Information and Communication Technology (ICT) penetrates every aspect of today's information society, and everyone is a stakeholder. Public interest can no longer be limited to public policy issues. It is impossible to foresee all the implications of the new technologies that are defined by standards and that shape our world. Such matters cannot be left to the sole domain of self-selected closed industrial groups that would prefer to manage innovation according to their own strategic business interests. A good example is IPR.

Abuses of IPR laws and practices today are harming the global standardization system, with a rush to attach IPR claims to otherwise open standards. Consortia have responded to this problem by forming patent pools and licensing authorities to control IPR claims and to cap royalty burdens on standards. But the cure may be worse that the affliction, since these consortia may then become essentially industrial cartels, harming competition and innovation even further. What is needed is not the simple legitimation of consortia practices, but the reform of IPR laws and the enforcement of anti-trust laws to create consistency with an open standardization system.

In any case, consortia and the needs that they claim to meet are being addressed by the formal SDO system through adaptation and accommodation. These include, hybrid standardization practices whereby SDOs and consortia are collaborating, and the introduction of new alternative processes such as PAS, fast-track, and workshop procedures. The formal process is being redefined, meeting the needs of industry through a graduated system of standardization methods, and these new methods obviate much of the need for consortia.

The cost to industry and to the world economy of the proliferation of consortia is becoming unmanageable. It is soaking up available resources and tilting the system unfairly toward the largest players—hardly the way to encourage competition and innovation. Consortia are becoming self-serving business ventures in themselves. Often they resemble a poker game dominated by the strongest players, who keep raising the stakes, while the other players are afraid not to stay in the game. Such participation is not motivated by trust or by a common goal of standardization, but is simply defensive.

It is necessary to see consortia standardization in the proper perspective. Consortia do meet a need, particularly in the role of commercialization and diffusion of new technologies into the market, but they should be limited and should not be legitimated by further recognition. On the contrary, they should be subject to antitrust enforcement, while the IPR system is reformed to alleviate some of the pressure that drives them. The formal system and consortia are learning to live and work together.

## Purpose of Principal Strawman Arguments

The previous two sections presented the principal strawman arguments for two conflicting sides of the consortia legitimation issue. This presentation attempted to make strong cases for each position that brought out the possible arguments in a single coherent text. These arguments can be made in many differing ways, and this text is only a guide. The next step will be to examine actual texts of cases where these arguments are made and to study how the arguments are constructed and applied.

# ANALYSIS OF CASES

The discourses considered in this study ultimately center on the issue of legitimacy. Specifically, the quest for legitimacy by consortia is driven by the desire for access to systems of government procurement and international trade. Two specific cases have been chosen for close examination because of the representative rhetorical arguments they advance. These cases are not isolated or anecdotal narratives. They have been selected because they are being advanced by influential individuals or bodies whose opinions were solicited by significant policy-making forums considering the issue of consortia legitimation for the specific purpose of legislation or administrative policy. This analysis does not claim to be a representative polling of opinion; it is an analysis of terms, arguments, discourse, and the discursive formation of meaning.

The first case is an industry position paper supporting testimony before a U.S. Congressional committee recommending changes to federal procurement policies (*i.e.*, to *OMB Circular A-119*) regarding the inclusion of consortia standards. The second case is a report on an academic study of consortia to the European Commission (E.C.) recommending similar consortia-including changes to European Union (E.U.) procurement policies. These two discourses are analyzed and compared with specific opposing discourses from academic, industrial, and governmental sources. What is found is the way in which all sides couch their arguments in various ideological conceptions of what is open, what is a consortium, and what is in the public interest and ought to be legitimated.

## U.S. Congressional Testimony Case

In support of his testimony before a Congressional committee,[20] Carl Cargill, provided a thirty-page position paper (2001a) [hereafter called the *Paper* and available in the Appendix of this work] that he summarized as follows:

*Standardization is essential to the growth of the IT industry. Within the IT industry, well-developed consensus consortia standards should be placed on an equal footing with standards developed by ANSI accredited organizations. The current Federal procurement practices—as mandated by OMB A-119—discourage the use of consortia specifications. The paper concludes with a proposal for a legislative change to permit and encourage Federal use of consortia-created standards in procurement.* (p. 2)

The *Paper* describes the importance of standards to the IT industry and to the broader environment of business, society, and culture. It then offers a three page evolutionary history of IT standardization including consortia, it attempts to define consortia and their strategic nature, and it offers criteria for a "good consortium." The *Paper* then discusses the role of national policy on standardization and its "significant consequences," and offers strictures about the language in *OMB Circular A-119*. The *Paper* calls for an expanded definition of *voluntary consensus standards body* that includes "good consortia," making them legitimate. Finally, the *Paper* provides an expanded history that places consortia at the end of the evolutionary chain of standards-setting organizations.

The *Paper* begins its first rhetorical arguments by tracing the historical role of IT standardization up to the formation of consortia, then offering the IETF as the role model:

*...in the later 1980s, a different form of standardization activity appeared... Providers began to move technology standardization away from the formal ANSI and ISO recognized SDOs to those of consortia, which did not have the intricate processes of the SDOs. ...[processes] which were both time consuming and often Byzantine, were necessary because "[m]ost delegates represent[ed] personal, professional, national, disciplinary, and industry goals..." and managing this vast and sometimes contradictory set of expectations forced these groups to create intricate rules to make sure that all voices were heard. Consortia, on the other hand, because they usually consisted of groups of like minded participants (either for technical or market reasons), did not need to have the lengthy discussions over the mission and intent of the proposed standardization activity—an organization's presence was, in many cases, proof of a general agreement. The archetypal consortium was the Internet Engineering Task Force (IETF), the group that manages the Internet. The success of this group in both keeping the Internet a leading-edge technical architecture leader as well as clear of greed, parochialism, and lethargy is a significant accomplishment.* (p. 4)

Here in this text the *Paper* builds a logical argument similar to Balto's overinclusive argument. But, then it makes an interesting rhetorical leap. It presents the IETF as the archetypal model of a consortium, and then devotes most of the remainder of this text to extolling the success of the IETF and of the Internet (and the Web) as an open medium for "...making decision-making more transparent" and for the creation of "highly open, highly visible specifications." The *Paper* praises the IETF's practices of open and free membership, document access, meeting participation, consensus building, and public comment. Although few would disagree that the IETF has been a highly successful and highly regarded consortium (certain arguments about its openness notwithstanding), the problem here is that the IETF is unique and bears almost no resemblance to the other membership-based consortia that the *Paper* really represents. This argument rhetorically shifts from consortia in *general* to the IETF as a *specific* example, but the IETF is completely unique and atypical of the sort of consortia (*e.g.,* paid membership, industry driven, etc.) that the writer advocates in the *Paper* and elsewhere.[21] IETF participation is completely free and open, and it is based on individuals, not organizations. The result is a diverse group of participants, hardly a like-minded group in the sense used by the writer in the aforementioned excerpt.

This part of the *Paper's* discourse seeks to create an identity between consortia as a class and the IETF (and the Internet itself). It presents an enthymematic argument, generalizing from a specific case to consortia as a class. Good things about *x* are extended to *y* as a member of the arbitrary class, without justification: *i.e.,* the Internet and the IETF that created and maintains it are open and effective, and the IETF is a consortium, therefore these merits may be attributed to other consortia. It is important to remember that the class, consortia, is a catch-all term for virtually all standards bodies that are not formal SDOs.

There is a certain irony in the *Paper's* valorization of the IETF, which is not structured at all like the other consortia for which the *Paper* advocates. Although the IETF has performed well as a standards body, there are certain things about it that make it not a generalizable model for other consortia. For one, it lets the "little guy" participate, who could not pay-to-play in other consortia that motivate the *Paper.* Additionally, the IETF has a unique governance structure that depends on the good will of its traditional leadership as individuals. It completed a major part of its standardization work long before the Internet became economically important. At that time, no one cared what the academics were doing in their labs. The IETF established its own set of cultural practices which became institutionalized. It may have resisted corruption by the wealth it has created—so far—largely because of the personalities of its leadership and its culture. It would likely be difficult or impossible to form such a body and collegial governance structure in the highly competitive and strategic atmosphere of the ICT industry today.

In its attempt to define consortia, the *Paper* emphasizes the term *strategic*, referring to the important "...systems, architectures, or new emerging markets where there is a need for a large number of interrelated and/or continuous specifications" that such consortia are standardizing (p. 7). The implication here seems to be that because consortia are strategic, they are important and should be accommodated (or legitimated). But strategic also means something more than important. *Strategic* often refers to actions taken in response to, or in anticipation of, the actions of others, taken in pursuit of some advantage and focused entirely on that end. Applying this meaning to consortia as strategic would seem to argue for even more public access and/or oversight, not less. Lessig (and the *enclosure* discourse) argues that system architectures in particular are not the place for exclusiveness. This line of reasoning leads to the issue of *balance*.

Perhaps the primary argument in the *Paper* is over the ANSI/SDO principle of balance. The *Paper* acknowledges that consortia can often meet essentially all other criteria in the voluntary consensus system except that of balance. Lack of balance seems to be an essential quality of consortia. The *Paper* argues:

*With respect to participation, ANSI-accredited SDOs cite "balance of participation" (parity between the various affected parties, usually providers, users, and others) as one of the criteria for judging whether an organization is legitimate. By definition, a consortium tends to be biased towards those who are interested enough to "pay to play", which may be enough to violate the ANSI rule of balance. What must be assured is that no party is denied the right to participate based upon the nature of the would-be participant, unless the participant is unwilling or unable to meet the common entrance requirements of the consortium.*

*The key to judging the "openness of the consortia" is one of the major differentiators between the consortia and the SDO forms of standardization. Openness has traditionally been viewed as the willingness to admit all concerned parties to the table. Consortia typically do not do this. Only consortium members may be allowed at the table to discuss specifications. This is why the members are willing to pay—they are trading money or other resources for the ability to determine the specification. This is not substantially different than the SDOs, where participants trade resources (time and travel budget) for the right to participate. Both groups traditionally charge fees—the difference is the amount of the fee charged. Therefore, it is necessary to create new criteria for "openness" among consortia.* (p. 8)

The *Paper's* argument for legitimacy is nowhere more clearly laid out. In the phrase, "interested enough to 'pay to play'," it equates interest directly with the willingness or ability to pay money, and, conversely, the lack of willingness or ability to pay money with lack of interest. This rationale could hardly be further from IETF standardization practice, which engages large numbers of small participants, has nothing to do with money, and serves to institutionalize principles of inclusiveness. The passage ends with an unconvincing attempt to draw a parallel between consortia and SDOs in respect to payment of fees, asserting that the "difference is the amount of the fee charged." In practice, that difference is measured in orders of magnitude.

The *Paper* contends specifically that *OMB Circular A-119* is overly restrictive about what it considers open and thus what is included in procurement policy. It maintains that the *Circular,*

*...seems to state...that the use of consortia based standards, which are open, consensus driven, and lack only the "balance" described in 4.a.(1)(ii) are the equivalent of proprietary or* ***de facto*** *standards, which they are not. Consortia standards represent standards that have been developed in an atmosphere that is as rigorous—if not more so—than most SDO standards, yet it is deprecated because it does not meet the five voluntary criteria (p. 11).*

So, if not balance, then what should be the criteria for openness? The *Paper* continues:

*The primary test for openness should be the outcome of the consortia—(1) the specification should provide an open (RAND minimum) reference implementation, (2) two or more competing implementations should exist, and (3) there should be, if appropriate, a testing regime to ensure interoperability among the various implementations. This approach focuses on the rationale for standardization—that is, there should be a mechanism by which the users have a choice of implementations from which to choose, providing guaranteed alternative sources for critical products. (p. 8)*

The argument here is that open refers not to the process, but to the outcome—*i.e.,* that the outcome (the standard) is open and free. But the reference to *RAND* (Reasonable and Non Discriminatory license terms) means that open outcome does not mean non-proprietary or free, merely available on a non-discriminatory basis. Thus, in the case that the consortia may enjoy an intellectual property right to the standard, it agrees to license the standard openly. Such would imply that the consortium is a licensing pool or cartel—a far different meaning of open than that intended in ANSI, OMB, ISO or other SDO practice.

The *Paper* acknowledges that all consortia may not deserve legitimation and it proposes a six-point criteria (p. 9) for legitimate consortia, or good consortia (p. 13). A good consortium must:

1. Produce "usable" technical specifications.
2. Have a legal basis—some form of legal entity (and some form of government oversight).
3. Have well-defined rules and processes (*i.e.,* charter, by-laws, *etc.*), assuring fair representation of members and anti-trust protection for members.
4. Have clear and legitimate IPR policy that requires, at a minimum, RAND licensing of all IPR included in its specifications.
5. Have membership not arbitrarily restricted (restricted only on economic basis).
6. Should create reference implementations, competing implementations, testing, and conformance.

A review of these criteria raises the question of how would it be known that a particular organization properly met them. Such is precisely the function of accreditation—the primary purpose for ANSI. The *Paper* alludes to acknowledging the appropriateness for "some form of governmental oversight" (p. 7) and specifically suggests that, "It may be appropriate to include a directive to NIST [National Institute of Standards and Technology]... if the private sector demands consortia accreditation" (p. 13). It appears obvious that some accreditation process would be necessary, with the wide range of organizational structures and practices involved among hundreds of consortia. Here, ironically, appears to be an invitation for more government regulation of exactly the form that was threatened by proposed legislation in the 1970s, as discussed earlier, and that became the prime impetus at that time for establishing ANSI's non-governmental accreditation role and for defining the voluntary consensus system in the *Circular* and in the *National Policy on Standards for the United States*. One conclusion that might be drawn is that the *Paper's* real complaint, at least in part, is with ANSI as an accreditor rather than with accreditation in principle. Such might be implied by the phrase, "...ANSI is focused on maintaining its hegemony..." (p. 13), a thread that seems to run through the *Paper.*

The *Paper's* concluding paragraph (p. 22) is interesting in its rhetorical arguments:

*All of the various forms of standardization can and do serve a purpose in the IT sector. There is the need for stability (provided by the formal arena), a need for defined and structured faster change (provided by consortia and alliances) and the need for complete community involvement (provided by open source.)*

This presents a plea for a place among equals. It would then follow that if SDOs and (some) consortia are equals, they ought to enjoy equal recognition and legitimacy. Speed is again emphasized as a primary value worthy of recognition. The need for community involvement and open source is also emphasized.[22] The *Paper* continues:

*The groups within each arena have not learned to work together for the good of "open systems". Rather than considering proprietary and closed systems to be the force to be changed, they have dissipated their energies arguing about which form of standardization is best, forgetting that the answer is that "Standardization is best, and non-standardization is less than optimal."*

This brings in the *common good* argument, which might be paraphrased as, "Consortia may not be as open as SDOs, but they are better than Microsoft, *etc.* If we fail, look what you will have to live with."

*ANSI is a necessary, but not sufficient, standardization component for the needs of the IT sector. Consortia are central to IT standardization success—but need the stability that the formal process can offer.*

Here is a vague complaint about ANSI followed by an appeal for stability and some accommodation with the formal process. It is not clear here what problem the *Paper* has with ANSI, but ANSI appears to have recognized a need for some accommodation. In the latest *National Standards Strategy for the United States*, policy 4 is to "Broaden the U.S. standards 'umbrella' to include all those organizations that are contributing to the standards system" (ANSI, 2000, p. 8). This policy includes the exhortation, "Non-traditional standards organizations should review their objectives to determine where closer interaction with the formal system will help add value to their efforts."

The foregoing critique of the *Paper* attempts to follow its logical, enthymematic, and ideological arguments. It has pointed out inconsistencies and identified the central argument and purpose. It has shown how the arguments identified earlier in the taxonomy are applied in practice.

A recurring and central argument for consortia standardization is the issue of speed—or the slowness and unresponsiveness of the traditional formal standardization process to market needs. This argument has been challenged by Sherif in several respects. He argues first that the preoccupation with speed is dangerous:

*If standardization is about new concepts, urgency can distract from aspects that may or may not be apparent at the onset. In addition, the self-selecting nature of [consortium] membership, lack of critical distance, and pressure to produce quick results could lead to "group think" or "tunnel vision." The adopted solution, while satisfactory on a small scale, may not be suitable for wider deployment.* (2001, p. 94)

Sherif further argues that the notion that the formal standards process is inherently slow is a myth that is unsupported by objective evidence and masks other agendas, primarily of manufacturers.

*It is widely believed that formal standard bodies are less responsive to market needs than industrial associations and consortia. ...[A]greements on shaky technical foundations for the sake of producing documents may not be very helpful in the long run. ...[T]he time horizon of manufacturers...is much shorter than of [users]. ...we propose...separating the technical issues from the immediate business needs and/or ideological persuasion. ...One cannot escape the conclusion that statements like "formal standardization is too slow," which are contradicted by objective facts, have political and ideological motivations.* (Sherif, 2002)

Sherif cites cases where formal standardization was fast and responsive, and cases where consortia became backlogged, delayed, and missteped. He attributes "political and ideological motivations" to such issues as intellectual property rights and business strategies among closed industrial groups (*e.g.*, manufacturers) for the control of certain technologies or innovations.[23]

A careful reading of the *Paper* does not dispel Sherif's concerns, and certain passages suggest that consortia are really about much more than *speed.*

*The consortia, responding to the pressure of "time is money, especially when the product life cycle was shrinking", wanted a faster system. The proponents and opponents of consortia have focused on this "speed issue", not realizing that increased speed was achieved in a consortium by changing the process. The argument has never been about speed; it has been about the process needed to achieve the speed necessary to satisfy the market needs of the members of the organization.* (p. 19)

Here, the *Paper* acknowledges that the argument has never been about speed, but about the market needs of the members—about changing the process. Here, the *Paper* implies that the project is not about technical issues but about market issues and business strategy for a self selected group—or, in terms of the *enclosure discourse*, about the enclosure and control of technical innovation.

Experience has shown that even though some consortia have produced excellent and useful stand-alone specifications, they have increasingly discovered the value of getting these adopted as formal standards (Jakobs, 2001, p. 137). This is evidenced by an increasing clamor around ISO and IEC "PAS submission"[24] to get consortia specifications incorporated into the formal ISO/IEC/ITU system as standards, so that they may obtain a level of legitimacy and respect. Some consortia are learning to work alongside the formal bodies and make their work compatible and acceptable from the beginning—not just as an afterthought (Schoechle, *et al.*, 2002; Vercoulen & van Wegberg, 1998). "[This] model also allows for a consortium to do the initial specification and subsequently have a formal body to transform it into a standard once it has sufficiently matured" (Jakobs, 2001, p. 141).

Many of the same questions and rhetorical arguments presented in the *Paper*, and some additional ones, can be found in standards discourse on the European side of the Atlantic, also focused on the issue of government procurement.

## European Commission Report Case

In a report to the European Commission[25] (E.C.), a European university[26] provided a sixty-nine page study (Egyedi, 2001) [hereafter called the *Report*] that is summarized as follows:

*Current standards policy appears to be caught up in a polarised discussion about what type of organisation best serves the market for democratic and timely standards: standards consortia or the traditional formal standards bodies. The general feeling is that standards consortia work more effectively, but that they have restrictive membership rules and are undemocratic. The latter is a cause of concern for the European Commission, which requires democratic accountability in the standards process if it is to refer to such standards in a regulatory context. The Commission's request for new input on how to deal with consortium standards is set against this background.* (p. 3)

The *Report* is organized in three parts. It proceeds by first presenting two case studies of specific consortia standardization efforts, Java™ standardization in European Computer Manufacturers Association (ECMA), and XML (Extended Markup Language) standardization in the World Wide Web Consortium (W3C). It then confronts dominant assumptions on consortia standardization (*i.e.,* that they lack openness and are undemocratic) with the case findings and critically examines the current basis for standards policy. Finally, the *Report* draws conclusions and recommends a differentiated European standards policy recognizing consortia standards, and a focus on compatibility as a goal rather than on standardization as a process.

The *Report* recognizes the European Commission's commitment to formal European and international standardization and the voluntary consensus process, based on essentially the same principles as ANSI. It elaborates on this commitment in what it terms a democratic ideology characterized by,

*decision making by consensus; voluntary application of standards; broad constituency of (national) delegations; well-balanced influence of national members in the management of international standards bodies; and impartial, politically and financially independent procedures.* (p. 11)

It is important to remember that the European system differs from that in the United States in that formal standardization is typically carried out by national bodies rather than by trade associations, and that representation in regional and international committees is still determined largely on the basis of national bodies.

The *Report* then frames the debate, and seeks to challenge this commitment, asking,

*...why sometimes consortium standardisation is preferred to formal standardisation, and whether consortia work in ways that will deliver open standards ... Does the way the problem of standards consortia is defined—i.e. that their procedures are restrictive and undemocratic, and that their standards are therefore unfit as an instrument of regulatory governance—accurately describe what is at stake?* (p. 3)

This way of asking the questions rhetorically constructs two arguments. The first is that the goal to "deliver open standards" is about the outcome, (*i.e.,* use of the term "delivered") rather than the process. This is a thread that runs through the *Report* and is consistent with the its later focus on compatibility (outcome) rather than standardization (process). Here, as in the *Paper*, open is left undefined but assumed in terms of outcome, and the process is thus made unimportant. The *Report*, in asking its central question—whether consortia can deliver open standards—fails to define what is meant by open, or even to recognize the problematic nature of the term.

The second argument is that the only proper realm of democratic interest is in standards that are to have regulatory governance application. Such applications might include standards related to health, safety, or other consumer issues. It would follow then that if there is no regulatory or public interest standard (a term used later on in the *Report*, p. 45) application involved, then exclusive, restrictive or undemocratic processes would be much less of a problem, especially if the outcome (*i.e.,* the final standard) were open. Here, as in the *Paper*, and as in the *overinclusiveness* argument described earlier, the *enclosure* discourse would argue that, who gets to define the code and the architecture *is* important, not simply who gets to *use* it—*i.e.,* that the issue is less about regulation or democracy, but rather more about access and innovation.

The *Report* sets about its study by focusing on two specific consortia standardization projects: ECMA standardizing the Java computer language, and W3C standardizing the XML language. It finds that, not surprisingly, these two bodies are fairly open (inclusive and consensus based) on the lower working group level, whether or not they are democratic on higher oversight governance levels (p. 40). The *Report* observes that in even formal democratically constituted bodies, "...formal procedures are often exploited in 'undemocratic' ways," (*i.e.,* manipulation of voting, tilting the program of work, stuffing delegations, *etc.*) and that, "Regional governments and formal standards bodies are well aware that in formal standardisation the objectives of democracy, diversity and openness often are not met" (p. 43). This observation may be correct, but here the *Report's* rhetoric argues further—that because the formal process suffers deficiencies in democratic application, it should not have preference in principle.

The *Report's* Conclusion and Recommendations section, proffers an answer to its central question, Do consortia deliver open standards?

*The two cases do not simply confirm the widely shared assumption that consortia are undemocratic. To the contrary, although there may be some practical exclusion mechanisms, in principle consortium membership is open. Indeed, in certain respects consortia appear more open than the formal bodies. For example, while the latter usually keep access to committee drafts restricted to participants—and, thus, seek consensus within a limited group—consortia more often post their drafts on the web and actively seek comments from outside.* (p. 57)

It is difficult to disagree with the point about access to committee drafts (and the like), and the shortcomings of many formal bodies, but the general assertion, "in principle, consortium membership is open," is deeply problematic. Also, there remains a fundamental difficulty with the *Report's* generalizations about *consortia* as a class from only two cases, ECMA and W3C. ECMA is a broad trade association, not unlike many of the ANSI accredited SDOs in the United States, and it would likely be so accredited if it were in the United States. W3C is operated by the Massachusetts Institute of Technology (MIT) and is structured more like a private club—perhaps more typical of consortia—yet it employs the "philosopher king" model of high level governance (under the leadership of Tim Berners-Lee, a distinguished inventor), much like the IETF. Here, the *Report* makes the same enthymematic argument as the *Paper*, generalizing from the specific. With over 255 consortia to look at, encompassing a wide diversity of structures, memberships, and governance arrangements, it is difficult to draw categorical conclusions about the general merits of consortia openness or their ability to deliver open standards, and make policy recommendations from these two case studies, especially when they are either atypical and/or do not support the argument.

Another central and problematic conclusion in the *Report* is that the primary aim of standardization is *compatibility*. This is asserted in the beginning of the *Report*, not concluded from the case studies, and is not supported by theory, citations, or arguments in the study. The term compatibility is left undefined, but carried forward as a mythological construction through the entire discourse.

*The cases further highlight that company and government policies overly emphasise the means of standardisation while largely bypassing its aim, namely technical compatibility. The latter can also be achieved by other means than standardisation. Among these are the proprietary and open source strategies to Information and Communication Technology (ICT) development. In certain circumstances, the latter strategies are more effective in achieving compatibility than standardisation.* (p. 4)

Here is an enthymematic argument that depends on the assumption that compatibility is the aim of standardization. It concludes that other more restrictive means could adequately achieve the aim.

However, others would argue that compatibility is only one of a number of aims of standardization. Other aims, particularly in the ICT arena include defining *unit* and *reference* standards (including registries and character sets), *similarity* standards, and *adaptability* standards (also called *flexibility* (Sherif, 2001), or *etiquette* standards) (Krechmer, 2000a; 2000b). Another aim of standardization is *interoperability*, which is not the same as compatibility.[27] In addition, a recent aim of standardization involves technology management systems (*e.g.*, ISO 9000 or ISO 14000) and most recently, accountability standards (*i.e.*, ISO 18000). Yet another aim of standardization, and possibly most important of all from the discourse perspective, is setting infrastructure standards that define basic system architectures (including gateway standards defining interoperability between systems). This last category is where the concerns of the *enclosure discourse* are primarily focused—on the infrastructure or platform technologies that constitute the commons for innovation (*e.g.*, the Internet), and which, it is contended, could be diminished by enclosure or privatization.

It is possible that the *Report* confuses the meaning of the term compatibility with *interoperability* or conflates the two. Such might be inferred from the pervasive use of the term *technical compatibility* and its manner of usage throughout the *Report*. Nevertheless, even if one takes "compatibility" in the *Report* to mean "compatibility and interoperability," the claim that such is the aim of standardization is still problematic.

To claim that the essential aim of standardization is compatibility, and that compatibility can as well be achieved in other more enclosed ways, appears to overlook, or to dismiss, any *public* character or interest that may exist beyond regulatory concerns in standards. In many cases, the *enclosure discourse* would argue that the process does matter. In framing the issue in terms of democracy or democratic ideology, the *Report* reveals another ideological perspective, that of the market. As mentioned earlier, the *enclosure discourse* would argue that certain technical realms ought to be kept open, and that the process is as important as the outcome—that in such cases, the determining principles should be access and accountability, in process and in outcome, because one constitutes the other.

In its conclusions and recommendations, the *Report* aspires to go "beyond standardization," hence its title, to what it regards as the real issue—a means to achieve technical compatibility—thus dispensing with the baggage of democratic ideology and the rivalry between consortia and the formal system. The *Report's* emphasis on compatibility as the aim of standardization is a reflection of its own ideological perspective—the ideology of the market. That perspective is evident as the *Report* argues that the European Commission should decide "what type of democracy is needed for what purpose"—that more democracy may be appropriate for standards of *regulatory* application, "...where democratic accountability is *still* important," [*emphasis* added] (p. 61), but not for market coordination:

*For market coordination, on the other hand, the democratic requirement of "balanced representation of interest groups" could be simplified to "multi-party participation".*

*A differentiated standards policy is recommended to better cater to the significance of standardisation as a means to coordinate the market and as an instrument of regulatory governance. Differentiation prevents a situation where democratic (or other political) ideals are diluted in order to be able to apply a market-oriented standards policy to* ***de jure*** *[regulatory] situations—or, as presently happens, vice versa.*

"Multi-party participation" appears to suggest that the process ought not be dominated by a single monopoly firm, but leaves the possibility that a cartel or other form of market power might be acceptable as long as it provides market coordination. The second paragraph attempts to separate standardization into realms of regulatory governance and market coordination—and to represent the market as depoliticized. Such ideological depoliticization has been addressed by Sherif:

*...criticism of established standards association, particularly those with government involvement parallel sustained attempts to replace political and democratic control of institutions with technocratic and financial procedures and to privatize the public space. The confluence of all these phenomena can be interpreted through the observation that unfettered markets must be engineered by weakening or dissolving all intermediary social institutions between individuals and economic entities that may challenge market-driven mechanisms.*

The *enclosure discourse* would likely concur here with Sherif. Its response to the *Report* would likely be that standardization is mostly not about regulation *vs.* market coordination, but should be more concerned with access and innovation; and that market coordination is inevitably and inherently strategic and political. This view is also supported by Mansell in regard to reliance on the forces of the market to produce appropriate and timely standards:

*...choices as to when to standardize a technical design or to encourage diversity do not necessarily reflect the relative superiority of alternative design innovations. They are often the result of oligopolistic competition, political bargaining processes and conflict resolution not just within, but among, a large number of institutions and actors with an interest in the outcomes.* (1995, p. 224)

The analyzed cases are parts of a discourse that focuses on bringing about institutional and policy changes—a discursive re-construction of the standardization system—specifically the legitimation of consortia for purposes of government procurement and trade. These cases are only two examples, but they have been chosen because they are influential and illustrative of the principal rhetorical arguments being made. Both the traditional standards institutions and industry have responded to such discourse, and to the changing political, economic, and technological landscape, in a variety of ways. Some of these responses will be considered.

## RESPONSES

In view of the analysis, it is interesting to examine briefly some of the institutional and societal responses to the discourse about standardization and to look at their adaptations to the changing environment for standardization. National standards bodies and international SDOs have been undergoing dramatic changes brought about by the challenge presented by the ascendancy of consortia, the dislocations resulting from privatization of the telecommunications industry, the general reduction of government support, and the loss of publishing revenue due to electronic distribution of documents. They have been forced to change their business models and to expand the scope of their memberships. They have had to adopt hybrid forms of standardization that include collaboration with each other and with consortia that introduce new practices such as the *workshop agreement*. The European Commission has launched major initiatives such as the *New Approach*[28] (EC, 2000) to reconcile the European standardization system with European integration and the global environment. The WTO has become an important factor in the equation of standardization, especially in emerging markets and economies.

Other forms of standardization have emerged, including, most significantly, the open source movement. The scramble over intellectual property rights has become a major issue in standardization, posing legal and political dilemmas for both formal standards bodies and consortia. This trend toward the *enclosure of ideas* has also inspired both source as well as other even less conventional responses such as the *Creative Commons*.

## European Institutional Responses

Generally speaking, the traditional national bodies, the British Standards Institution (BSI) and, to a lesser extent, the *Deutsche Institute für Normung* (DIN), have responded by privatizing and internationalizing themselves. They are transforming from arms of their national governments to essentially quasi-private industry associations, expanding their clientele beyond their national borders and deriving revenues from an array of activities including publishing, training, consulting, testing, and certification. The European Telecommunications Standards Institute (ETSI), a quasi-consortium,[29] has endeavored to do the same; it is active globally with members in 54 countries. DIN is a central contributor to European standardization and has taken a lead in establishing standardization policy for the European Commission (E.C.).

A recent policy paper has been prepared by DIN setting forth the German position on current issues in standardization, including consortia, titled, *Strategy for Standardizing Information and Communications Technology* (DIN, 2002). This paper [hereafter called the *DIN paper*] provides an interesting response to the consortia discourse, including a number of the positions voiced in the *Report* discussed earlier. On the issue of compatibility/interoperability, The *DIN paper* remarks:

*...interoperability is the main standardisation objective for ICT. ...However, the standardisation objective of ICT is not limited to interoperability. Much rather, the penetration of ICT into all areas of society, especially the consumer area, means that issues of relevance to society are also gaining in importance. Security including biometry and data protection are in great need of standardisation, as are ergo-*

*nomic aspects, especially with regard to allowing the disabled barrier-free access to new technology ('disability access', 'design for all'). These issues require the participation of the general public and are consequently predestined for treatment in standardisation organisations instead of other bodies.* (DIN, 2002, p. 4-1)

Then the *DIN paper* comments on the relationship between consortia *specifications* and SDO *standardization*:

*The development of specifications and standardisation are methods for technical harmonisation called into being and supported by industry – they all have their own justification. This leads to industry's interest in combining these various methods and their products in accordance with their specific strengths and making the competent organisations and bodies cooperate purposefully, where this is necessary.*

This is a call for hybrid models of standardization and collaboration between consortia and SDOs. Then the *DIN paper* begins to take a harder line toward consortia activity:

*The proliferation of consortia in the ICT field that can be seen has considerable cost consequences for the companies affected: participation in many of these organisations requires high expenditure of personnel and financial resources, irrespective of the duplication of effort resulting from the lack of coordination between the consortia. It is therefore in the interests of the industry concerned to limit the number of consortia.*

*[this paper] therefore expressly welcomes the fact that the international (ISO, IEC) and European (CEN, ETSI; CENELEC under consideration) standardisation organisations, as well as DIN, make the establishment of fora and consortia superfluous to a certain extent by meeting the needs of industry by introducing alternative standardisation processes (workshops, etc.) and products (Workshop Agreements, etc.) and by a graduated system of standardisation products.* (p. 4-2)

The *DIN paper* proposes to pre-empt consortia activity by further developing "alternative standardization processes" such as the *workshop* process that operates under a much less formal set of procedures and consensus requirements, yet still remains under the SDOs "tent." The *enclosure discourse* might suggest the argument that these bodies are trading some of their legitimacy for "market relevance" by compromising their processes. The *DIN paper* goes on to give its perspective on market needs, or what the *Paper* and the *Report* called *market relevance*:

*Standards that do not comply with market needs do not develop an effect; developing them is a waste of resources. The difficulty is recognising market needs when initiating a project with a certain degree of reliability or recognising declining market interest on time. If the players in a standardisation project are only economically interested parties, it can be assumed that a project meets the market needs. However, the situation is different in standardisation with a potentially broad spectrum of players.* (p. 4-4)

Here, both sides of the debate seem to agree that market needs, or market relevance, is important. The problem is how market needs are determined and how interpreted. It could be suggested that the market is not always good at knowing its own needs, much less the needs of society. The last sentence in the previous quote seems to admit that there are broader needs to be considered than simply market needs. For instance, as noted earlier in this study, the market has not usually been a good indicator of infrastructure needs or the needs for public goods.

## European Union Responses

With the objective of creating a single European market beginning in 1985, the challenge has been to move from a system of mandatory or quasi-legislative-based disparate national technical standards (some intended to impede trade) to a unified and integrated voluntary consensus-based system. The *New Approach* is a set of policies to facilitate this transition through directives setting *essential requirements* that then are incorporated into voluntary *harmonized standards.*

*A fundamental principle of the New Approach is that the legislator limits harmonisation to the essential requirements that are of public interest. These requirements deal in particular with the protection of health and safety of users (e.g., consumers and workers) and sometimes cover other fundamental requirements (e.g., protection of the environment). By setting the technical specifications for the products to meet those essential requirements, European standardisation completes the picture: this makes the New Approach a good example of co-regulation.* (Bilalis & Herbert, 2003, p. 47-48)

The *New Approach* emphasizes the need for collaboration between SDOs and consortia, calling for consortia to input specifications into the formal process, and states that "a 'good' standardization system must satisfy market needs, has to be used in practice and be compatible with public interests" (p. 49).

## International Institutional Responses

The international SDOs have also adapted to the changing environment, with varying success. One response by ISO has been the introduction of the PAS (Publicly Available Specification) and the "fast-track" processes. PAS is a shortcut endorsement process for conferring "international standard" status on selected consortia-developed specifications. It essentially allows the ISO to trade some measure of its legitimacy for a consortium's market relevance.

The ITU has had a different and more difficult problem with its legacy as a clearinghouse for monopoly telecommunications carriers and with representation only by national governments under treaty obligations.[30] Its processes were convoluted, bureaucratic, and not at all public or open by most measures. In recent years, the ITU-T (ITU Telecommunication Sector) has been striving under new leadership[31] to partially re-invent itself, and has been initiating processes for collaboration with SDOs and various other standards organizations and private entities that it did not work with in the past. These reforms have included granting membership status to groups such as the IETF, IEEE, ETSI, ISO, IEC and even some corporations known as Scientific and Industrial Organizations (SIOs).

## Other Responses

From the U.S. perspective, another interesting response to consortia has emerged. Krechmer considers consortia partly a response to what had become a "lengthy and expensive two-stage process," whereby regional SDOs would develop standards and then bring them to the international SDOs. He sees this process evolving, presumably under pressure of consortia competition, "...as the regional SDOs become caucuses for the international SDOs. For example, the rapid completion of Digital Subscriber Line (DSL) standards in ITU Study Group 15 was in large part due to the extensive standardization work that had already taken place in ATIS Committee T1E1.4" (Krechmer, 2000, p. 4). Another closely related response is that in recent times, ANSI has been encouraging the initiation of work directly in international committees (*e.g.,* ISO, IEC, and JTC1), aided by its system of TAGs (Technical Advisory Groups) for each international committee.

Another extremely important response to the perception of enclosure of standards and to the ideology of intellectual property has been the rise of the open source movement—an outgrowth of the free software movement that began in the early days of personal computers. A seminal paper on the topic is *The Cathedral and the Bazaar* (Raymond, 1998), and the phenomenon of open source has been examined in some detail in the *enclosure* discourse by Benkler (2001). Basically, open source can be seen as a form of standardization entirely outside the traditional institutions, either SDO or consortia—in fact, outside the market paradigm altogether—and more akin to the commons of the *enclosure discourse*. Yet it can create platforms upon which new markets can be built. Open source has learned to protect itself from enclosure or appropriation using a novel mechanism of licensing that perpetuates itself, known as General Public License (GPL).

Another response to *enclosure* and the ideology of intellectual property is the Creative Commons, a recent initiative based at the Stanford University Law School and inspired by some of the contributors to the *enclosure* discourse, including Boyle, Lessig, and van Houweiling. This approach seeks to apply the land trust conservancy model of preservation from enclosure and GPL concepts to the realm of ideas and creative works—presumably including standards.

## CONCLUSION

This chapter set out to explore the issue of the *privatization* or *enclosure* of standardization by proposing the hypothesis that the process is being enclosed, and then asking why and how? Informed by a conception of enclosure from another discourse on ideas and intellectual property, this inquiry proceeded by examining the general discourse from the standardization field around this proposition, and then it focused on the specific discourse about the legitimation of consortia standardization. The inquiry converged on two specific documents, which were examined in detail to find the meanings of the terms in the discourse, the rhetorical devices used, and the way in which the practice of standardization was being discursively re-constructed. An array of societal and institutional responses to the challenges of enclosure were also examined.

This study found that the principle issues addressed revolved around the use of such terms as *open* and *balanced*, and around the ideological and mythological constructions the words convey. It is hoped that this analysis will contribute to a better understanding of these terms and the ideas they represent. It is also hoped that this analysis can provide a framework that can inform policy discourse dealing with important issues that relate to technological innovation, global commerce, economic growth, and the stability of social institutions. The next and final chapter will attempt to draw broader conclusions from this analysis, make observations, provide recommendations, and suggest areas for further research.

## REFERENCES

ANSI (2002b). *Breaking Down Borders: business, standards and trade: ANSI Annual Conference 2002.* Washington, DC. October 15-16, 2002.

Baker, Mark (2002). *Europe: Official Says Accounting Standards are Sound.* Prague: Radio Free Europe/Radio Liberty. August 7, 2002. Available at <http://www.rferl.org/nca/features/2002/08/07082002144442.asp>

Balto, David (2001 June). Standard Setting in the 21st Century Network Economy. *The Computer & Internet Lawyer, 18*(6), 1-18.

Benkler, Yochai (2001). Coase's Penguin, or, Linux and the Nature of the Firm. *29th Telecom Policy Research Conference.* Alexandria, VA. September 24, 2001. Available at <http://www.arxiv.org/abs/cs.CY/0109077>

Bilalis, Zacharias, & Didier, Herbert (2003, January-March). (IT) Standardisation from a European Point of View. *International Journal of IT Standards & Standardization Research, 1*(1), 46-49. Hershey, PA: Idea Group Publishing.

Cargill, Carl (1997). *Open Systems Standardization: A Business Approach* (p. 327). Upper Saddle River, NJ: Prentice-Hall.

Cargill, Carl (1999). Consortia and the Evolution of Information Technology Standardization. *SIIT '99 Proceedings: 1st IEEE Conference on Standardization and Innovation in Information Technology.* University of Aachen. September 15-16, 1999. (pp. 37-42).

Cargill, Carl (2001a). *The Role of Consortia Standards in Federal Procurements in the Information Technology Sector: Towards a Re-Definition of a Voluntary Standards Organization.* Submitted to the House of Representatives, Sub-Committee On Technology, Environment and Standards. Palo Alto: Sun Microsystems. June 28, 2001. 30 pp. Available at <http://www.house.gov/science/ets/Jun28/Cargill.pdf/>

CEN (2002). *CEN/ISSS Survey of Standards Related Fora and Consortia.* Brussels: CEN/ISSS Secretariat. October, 2002. Available at <http://www.cenorm.be/isss/Consortia2/>

Cerni, Dorothy M. (1984, December). *Standards in Process: Foundations and Profiles of ISDN and OSI Studies.* NTIA Report 84-170, Washington, DC: U.S. Department of Commerce. 247 pp.

David, Paul A., & Monroe, H. K. (1994). *Telecommunications Policy Research Conference*, held 1-3 October 1994 at Solomon's Island, MD.

De Vries, Henk J. (1999). *Standards for the Nation.* Doctoral dissertation, Erasmus University Rotterdam. published as *Standardization—A Business Approach to the Role of National Standardization Organizations.* Boston: Kluwer Academic Publishers.

DIN (2002). *Strategie für die Standardisierung der Informations und Kommunikationstechnik (ICT) - Deutsche Positionen,* Version 1.0. Berlin: DIN Deutsches Institut für Normung, eV. Available at <http://www2.din.de/sixcms/detail.php?id=3871>

EC (2000). *Guide to Implementation of Directives Based on the New Aporoach and the Global Approach.* Brussels: European Commission, DG Enterprise. 118 pp. Available at <http://www.newapproach.org/>

Egyedi, Tineke M. (1996). *Shaping Standardization: A Study of Standards Processes and Standards Policies in the Field of Telematic Services.* (Doctoral dissertation) Delft University of Technology. 329 pp.

Egyedi, Tineke M. (2001). *Beyond Consortia, Beyond Standardization? New Case Material for the European Commission. Final Report to the European Commission.* Delft: Faculty of Technology, Policy and Management, Delft University of Technology. October 2001. 69 pp.

Flood, Tony (2002). *Presentation and subsequent personal conversation.* ANSI Annual Meeting, October 15, 2002.

Gifford, Jonathan L. (1997). ITS Standardization: Assessing the Value of a Consortium Approach. In *Proceedings of the ITS Standards Review and Interoperability Workshop.* George Mason University Dec. 17-18, 1997. (pp 1-7). Available at <http://www.itsdocs.fhwa.dot.gov/jpodocs/proceedn/2lz1!.pdf>

Gray, John (1998). *False Dawn: The Delusions of Global Capitalism* (p. 262). London: Granta Books.

Jakobs, Kai (2000). *Standardisation Processes in IT: Impact, Problems and Benefits of User Participation.* Braunschweig/Wiesbaden: Vieweg & Sohn Verlagsgesellschaft mbH.

Jakobs, Kai (2001). Broader View on Some Forces Shaping Standardization. *SIIT 2001 Proceedings: 2nd IEEE Conference on Standardization and Innovation in Information Technology.* Boulder, CO, October 5, 2001. (pp. 133-143).

Krechmer, Ken (1998, November/December). The Principles of Open Standards. *Standards Engineering, 50*(6), 1. Available at <http://www.crstds.com/openstds.html>

Krechmer, Ken (2000, July/August). Market Driven Standardization: Everyone Can Win. *Standards Engineering, 52*(4), 15-19. Available at <http://www.crstds.com/fundeco.html>

Krechmer, Ken (2000a, June). The Fundamental Nature of Standards: Technical Perspective. *IEEE Communications Magazine, 38*(6), 70. Available at <http://www.crstds.com/fora.html>

Krechmer, Ken (2000b). The Fundamental Nature of Standards: Economics Perspective. *Schumpeter 2000: Eighth International Joseph A. Schumpeter Society Conference*, Manchester, UK, June 28-July 1, 2000. (p. 70). Available at <http://www.crstds.com/fundeco.html>

Lash, William, III (2002). Keynote speech at ANSI Annual Meeting, Washington, DC. October 15, 2002.

Lim, Andrew S. (2002). *Standards Setting Processes in ICT: The Negotiations Approach.* Unpublished working paper 02.19. Eindhoven Centre for Innovation Studies. Technische Universiteit Eindhoven, Faculteit Technologie Management. 21 pp. Available at <http://www.tm.tue.nl/ecis/>

Lipset, Seymour Martin (1996). *American Exceptionalism: A Double Edged Sword.* (p. 348). New York: W.W. Norton & Co.

Mansell, Robin (1995). Standards, Industrial Policy and Innovation. In R. Hawkins, R. Mansell, & J. Skea (Eds.), *Standards, Innovation and Competitiveness: The Politics and Economics of Standards in Natural and Technical Environments* (pp. 213-227). Brookfield, VT: Edward Elgar Publishing Ltd.

NSPAC (1979). *National Policy on Standards for The United States and a Recommended Implementation Plan.* Washington, DC: National Standards Policy Advisory Committee. (reproduced in Cerni (1984), appendix c.2: pp. 223-226).

Raymond, Eric (1998). *The Cathedral and the Bazaar.* Available at <http://www.tuxedo.org/~esr/writings/cathedral-bazaar/cathedral-bazaar/>

Ritterbusch, Gerald. H. (2001). *Standards-Setting and United States Competitiveness.* Testimony before the House of Representatives, Sub-Committee On Technology, Environment and Standards on Behalf of Caterpillar Inc., June 28, 2001.

Scannell,Kara, & Slater, Joanna. (2008, August). SEC Moves to Pull Plug On U.S. Accounting Standards. *The Wall Street Journal, CCLII,* 28(70)August 2008, A1.

Schoechle, T. (1998). *A Public Policy Debate: Standardsmaking Practice and the Discourse on International Privacy Standards.* Standards Policy Research Paper

ICSR98-211. Boulder: International Center for Standards Research. Available at <http://www.standardsresearch.org>

Schoechle, Timothy, Shapiro, Stephen, Rinow, Michael., & Richards, Barnaby (2002). Evolving Approaches to Technical Standardization: Hybrid Standards Setting. *EASST 2002: Conference of the European Association for the Study of Science and Technology*, York, UK, August 1.

Shapiro, S., Richards, B., Rinow, M., & Schoechle, T. (2001). Hybrid Standards Setting Solutions For Today's Convergent Telecommunications Market. *SIIT 2001Proceedings: 2nd IEEE Conference on Standardization and Innovation in Information Technology* (pp. 348-351). Boulder, CO, October 5, 2001.

Sherif, Mostafa (2001, April). A Framework for Standardization in Telecommunications and Information Technology. *IEEE Communications Magazine, 39*(4), 94-100.

Sherif, Mostafa (2002). *When Is Standardization Slow?* Conference of the European Association for the Study of Science and Technology, York, UK, August 1, 2002. Revised version published in *International Journal of IT Standards and Standardization Research, 1*, (2003, spring), 19-32.

Smoot, Oliver R. (2001). *Standards-Setting and United States Competitiveness.* Statement of Oliver R. Smoot, Chairman of the Board of Directors, American National Standards Institute before the House Science Committee, Subcommittee on Technology, Environment and Standards, June 28, 2001. Available at <http://www.house.gov/science/ets/Jun28/Smoot.htm/>

Thomas, James (2002). Remarks of James Thomas, President, ASTM International at *Standards Activity Promotion Workshop: How to Promote Voluntary Consensus Standards Activities in Korea.* Invited panel expert at Korean Standards Association, Intercontinental Hotel, Seoul, Korea, December 4, 2002.

Vercoulen, Frank, & van Wegberg, Marc (1998). *Standard Selection Modes in Dynamic, Complex Industries: Creating Hybrids between Market Selection and Negotiated Selection of Standards,* Maastricht: Universiteit Maastricht, (pp. 1-14).

Warshaw, Stanley I., & Saunders, Mary H. (1995). International Challenges in Defining the Public and Private Interest in Standards. In R. Hawkins, R. Mansell, & J. Skea (Eds.), *Standards, Innovation and Competitiveness: The Politics and Economics of Standards in Natural and Technical Environments.* Brookfield, VT: Edward Elgar Publishing Ltd. (pp. 67-74).

Zhao, Houlin. (2001). The IT Standardization and ITU. Speech at *SIIT 2001: 2nd IEEE Conference on Standardization and Innovation in Information Technology.* Boulder, CO, October 5.

## ENDNOTES

[1] It is interesting to note that, in some cases (*e.g.*, IEEE, ASTM), ANSI accredited bodies may be reluctant to emphasize this SDO/SRO distinction because they are endeavoring to extend their role into the international arena. They are, in a sense, competing with their accrediting body for authority and recognition.

[2] This criticism might seem somewhat ironic, at least in the case described earlier of Global System Mobile (GSM) (clearly in the negotiated category) and its unquestionable global market success *vis a vis* the fragmentation of the U.S. mobile telephony market, which pursed the market-based bottom-up approach.

[3] IETF (Internet Engineering Task Force), not a formal SDO, is considered a consortium in this context, albeit highly atypical of other consortia.

[4] From legal usage, the difference between *de facto* and *de jure* can be inferred to hinge on the presence of some formal process or methodology.

[5] In this context, *formal* merely means *procedural*, not *institutional.*

[6] IAS has been endorsed by The World Bank, United Nations and the European Commission. (EC, 2000)

[7] It is interesting to note that in 2008, the U.S. Securities and Exchange Commission released a proposed timetable for converting U.S. publicly traded companies to the IAS and abandoning the GAAP in the 2011–2014 timeframe (Scannell, 2008).

[8] Flood further commented that, "The CENELEC mantra overpowers everything else...has a program to sell CENELEC standards throughout the world...a goal stated in the Portugal [plenary] meeting...to expand its influence worldwide" (Flood, 2002).

[9] For comparison, ANSI's entire current annual budget is only about $16 million.

[10] The IT industry grew up largely under the leadership of U.S. firms. IT standardization was thus largely a U.S. centric entrepreneurial culture.

[11] Carl Cargill has for over a decade served as the Director of Standards for Sun Microsystems and is the author of numerous books, journal articles, and conference papers on the topic of Standards and Standardization.

[12] *Comité Européen de Normalisation*, a formal European regional standards body

[13] Open Systems Interconnection (OSI) was a large, formal standardization undertaking that was later largely pre-empted by the Internet, a system of standards that did not come out of the formal process.

[14] Apparently this is a reference to telecom standardization in the ITU.

[15] Such as the proposed *Voluntary Standards and Certification Act of 1976* (S.3555)' and the *Voluntary Standards Accreditation Act of 1977* (S.825). A detailed account of this crisis period is provided by Cerni (1984, pp. 49-61).

[16] It is interesting to note that Carl Cargill, a vocal proponent of revising *OMB Circular A-119* to recognize consortia standards, presented testimony at the same hearing.

[17] For example, the *1394 TA—The 1394 High Performance Serial Bus Association* (*TA* means Trade Association) formed to commercialize support for the IEEE 1394 standard. The Institute for Electrical and Electronic Engineers (IEEE) is an ANSI accredited SDO but the 1394 TA is not.

[18] *Topoi* is a term borrowed from Aristotle. It is his theory and method of identifying the lines of reasoning used to develop an argument. Here, it refers to a set of stock arguments and issues that can be found in the standards discourse.

[19] *Structural* here means about the relative relationship between component parts.

[20] U.S. House of Representatives, Committee on Science, Sub-Committee on Technology, Environment, and Standards, June 28, 2001.

[21] The IETF could not qualify as an SDO for another reason. Although regarded as extremely inclusive and consensus-based on its lowest working level, the IETF is governed at the highest technical levels by a "philosopher king" model of governance, with ultimate authority resting entirely with its Area Directors.

[22] The reference to community involvement and to open source is ironic. Elsewhere, the *Paper's* author has characterized open source as *viral*, because of the self-propagating nature of its General Public License (GPL) tradition and its resistance to control or enclosure.

[23] Consortia may be seen as business ventures unto themselves. A case in point might be the Bluetooth Consortium. Substantial financial resources and an impressive list of corporate members were committed to what turned out to be a technically weak network protocol and a premature marketing, licensing and promotional strategy. Technical incompatibility problems with other standards and products emerged.

[24] The "Publicly Available Specification (PAS)" process is a formal procedure within the ISO directives that provides for a consortium to qualify as "PAS Submitters" and then feed their specifications into the formal committee adoption (voting) and maintenance system.

[25] To the Standardisation Unit of Directorate General Enterprise.

[26] Department of ICT, Faculty of Technology, Policy and Management, Delft University of Technology, Delft, The Netherlands.

[27] Compatibility refers to co-existence or non-interference of elements in a system, whether or not they are interoperable (*e.g.,* electromagnetic compatibility, such as concerns IEC Technical Committee 84).

[28] *New Approach* to technical harmonization and standardization (OJ85/C136/01) and the *Global Approach.*

[29] ETSI, is a hybrid standards body, created by the E.C., but organized like a consortium or trade association, it enjoys a exceptional, but not uncontested status or legitimacy, more like an SDO.

[30] The ITU is an international treaty organization and has traditionally been organized on a national body basis. Submissions come through national committees. In the case of the United States, the State Department officially coordinates and represents the U.S. position. The part of the ITU concerned with telecommunications standardization is called the ITU-T (formerly called the CCITT). The various ITU-T committees are called Study Groups and attendance is authorized through the national bodies. Working documents, contributions, *etc.*, including meeting minutes, are not generally publicly available (they are on the Web site but are password protected).

[31] Houlin Zhao, Director, Telecommunication Standardization Bureau. (Zhao, 2001)

# Chapter VIII
# Conclusion

*[in regard to the caucus process] ...rhetoric of inclusiveness is actually exclusive; who has the time to kill—all the action is after midnight.*

—John Durham Peters, 2002

## INTRODUCTION

This chapter offers the conclusions of the study, briefly summarizes the entire study, and then presents the results and their relevance to the study's theoretical perspectives. Recommendations are also provided about how the discourse on standardization might be clarified and employed more fruitfully by those exploring policy alternatives, including academia, government, and industry. The limitations of the study are assessed and some suggestions are made about possible areas for further research, both theoretical and practical. Finally, observations are included that relate this work to the larger context of global economic, political, and social change.

## RESEARCH QUESTIONS AND CONCLUSIONS

### Research Questions Revisited

This study began by establishing the following research questions: *How public has standardization been in the past and how does that compare with the standardiza-*

*tion process of today? What is meant by terms such as "public," "private," and "open," and how are their meanings constructed and applied? If the process is now undergoing increasing enclosure, what are the roots of such enclosure? What are the responses to arguments that enclosure is occurring? What are the institutional responses to the perception of enclosure?*

The study has sought to explain the concept of standards in the context of American, European, and global industrial policy (*e.g.,* what are "standards"? What is "open"?). Initially, the study examined a discourse about the enclosure of ideas to establish a vocabulary and basic framework for its hypothesis of enclosure. Also included was an analysis of the discourse on standards and standardization. An historical review of several of the most established standardization institutions followed with attention to how terms and concepts of *openness* and the *public* were applied, and what were their structural basis (*i.e.,* what interests were they structured to serve). The study then an examined the discourse on standards and standardization and considered what enclosure might imply for the future of *open standards.* A taxonomy of arguments was established to aid in the analysis. Next, the study considered specific case studies to find the application of the arguments and their relevance to the enclosure hypothesis. Finally, it examined institutional responses, and proposed policies and remedies to enclosure.

As the study proceeded, the basic research questions, revisited previously, suggested the following operational questions: What is the meaning of "open" and how is its meaning constructed and applied? Can a standard developed in a closed committee be an open standard? What are the views on the effects of open/closed processes of standards-making?

## Conclusions

The conclusion reached is that *open* is a multidimensional term universally embraced, but never adequately defined. This ambiguity is used to rhetorically construct and legitimate a wide variety of activities. The most common problem in the use of the term is confusion between *open process* and *open result.* Claims for legitimacy of consortia involved in standards-setting are frequently based on defining an *open standard* as a completed standard that is then *openly* available, whether or not it was developed in an *open process.* The claim is made, on behalf of consortia, that process is irrelevant and the defining criterion for establishing legitimacy is market relevance. Traditional SDOs claim superior legitimacy because of *open processes,* even though they often fail to meet many of their own criteria for *openness.* This question of *open* is entwined with similar confusion over the meaning of *public* and *private.* As a result, standardization, as a quasi-public function, lacks recognition and support from both government and industry, in spite of its economic importance, as is often the case with "public goods". This is particularly the case in the United States. Another problematic element is the use of the term *sector* and *sectoral* because confused and contradictory application of the term it reveals deep structural conflicts and contradictions within the traditional standards system that leave the standards-setting process vulnerable to enclosure by private interests.

Views on the effects of such *open* or *closed* (or rather *open* and *not-so-open*) processes have been presented. These views relate mainly to the issue of *legitimacy*, particularly in regard to government procurement regulations and trade policies. The study concludes that enclosure of standardization is driven by market-based political and economic trends that do not adequately consider issues of public good. Furthermore, governments and institutions face significant questions about how to structure and support the standardization process in the future.

The presented research and analysis offer the following summary of conclusions:

1. **Conflicting systems:** A conflict exists between prevailing U.S. and European views pertaining to the structure of the global standardization system, based on differences in cultural meaning and institutional practices. Globalization exacerbates this problem.
2. **Unclear terms and assumptions:** Standards discourse is based on problematic and ill-defined terms and unclear assumptions on which conflicting policy decisions are being advocated.
3. **Legitimacy:** The central issue in the conflicting discourse, raised by the perception of enclosure, is the *legitimacy* of emerging consortia and of older institutions.
4. **Vocabulary:** A mutually acceptable vocabulary can be established for defining the problem of enclosure and what a reasonable solution might look like.
5. **Hybrid Standardization:** Hybrid standardization is a pragmatic solution that offers a mutually acceptable mode of cooperative action in spite of ideological differences.

## ELABORATION OF CONCLUSIONS

### Conflicting Systems: United States and Europe

The discourse about the U.S. *sectoral approach* to its system of standardization is colliding head-on with the European centralized system based on a more generalized approach. The U.S. system reflects a unique political and economic culture. The rest of the participants, meanwhile, either follow the European system or find themselves in a formative stage. This gulf was less a problem before the demands of economic and market globalization began to bring pressure for harmonizing national standards and internationalizing standardization. The United States is disadvantaged because it relies heavily on other nations for manufacturing, yet largely abjures the international standardization process. It even fails to adequately support its own national process. Present U.S. policy, as reflected in the standards discourse, appears to largely focus on advocating the country's *privatized* and *sectoral* system rather than reaching an accommodation with Europe and the international system. Europe, meanwhile, is moving proactively to fully adapt its system and to harmonize it inside the European Union, to expand the European Union itself, and to internationalize its system. The American National Standards Institute (ANSI) functions as a *privatized regulatory body* and represents a model uniquely adapted to U.S. industrial and political culture. However the model is not one that is likely to be adopted by the rest of the world.

The question remains: Why won't U.S. industry better support its standardization system? The entire ANSI annual budget is only about $16 million—a fraction of what major corporations spend on their own standardization activities, and miniscule compared to what is spent on other government priorities such as research and development. Part of the answer is suggested by former ANSI Board Member and former ISO President, Oliver Smoot.

> *A significant national need with regard to standards-setting and U.S. competitiveness is a* ***much higher level of executive awareness and understanding*** *both within industry and within government of the strategic significance of standardization.* (Smoot, 2001, p. 5) [emphasis added]

One conclusion is that there is a lack of *awareness* and *understanding*—at least there is not enough to overcome the inherent difficulty in supporting a *privatized, voluntary* system devoted to a *common good.* However, it is also likely that the problem reaches much deeper.

## Unclear Terms and Assumptions

The research in this study has shown that the discourse around standards and standardization is to a large extent based on problematic or undefined terms, thus creating a discursive space that is fluid or mobile and unclear. In particular, the term *open,* while universally used, is virtually never defined, except contextually. Virtually all claims are based on *openness.* Everyone claims to be *open* and no one will admit to being closed. The more that standards policy discourse is based on such fluid and unclear terminology, the more it is subject to manipulation and rhetorical constructions to achieve the policy goals of special interests. In particular, the terms *open*, *public*, *sectoral,* and *compatibility* lend themselves to contextual and enthemymatic application and to the construction of mythological and ideological frames of reference.

The discourse on standardization, as with all political discourses, is inherently rhetorical. It is also founded, however, on a long-standing commitment to reasonable deliberation and argumentation. Clarification of the terms of discourse and their related concepts offers to provide a firmer grounding for public policy decisions. This is important because such decisions have significant material consequences in many vital national areas including commerce, government procurement, access to resources, economic and social equity, and technical innovation.

The term *public* is vague and used in a fluid and contradictory manner. The dichotomy between *public sector* and *private sector* is particularly confusing. It effectively excludes the most important discursive space, the *public sphere* or *commons*, from the discourse about standardization, rendering it invisible and de-legitimized. The term *compatibility*, meanwhile, is used in certain discourse to bring standardization practice into a discursively depoliticized,[1] technocratic, and ideological framework that justifies the legitimation of enclosed practices.

In the past, standardization has been an arcane world. Essentially, it was the realm of technical experts—an epistemic elite—who know the practices and the vocabulary of the standardization process as well as the highly technical subject matter of the standards themselves. These elites have been characterized by Streeter (1986) as technocrats. In terms of rhetorical theory, Farrell (1976, p. 4) has termed this sort of discourse as "technical knowledge," in contrast to "social knowledge," which is accessible to the public. Cadres of experts are extremely expensive to maintain. Thus, it is no surprise that this discourse has been largely the exclusive domain of corporations and, to some extent, governments. This is the case, even though, as this study has shown, the discourse has, in theory, been largely open to public participation. Now, enclosure threatens to remove even the possibility of public access. This study proposes that the examination and clarification of the vocabulary of standardization will help in bringing standards discourse into the public domain.

This study has attempted to gain an understanding of the enclosure of standards and standardization by examining its discourse and practices. In particular, the study has focused on the issue of consortia standardization and its relationship with notions of *private* and *public*. What has been found is that there is no clear-cut delineation between these concepts, but a spectrum of relations and institutional structures. This spectrum is shaped by the language and the discursive practices used to talk about it. The significance of these practices is reflected in the discourse itself:

*Pure public goods will not be produced privately. There are only a few pure public goods, one example being national defense. Other goods, like education and standards, are impure public goods. These combine aspects of both public and private goods. Although they serve a private function, there are also public benefits associated with them. Impure public goods may be produced and distributed privately in the market or collectively through government.* ***How they are produced is a societal choice of significant consequence****. If decisions about impure public goods are made in the market, on the basis of personal preferences alone, then the public benefits associated with them may not be efficiently produced or equitably distributed.* (Congress, 1992, p. 9, footnote 23)[2] [emphasis added]

This study has found that there is a significant divergence between the approach taken toward standardization in the United States and that taken in Europe and the rest of the world. This divergence is a reflection of the much different ways in which the public interest is defined and served in the two political economies. It is expressed in basic attitudes and assumptions about the proper role of government and of private initiative and commerce in society. This study has attempted to explore this divergence through the discourse surrounding standardization and its terminology, and through the way the social practice of standardization is institutionalized and supported. It is hoped that this research has shed some light on how these societal choices are made. It is also hoped that this research has helped to clarify the issues and terms for those concerned with policy-making in the standards arena.

## Legitimacy

The research has shown that the central issue in the standards discourse related to enclosure is *legitimation*. This issue raises questions regarding what constitutes *legitimacy*, who confers it, and who shall enjoy it. For instance, the International Standards Organization (ISO) and the International Electrotechnical Commission (IEC) are generally regarded as legitimate international standards bodies because of their historical role, general recognition by other national and international bodies, institutional commitment to practices of openness, due process, and balanced participation. However, to cite another case, the International Telecommunication Union (ITU) is also regarded as a legitimate international standards institution, even though its practices would not meet the same test because of a conferral of legitimacy by national governments via an international treaty, and now by the United Nations. ANSI is regarded as legitimate, even though it is a *private* membership organization, because of its history of industrial practice and public service, its government endorsed privatized regulatory function, and its adherence to, and enforcement of, practices of *openness*, *due process,* and *balanced* participation. Standards discourse currently reflects a debate over an appeal for consortia legitimation by government on the basis of a redefinition of *openness* and *balance*. Conflicts between U.S. and European practices also reflect discourse over issues of *legitimacy*.

## Vocabulary

This study concludes that a mutually acceptable vocabulary can be established for defining the problem of how the standardization system should be structured and what a "reasonable" solution might look like. The relationship between rational and rhetorical elements of standardization discourse and practice were examined in this study. It was found that any particular situation, any word, argument, or utterance, might function simultaneously in each of three separate categories: locutionary (communicative action), illocutionary (symbolic action), and perlocutionary (strategic action). In any particular situation, the dominant meaning or mode of action is highly contextual and depends on the circumstances and the audience or participants in the discourse. In this manner, different speakers and different audiences might interpret the same words in different ways. In this process, a certain level of ambiguity allows adversaries to believe that their claims have been accepted without accepting their opponents' assumptions or beliefs. In other words, it opens the door to a covert ideological rationale contained in the vocabulary shared by the contesting parties. The clarification of some of the most basic terms proposed in this study will allow the establishment of a mutually acceptable vocabulary upon which further agreement and an acceptable mode of cooperative action can be built. One manifestation of such a mode of cooperative action that has been identified in the present study is *hybrid* standardization.

## Hybrid Standardization

An additional conclusion of this study is that a great potential may rest in the *hybrid* process that is currently evolving out of necessity and expediency. For instance, it has been noted earlier that the European *workshop process* is a hybrid response to consortia. If this trend progresses, it may alleviate the need to legitimize consortia. Another example is the recently adopted ISO/IEC PAS process, whereby an existing (consortia) standard, known as a Publicly Available Specification (PAS), can be submitted to the ISO or IEC ballot process by a pre-qualified "PAS submitter" organization that has met certain requirements. A variant on this is the "fast track" process whereby a national body member of ISO/IEC may submit an existing national standard to the ISO/IEC ballot process. By these methods, the PAS or "fast track" specification obtains the ISO/IEC imprimatur even though it did not originate within, or endure, the usual rigors of the ISO/IEC committee process. Experience has shown that even though some consortia have produced excellent and useful stand-alone specifications, they now are increasingly discovering the value of getting these adopted as formal standards. Today there is an rising clamor around ISO and IEC "PAS submission" to get consortia specifications incorporated into the formal ISO/IEC/ITU system as standards (they seek the legitimacy and respect that comes with the ISO/IEC logo). Some consortia are learning to work alongside the formal bodies and make their work compatible and acceptable from the beginning—not just as an afterthought.

On the down side, some see the PAS process as ISO and IEC selling their legitimacy in exchange for some perception of "market relevance" and/or for publishing revenues. Most recently, certain PAS ballots have been highly contentious, with appeals to the ballot initiation and/or ballot resolution process being brought by developing nations. There may be a growing perception that the PAS process is lacking in recourse or due-process and is being "gamed" by dominant nations and/or corporations to advantage their proprietary interests by employing proxies or surrogates in voting and committee deliberations.

In any case, the *open* argument cannot be easily resolved, but may be avoided by accommodation and collaboration between consortia and SDOs within a framework of antitrust law application. From a public policy standpoint, there is no advantage in conferring legitimacy on consortia because one consequence may be a further disadvantaging, or even the destruction, of the formal system. There exists at present a sort of balance of power and a creative tension. Persuasive arguments have been made on both sides. Often consortia represent a response to real needs, but may not be viewed favorably. Market-driven standards have limits; markets are not typically good at picking long-term needs or building basic infrastructure. For instance, the Internet was built by a government initiative. It is difficult to imagine it having been constructed by market forces. Standardization arenas are needed that can focus on longer-term tasks, rather than those to which consortia are best suited. Consortia are diverse and are not easily assessed and classified with regard to *openness* and other characteristics. Also, many consortia are ephemeral in nature.

In any case, before policies can be established through legislation or by changing governmental administrative procurement or regulatory rules, some criteria and accreditation scheme would need to be developed. The criteria proposed in a policy paper referred to throughout the present work as the *Paper* (Cargill, 2001a), is only conceptual and incomplete, and lacking on major points (*e.g., balance*). To embark on such a project without taking into account the existing criteria and accreditation system could be a big mistake. The inevitable question that arises then is: Can the need for some level of consortia legitimation or inclusion be accommodated within the existing structure, *i.e.,* ANSI in the United States? Many consortia are really marketing associations and the possibilities of collaboration with SDOs would seem to have obvious potential.

## RECOMMENDATIONS

### Eliminate the "Public Sector"

The single most important recommendation this research can make is the banishment of the term *public sector* from standards discourse—and from all policy discourse in general. A new space between *private* and *government* should be carved out for the *public*. For example, ANSI committees hold public meetings, but are *not part of government*. They need a discursive space in which to live. At a minimum, the term *government sector* should be revived from older policy documents. Ideally, the term *sector* should be replaced because it perpetuates the misleading illusion of a precise division between *private*, *public*, and *government*. *Sector* also is confusingly assigned in an alternative meaning when used to refer to an *industrial sector*, as in *sectoral approach*. The *sectoral approach* has enough confusion associated with it already. These problems only get worse when they are introduced into the international discourse and are translated into other languages.

### Support Standards Research and Education

Another recommendation is to advance the emerging field of standards research and education. This may be achieved by engaging academic institutions, government, and industry in collaborative forums to clarify and solidify the research agenda and define academic curricula. Specifically, a goal should be to clarify the terms and identify the issues pertaining to the discourse on standards policy. Such initiatives have been started, in Europe, Asia, and in the United States,[3] but they need support, reinforcement, and participation. [4]

## SUMMARY OF THE STUDY

This study was initiated with the purpose of exploring and gaining greater understanding of an important and contentious issue in the increasingly essential global system of technical standardization—the rise of the consortia movement. It sought to do so by analyzing the discourse surrounding this issue to ascertain the cultural meanings that drive related policy decisions by corporations, governments, and institutions. It proceeded by establishing a perspective of analysis and a hypothetical problem statement based on discourse about the *enclosure of ideas* and intellectual property, and approached the enterprise of standardization as a form of idea generation in an *intellectual commons.*

This study then reviewed the historical roots of the present global standardization system and its institutions, with specific attention to the national, institutional, political, economic, and cultural factors that led to the present structure. It also examined this historical perspective with regard to the *discursive construction of standards practice* and the terms and assumptions of policy discourse surrounding these practices and institutions. Of particular significance was the divergence between U.S. and European practices and institutions. Also of consequence was the emergence of the IT industry, the deregulation and privatization of the telecommunications industry, and the ongoing convergence of these two industries.

This study then reviewed the relevant literature and research and attempted to situate the study in a broader intellectual and historical context. First, the discourse about *enclosure* was reviewed in detail, including concepts and terms of discourse relating to the nature of technical innovation and markets. This review also included the relationship between intellectual property rights and the intellectual *commons*, the derivative nature of intellectual creation and authorship (*i.e.*, of *ideas*), the notion of the *2nd enclosure movement*, alternative modes of production (*i.e.*, of *peer production*), and specific terms such as *public*, *private*, *free*, and *open.* Then literature and research on the discourse on standardization was reviewed with particular attention to conceptions of *private* and *public*, *sectoral,* and the rise of *consortia* and their claims to *legitimacy.*

The study then considered and defined relevant theoretical perspectives and its methodological approach—that of Habermas' *public sphere* theory, *political economy,* and *discourse as a social practice. Discourse analysis* was established as the primary methodology. The study then presented research and analysis by focusing first on some basic terms of the discourse, namely *public, private, sectoral, and open.* Then the study focused on the discourse on *negotiated vs. market driven* standardization, *hybrid* approaches, international institutions, *sectoral* discourse, and on the motivation for *consortia* standardization practices. Finally, the study deconstructed two specific policy position documents, in both U.S. and E.U. discourses, and the claims they make for the legitimacy of consortia. It examined counter arguments from other related policy position documents and discourses. Finally, the research and analysis considered societal and institutional responses including related discourses and practices.

## RESULTS OF THE RESEARCH AND ANALYSIS

In order to show the relevance to standards practice of the theoretical perspectives that have been employed in this study, the research results will be reviewed as follows. Early in this study, in the review of the literature, basic concepts of *public* and *private*, including those theorized by Habermas, were introduced. Introduction of the *enclosure discourse* then followed. The terms *public* and *private*, and others such as the *commons*, *open*, *free*, and of *property talk* were also explored in relation to their mythological and ideological implications. This preview was provided to show the significance of the terms of discourse. Later, the full theoretical framework for the study was developed and, in that context, the research and analysis of specific examples of the Standards Discourse on policy were presented.

### Basic Terms of the Discourse

Some of the most basic terms of the standards discourse, including the problematic nature of the *public sector vs. private sector* dichotomy, were explored by seimotically mapping the signifiers to what was being signified, and attempting to establish the range of meanings being used in the discourse. This was done by applying Habermas's theoretical perspectives of *public sphere*, *communicative action,* and *practical discourse* to their usage in standards practice. Missing meanings that were being excluded by such usage were revealed and mapped back to the Enclosure Discourse. The problematic use of the term *sectoral* was similarly explored. This exploration showed the resultant conflicting meanings and illustrated *semiotic* (*i.e.,* mythological) and *enthymematic* usages that rely on reader/listener/user assumptions about the proper role of standardization in a market economy. This role includes the proper political economic distributions of power among industries and public/private sector roles. Finally, the important and problematic term *open* was similarly explored. A connection was drawn with the *enclosure* discourse, with Habermasian *discursive deliberation* and *public sphere* theory, and with Althusserian "interpellation" as a constitutive discourse. An attempt was made to narrow the definition by tying these and other theoretical constructs with the actual use of the term *open* in standards discourse and practice. An effort was made to trace the historical migration of the meaning of *open* for political economic reasons having to do with dominant interests and institutional priorities.

### Discourse on Standards

The standards discourse was explored, using relevant documents and conference presentations and discussions. Additional terms of discourse were identified and explored. The discourse about *negotiated vs. market-based* standardization was studied. Its use was defined in terms of political economic power relations between industries and different forms of standardization institutions, (*i.e.,* SDOs, consortia and hybrid forms of standardization). These terms were defined according to institutional cultural values and of negotiated authority among institutions. Possible implications were observed for these values in the context of both the *enclosure* discourse the *public sphere.* The problem of democratic *legitimacy* was introduced and explored in terms of openness, stakeholder participation, and market power.

The discourse on openness of the international standardization system was then examined, including the different concepts associated with many of the terms examined earlier, and their mythological and ideological meanings. In particular, a diverse discourse taking place at the *2002 ANSI [A]nnual [C]onference* was studied, providing examples of rhetorical statements and constructions by high ranking players from the United States, the European Union, Asia, the Middle East, and South America. From this discourse, it was possible to identify key discursive practices and terminology that revealed regional, national, and cultural differences, including conflicting ideological and mythological constructions about *private* and *public* roles, *trade*, *regulation*, *harmonization*, and *strategic behaviors*—all of which tied back to theories of the *public sphere* and *political economic* motivations. One such conflict was identified in the discourse about *principles vs. rules*, or *horizontal vs. vertical* standardization. This conflict revealed basic ideological differences between U.S. and European culture and practice of standardization.

The driving forces for consortia standardization and its historical roots were then examined. Through analysis of historical policy documents, it was possible to extract the arguments being made for and against the *legitimation* of consortia standardization for purposes of government procurement and regulation. This examination also revealed the underlying mythological and ideological assumptions and the differing ways in which meaning is attributed in the discourse. Such terms as *balance* and differing mythological and enthymematic uses of *openness* and *consensus process* were highlighted and tied back to notions of the *public,* to the *public sphere,* and to *enclosure.* The *overinclusiveness* argument was then identified and explored. This argument, asserting a claim for consortia *legitimacy* based on instrumental rationality, helped prepare research and analysis for a later exploration of the discourse about *compatibility* and the de-politicization of standardization. Possible political economic institutional motivations were also considered.

## CLAIMS FOR LEGITIMACY

A further examination of the rhetorical and constitutive discourse about legitimacy prepared the way for the analysis of two specific policy documents. One of these documents was the policy paper referred to throughout the present study as the *Paper*, supporting testimony before the U.S. Congress. The other was an academic report, the *Report* (Egyedi, 2001), to the European Commission. Both advocated the legitimation of consortia standardization. The study deconstructed these documents in some detail to extract the salient messages and claims embedded in them. Several specific mythological, ideological, and enthymematic arguments were identified. The *Paper's* claim for legitimacy relied on mythological and enthymematic re-constructions of *consortia*. The claim was based on a instrumental argument for re-defining *open* and discarding *balance*. The *Report's* assertion was based on an *ideological* shift to an instrumental and depoliticized technocratic meaning for *standardization*, that of *compatibility*. Both of these documents were examined for their implications for public sphere theory and political economic considerations. For example, the mythology of *speed* as a motive for consortia was critically examined and compared with other discourses and arguments that traced the issue of *speed* to other possible political/economic agendas. Other counter arguments were examined including those put forth in a recent policy paper from DIN, refuting some discourse similar to that found in the *Paper* and the *Report,* and proposing alternative meanings and institutional structures.

### European and International Responses

Finally, European and international institutional responses to the consortia challenge were considered as part of the research. Various discourses were examined that proposed certain institutional adjustments, both toward and away from more democratic, public, or more enclosed discursive standardization processes. These were compared with the Habermasian and rhetorical notions of discourse, of the *public sphere,* democratic *legitimacy*, and with the *enclosure* discourse.

## LIMITATIONS OF THE STUDY

As in any study of this nature, it is important to recognize the present study's limitations in theory, method, and findings. Sometimes this recognition may be more valuable than the study itself. It is hoped that such is not the case with this study. But limitations clearly identify opportunities for further knowledge and understanding of an important, yet contentious, topic.

What has been presented in this study is a way of thinking about standards. The study provides an understanding of the meaning of some basic terms of discourse about standards and standardization, framed by the main terms and meanings of another discourse about *enclosure*. This choice of framework provided a set of research questions that guided the research and analysis. A different framework could have been chosen that would have yielded different research questions and different results—perhaps even a different way of thinking about standards. There is probably no unified theory of standardization—no grand narrative approach to understanding standards—for standardization is inherently a social practice.

## Research Questions

Two specific limitations are suggested by the choice of research questions, one of scope and the other of direction. One specific area of inquiry was a form of enclosure that was touched upon but not explored. This area is the enclosure of standards and standardization by private corporations as strategic action. It is an area that overlaps consortia enclosure, and while it was noted in the study, it was not explored. A second issue was the possibility of the seeming opposite of enclosure—the moving of privately or consortia-developed standards or technologies into the open accredited process. Instances of such, or at least attempts of such, were encountered in the study (*e.g.*, PAS process, hybrid standardization, *etc.*), but not pursued. The PAS process specifically need further detailed study.

## Anecdotal Sample

Another limitation of this study is the limited and somewhat anecdotal nature of its sample. Certainly, much more discourse is available for study, and more is being developed every day, but it was necessary to choose documents and discourses that were available and representative of certain general arguments. However, there is no assurance that these were truly representative, unless much larger surveys are conducted. This researcher relied on a personal knowledge of the practice and the discourse—introducing a subjectivity, as Habermas has noted, that lies between theory and practice. If other documents had been chosen, other meanings or interpretations may have been found and other terms may have been identified. Also, other practices might have been encountered. In this study, one of the criticisms offered was of the *Report*'s reliance on two case studies of consortia. The same problem should be recognized with this research. As noted earlier, one list of consortia contains 255 organizations. More case studies of consortia should be conducted in order to make generalizations about their practices and to possibly classify them into some useful taxonomy. This study was only able to address a few terms of discourse. In a broader study, others may have been discovered.

## Perspective of Researcher

An additional limitation is the perspective of the researcher. As an involved participant in the practice being studied, especially as contentious as it is, a potential for bias is introduced. The other side of this issue, however, is that an involved researcher knows where to look for the arguments. The exploration of this topic did provide an appreciation of alternative arguments that was not initially present on the part of this investigator and it influenced the direction of the research in a fruitful way and tempered its conclusions. Another subjectivity of the study is that the interpretation of the discourse and its terms, myths, and ideologies depends on finding patterns which may be illusionary. For instance, the generalization made earlier about *rules vs. principles* may not be justified in light of a more comprehensive cultural study, or, at least, counter-examples might be found. In a much broader example, the entire study was based on the *enclosure* metaphor—a conception that is not without challenge. In any case, the foregoing discussion of limitations also suggests many opportunities for further study.

## IPR and Antitrust

Finally, two additional areas that the study encountered, but did not delve into, should be noted. One was standards incorporating intellectual property rights (IPR), and the other was standards as intellectual property *per se.* These topics have become among the most significant issues in standardization over the past decade. In particular, the issue of *ex ante* patent disclosure requirements and even *ex ante* license negotiation within standards bodies have emerged in the last few years as a major patent policy debate within standards bodies and among legal scholars and policymakers, due to litigation, court decisions and FTC rulings in the United States and in Europe. These topics pose thorny legal and policy problems and could provide rich and important opportunities for further research.

In certain areas such as wireless telephony, a sort of "feeding frenzy" has developed as the financial stakes rise and participants in the standardization process seek to embed their own IPR in the standards and to position themselves to collect royalties from users of the standard or, conversely, to fend off royalty payments by cross licensing strategies.[5] Often, as a defensive strategy, firms seek to develop patent portfolios that can be used in cross-license negotiations. As mentioned earlier in this study, consortia may be in a better position to deal with or enforce more elaborate IPR policies. SDOs, at least at the working group level, are largely composed of engineers and not lawyers. Defensive cross-license strategies may work for large corporate participants, but smaller participants without patent portfolios are disadvantaged, thus tilting the process and making it more exclusive and enclosed. A patent régime that allows claims on algorithms and business practices exacerbates this problem.

Discussion of license terms *ex ante* has long been strictly avoided in open standards bodies, due to the risk of anti-trust litigation. But without such negotiation these bodies have found themselves exposed to "submarine" patent practices wherein members or participants have filed patent applications on the material being standardized and later demanded royalty payments by users of the standard. Royalty-free or royalty-cap policies, in response, may go even further toward positioning the consortium as a cartel. Partially in response to such cartels, China, with its a large domestic market, has begun to challenge the system by forming its own competing standards (Stuttmeier, Yao & Tan, 2006) in several areas (*e.g.*, 3G, DVD, MPEG, digital TV, Wi-Fi, *etc.*) and thus threaten, to further fragment the international system

A second but related problem is that traditional standards bodies have customarily relied on copyrights and their attendant publishing revenues as a primary source of income for their host organizations or trade associations. Such reliance has become a problem as demand for free or low-cost electronic dissemination of standards increases, and consortia do not typically rely on such publishing revenue. This reliance has also long posed a dilemma since it conflicts with the basic purpose of SDO standards—their widest possible dissemination and use. This property right also conflicts with the quasi-public goods nature of standards, particularly in the case of standards that become mandated by law or are incorporated into regulatory or administrative rules.[6] Such cases pose legal dilemmas because they are essentially a form of privatized law.

## DIRECTIONS FOR FURTHER RESEARCH

The field of standards research is rich in subject matter for theoretical and practical studies in standardization discourse and practice. Five areas in particular are recommended for further research:

1. Studies of other forms of Digital Enclosure, including IPR and antitrust issues
2. Construction of taxonomy of consortia, based on case studies
3. Case studies of standards-related technical collaboration (*e.g.,* open source).
4. Theorization of the rhetoric of standards

5. Theorizing about the public, democratic deliberation, and institutional governance

## Studies of Other Forms of Digital Enclosure including IPR and Antitrust Issues

Other forms of enclosure of standards and standardization require study and better understanding. This study has focused on the transfer of standardization from public processes to private consortia. Another form of enclosure is the transfer from public to proprietary ownership, or the appropriation of established standards by specific firms. Van Howeiling (2002) has described the standards "pollution problem," where public domain standard protocols are enclosed by a process she characterizes as "embrace, extend, extinguish." Lemley (1998) has discussed the notion of standards "pollution" and Lemley and McGowan (1998) have referred to "intellectual property ambush," particularly, but not solely, regarding the practices of Microsoft Corporation. Forms of un-enclosure may also be found and would provide valuable insight into the process of standardization.

Two areas of increasing importance in need of research are 1) the relationship between intellectual property issues in standards bodies that may be driving consortia and cartel formation, and 2) the relationship between intellectual property rights (particularly copyright) and the dependence of standards bodies on publishing revenues, on the one hand, and the problem of privatized regulation or privatized governance, on the other.

The fundamental rationale for the patent system is that patents play a role in technical innovation for the benefit of the public. This notion is based on the presumption that inventors need an incentive to invent and to offer their inventions to the public. Some investigators challenge that basic premise. Benkler (2001), writing from a law perspective, has studied alternative forms of production, both historical and recent, that are highly successful and do not rely on such a presumption (*e.g.*, open source). Boldrin and Levine (2002), from an economic perspective, challenge the conventional wisdom, asserting that, "there is nothing either natural or socially useful in the monopoly power the state confers upon innovators…that, from a viewpoint of social welfare, current legislation on copyrights, licensing, and patents plays a harmful role in the innovation process" (p. 1). These studies and others suggest that the patent system may be afflicting the standards process for no good reason. In any case, it is suggested here that the IPR system has diverged far from its original Jeffersonian mission. This line of research needs to be further pursued—extending the realm of Enclosure Theory.

## Construction of Taxonomy of Consortia

Case studies of consortia standardization need to be carried out in a consistent manner in order to establish classification criteria and to compare their governance mechanisms, rules, procedures, memberships, and openness with the formal standardization and accreditation systems. A classification of openness was suggested in this research in Chapter VI. A basic classification of governance systems has been suggested by Choh (1999), but needs to be developed and expanded. Interesting work has been done on the topic of strategic behavior of participants in standardization activities such as Global System Mobile (GSM) standardization (Bekkers, Verspagen, & Smits, 2002). More such work needs to be done in other areas of standardization. The topic of IPR is also entwined with standardization, in both consortia and in SDOs, and needs further study. Hybrid standardization needs more case studies to increase understanding of how consortia and SDOs work together to inform policy discourse.

## Case Studies of Standards-Related Technical Collaboration

Case studies of open source and other alternative standardization processes would also contribute to an understanding of standardization as it is now practiced. The area of "peer production," which has been described by Benkler (2001), is closely related to the study of innovation, and needs further study.

## Theorizing the Rhetoric of Standards

This study has been a study of the rhetoric of the discourse of standardization. It is a study of a meta-discourse, or the discourse about the discourse of standards. The analytical theory and methodology used in this study could be applied to the actual discourse of standards-making itself. Such research could reveal much about the process of group collaboration and consensus. It could also explore the role of the technical expert and tendencies toward technocracy, an issue that has been noted in this study, but not thoroughly explored. Such an exploration could provide a better understanding of how the process may be made more public.

## Theorizing the Public, Deliberation, and Institutional Governance

It has been noted earlier in this study that the theorization of the *public* is incomplete and needs to be developed—it is the essence of Enclosure Theory. Some of this research is beginning. In particular, processes of negotiation are beginning to be studied (Lim, 2002) in relation to standardization. The subject of deliberative democracy has been studied extensively, but not specifically in relation to the standardization processes. The topic of "private governance" is beginning to be seen in public policy discourse (Mueller, 2002), but it needs to be applied specifically to standardization practice and institutions. However, in a broader sense, there is an urgent need for more fundamental research and policy discourse; the world is changing rapidly.

This book has been largely about Enlightenment concepts of freedom, democracy, and the public. It has relied on democratic theorists such as Jürgen Habermas and others. An underlying assumption of this work has been that these ideas offer universal value for human societies. However, critical thinking[7] demands that such assumptions must be questioned and not simply taken for granted. Globalization, the re-emergence of geopolitics, and rapid economic development over the past decade—and specifically the advancement of new economies from the third-world into the second-world, or perhaps soon into the first-world—has raised fundamental questions about the value and appropriateness of Western notions of democracy, freedom, and capitalism and how they are being interpreted and applied in non-Western cultures. In a recent detailed analysis of such development, Khanna (2008) observes,

*East Asian communitarian traditions also challenge American notions of human rights by prioritizing social and economic rights over civic and political rights, justifying the denial of constitutional protections of individual liberty and free speech. The early Confucian scholar Mencius argued that violating the right to food and material well-being is a greater crime than denying political rights. Humility and compassion (ren), not flamboyance and egoism, are cherished virtues. ...Like Europeans, East Asians focus more on subsistence and economic equality, rights that are far more enshrined in the European legal tradition than in the American.* (p. 267)

Barely a decade and a half ago the triumph of western liberalism was celebrated —the victory of democracy and capitalism was seen as final, and was heralded as the "end of history" (Fukuyama, 1992). Today, the picture is quite different. The case for the value of democratic discourse and processes, and how they are applied, needs to be made far more clearly. This case will be particularly important in the realm of international standards and its institutions that stake so much on such discourse and such processes for their validity and legitimacy.

## OBSERVATIONS

The process of *enclosure* of standardization, and of *ideas* in general, may be seen as part of a larger global trend over the past two decades of deregulation, privatization, and commoditization. These trends were propelled in most part by the hegemony and seemingly endless success of the free market ideology and high technology industry springing from the United States during the 1990s. With the sudden decline of financial markets in 2001 followed by relevations of severe shortcomings in economic institutions, particularly in the United States, and subsequent collapse of financial markets and associated recession beginning in 2008, there is an opportunity for reevaluation of many of the assumptions that dominated policy and standards discourse for more than a decade.

In the late 1990s, things were different. Free market ideology was at its zenith. Technology was moving on "Internet time." There was an air of hubris with little patience for long-term planning, accountability, and talk of the public interest. As long as corporations were making many stockholders wealthy, it was easy to overlook the consequences. Now, it is suggested, there is room for a new and healthy skepticism about the omnipotence of unfettered markets and of science and technology, and a new appreciation of the value of long-term institutions of oversight and continuity. The mantra of "build shareholder value" has shown to be largely self-serving rhetoric, masking the systematic exaggeration of productivity and profits, and also masking the underlying motivation of management executive compensation. It is proposed that the economics of hypercompetitive consumerism and the destruction of intermediary social institutions may now have slowed, or are being reevaluated in both established and emerging societies.

Writing in the Forward, revised in 2002, Gray (1998, p. xiv)[8] suggested that, "...global laissez-faire is visibly unraveling today," and, "the events of 11 September, 2001, confirm [my] central thesis: globalization works to undermine the global free market." He predicted that "America's infatuation with the global free market is all but over," (p. xix) and the world economy will see a return of trade protectionism and to the geopolitical priorities of an earlier age of *realpolitik*. Gray further predicted that global markets will continue to grow, but suggested that American business culture will not be copied throughout the world. Rather, he suggested that capitalism will be re-interpreted into the various cultural idioms of each developing country and region, representing a fundamental separation of *capitalism* from the *free market* ideology (*e.g.*, China). Particularly in light of events subsequent to Gray's comments, the implications of his perspective for the role and practice of standardization are profound:

*A régime of global governance is needed in which world markets are managed so as to promote the cohesion of societies and the integrity of states. Only a framework of global regulation—of currencies, capital movements, trade and environmental conservation—can enable the creativity of the world economy to be harnessed in the service of human needs.* (p. 199)

*The task of transnational organizations should be to fashion a framework of regulation within which diverse market economies can flourish.* (p. 235)

It is proposed that the global technical standardization system is one of these *transnational organizations* to which Gray refers. It is one that elsewhere Gray calls, an *intermediary social institution* between human beings and the market.

Noble Prize economist Joseph Stiglitz (2003) has echoed much of Gray's position about the fallacies of free market ideology, and in particular Gray's call for adequate global financial and economic institutions that truly serve public needs and not simply those of dominant economic interests.

It is interesting to note that the early market theorist, Adam Smith (1776), would probably not disagree with Gray's assessment. Smith was deeply concerned about moral obligations, social justice, and stability, and about the potential for monopoly and for government policy to favor special interests. The implication for this research is that the *enclosure of standardization* may be slowing, or at least may have some prospect of establishing counter-balancing elements.

The new millennium has brought many changes. Gray has suggested that the hegemony of the U.S. economy in the world may be drawing to an end. It has been further suggested that any future clash of civilizations will not be "between the West and the rest, but between the U.S. and Europe." Kupchan observes that,

*Europe is strengthening its collective consciousness and character and forging a clearer sense of interests and values that are quite distinct from those in the United States. ...the EU will surely test its muscle against America if the unilateralist bent in U.S. foreign policy continues. ...Despite recent deregulation across Europe, America's laissez-faire capitalism still contrasts sharply with Europe's more centralized approach. Whereas Americans decry the constraints on growth that stem from the European model, Europeans look askance at America's income inequalities, its consumerism, and its readiness to sacrifice social capital for material gain. ...At root, America and Europe are driven by different political cultures. And the cultural distance appears to be widening, not closing, putting the two sides of the Atlantic on divergent social paths.* (2002, pp. 42-43; 2002a)

The observations have significant implications for the U.S. *vs.* E.U. discourse on standardization mentioned earlier in this research and in the aforementioned conclusions. These implications lend an additional urgency to clarifying the discourse and advancing policy initiatives that may then flow from such improved clarity.

## REFERENCES

Bekkers, Rudi, Verspagen, Bart, & Smits, Jan (2002). Intellectual Property Rights and Standardization: the Case of GSM. *Telecommunications Policy, 26*, 171-188.

Benkler, Yochai (2001). Coase's Penguin, or, Linux and the Nature of the Firm. *29th Telecom Policy Research Conference.* Alexandria, VA. September 24, 2001. Avaiable at <http://www.arxiv.org/abs/cs.CY/0109077>

Boldrin, Michelle, & Levine, David (2002, March). *Perfectly Competitive Innovation.* Federal Reserve Bank of Minneapolis, Research Department Staff Report 303.

Choh, Kwonjoong (1999). Governance Mechanisms of Standard-Making in Information Technology. *SIIT '99 Proceedings: 1st IEEE Conference on Standardization and Innovation in Information Technology* (pp. 49-54). University of Aachen. September 15-16, 1999.

Congress of the United States, Office of Technology Assessment. (1992). *Global Standards: Building Blocks for the Future.* TCT-512. Washington, DC: U.S. Government Printing Office. March 1992. 114 pp.

Farrell, Thomas B. (1976, February). Knowledge, Consensus, and Rhetorical Theory. *The Quarterly Journal of Speech, 62*(1), 1-14.

Fukuyama, Francis (1992). *The End of History and the Last Man.* New York: Avon Books.

Gray, John (1998). *False Dawn: The Delusions of Global Capitalism* (p. 262). London: Granta Books.

Khanna, Parag (2008). *The Second World: Empires and Influence in the New Global Order* (p. 466). New York: Random House.

Kupchan, Charles A. (2002, November). The End of the West. *The Atlantic Monthly, 290*(4), 42-44.

Kupchan, Charles A. (2002a). *The End of the American Era: U.S. Foreign Policy and the Geopolitics of the Twenty-first Century* (p. 391). New York: Alfred A. Knopf.

Lemley, Mark, & McGowan, D. (1998). Could Java Change Everything? The Competitive Propriety of a Proprietary Standard. *Antitrust Bulletin, 43*, 715

Lemley, Mark A. (1998). The Law and Economics of Internet Norms. *Kent Law Review, 73*, 1257-1288.

Lemley, M., & McGowan, David. (1998). Could Java Change Everything? The Competitive Propriety of a Proprietary Standard. *Antitrust Bulletin, 43*, 715

Lim, Andriew S. (2002). *Standards Setting Processes in ICT: The Negotiations Approach.* Unpublished working paper 02.19. Eindhoven Centre for Innovation Studies. Technische Universiteit Eindhoven, Faculteit Technologie Management. 21 pp. <http://www.tm.tue.nl/ecis/>.

Mansfield, Edwin (1970). *Microeconomic Theory and Application.* New York: W.W. Norton

Mueller, Milton (2002). Interest Groups and the Public Interest: Civil Society Action and the Milton Globalization of Communications Policy. *TPRC 2002: The 30th Annual Research Conference on Communication, Information and Internet Policy.* September 28-30, 2002. Alexandria, VA.

Smith, Adam (1776) *(1976). An Inquiry into the Nature and Causes of the Wealth of Nations.* Inn R. H. Campbell, *et al* (Eds.), Oxford & New York: Oxford University Press.

Smoot, Oliver R. (2001). *Standards-Setting and United States Competitiveness.* Statement of Oliver R. Smoot, Chairman of the Board of Directors, American National Standards Institute before the House Science Committee, Subcommittee on Technology, Environment and Standards, June 28, 2001. Available at <http://www.house.gov/science/ets/Jun28/Smoot.htm/>

Stiglitz, Joseph E. (2003). *Globalization and its Discontents* (p. 288). New York: W.W. Norton..

Streeter, Thomas (1986). *Technocracy and Television: Discourse, Policy, Politics and the Making of Cable Television.* Unpublished doctoral dissertation. University of Illinois at Urbana-Champaign.

Stuttmeier, Richard P., Xiangkui, Yao, & Tan, Alex Zixiang (2006). *Standards of Power? Technologies, Institutions and Politics in the Development of China's National Standards Strategy* (p. 52). Seattle: The National Bureau of Asian Research.

Van Houweiling, Molly S. (2002). Cultivating Open Information Platforms: A Land Trust Model. *The Journal on Telecommunications Law and High Technology.* Vol.

1. Also presented at the Silicon Flatirons Conference. "Regulation of Information Platforms." University of Colorado, Boulder, CO. January 28-29.

## ENDNOTES

1. To be de-politicized means to be rendered non-political or to exist as merely a technical issue—assumed to carry no consequences for policy.
2. Citing Edwin Mansfield (1970). *Microeconomic Theory and Application.* (New York, NY: W.W. Norton)
3. For example a special panel: "University Education and Research on Technical Standards," was organized by the International Center for Standards Research (ICSR), supported by ANSI and NIST and hosted by the Columbia Institute for Tele-Information (CITI), at Columbia University September 9, 2002. Further examples include a panel on "Academic Outreach," conducted at the ANSI Annual meeting, Washington DC, September 15-16, 2002 and the subsequent establishment of the ANSI Committee on Education as an ongoing activity.
4. To this purpose, an informal international conference known as the International Committee for Education about Standardization (ICES) was organized and hosted in Tokyo during February 2006, in Delft during February 2007, and in the Washington DC area during February 2008.
5. Such participant firm's strategies with regard to GSM (Global System Mobile) patents are described by Bekkers, Verspagen and Smits (2002).
6. *Regulatory* standards might include pharmaceuticals, auto safety regulations, food classification standards, and agricultural inspection standards, electromagnetic emission limits, building codes, fire protection standards, and spectrum allocations. The term *regulatory* actually refers a distinction between *voluntary* and *involuntary* compliance requirements. *Administrative* rules might include purchasing or procurement requirements.
7. By *critical thinking*, it is meant the discipline of routinely identifying and questioning the unstated assumptions that underlie one's own, as well as others, ideas, narratives, and truth claims.
8. In the "Forward to 2002 Edition" of *False Dawn* (Gray, 1998).

# Compilation of References

AFU (1996). The Decline and Fall of Betamax. *AFU White Paper*. Available <http://urbanlegends.com/products/beta_vs_vhs.html>

Albert, M. (1993). *Capitalism vs. Capitalism: How America's Obsession With Individual Achievement and Short-Term Profit Has Led It To The Brink of Collapse*. New York: Four Walls Eight Windows. 259 pp.

Althusser, Louis (1971). Ideology and Ideological State Apparatuses. In *Lenin and Philosophy and Other Essays*. London: New Left. , p. 170)

ANSI (2000). *National Standards Strategy for the United States*. American National Standards Institute. New York: ANSI. August 31, 2000, 14 pp.

ANSI (2001). *American National Standards Institute: 2000 Annual Report*. New York: ANSI.

ANSI (2001a). *ANSI Procedures for U.S. Participation in the International Standards Activities of ISO*. New York: ANSI. January 2001.

ANSI (2002). *American National Standards Institute: Annual Report Two Thousand and One*. New York: ANSI.

ANSI (2002). *ANSI Procedures for the Development and Coordination of American National Standards*. New York: ANSI. Availiable at <www.ansi.org/rooms/room_16/public/gov_proc.html>

ANSI (2002b). *Breaking Down Borders: business, standards and trade: ANSI Annual Conference 2002*. Washington, DC. October 15-16, 2002.

Arendt, Hannah (1958). *The Human Condition*. Chicago: Chicago: University of Chicago Press.

Baker, Mark (2002). *Europe: Official Says Accounting Standards are Sound.* Prague: Radio Free Europe/Radio Liberty. August 7, 2002. Available at <http://www.rferl.org/nca/features/2002/08/07082002144442.asp>

Balto, David A. (2000). Standard Setting in a Network Economy. Speech, *Cutting Edge Antitrust Law Seminars International.* New York, NY. February 17, 2000. Available at <http://www.ftc.gov/speeches/other/standardsetting.htm>

Balto, David A. (2001, June). Standard Setting in the 21st Century Network Economy. *The Computer & Internet Lawyer, 18*(6), 1-18.

Barthes, Roland (1972). *Mythologies.* New York: Hill & Wang (c. 1957, reprint 1972).

Bekkers, R., Bart V., & Jan, S. (2002). Intellectual Property Rights and Standardization: the Case of GSM. *Telecommunications Policy, 26*, 171-188.

Bekkers, Rudi, Verspagen, Bart, & Smits, Jan (2002). Intellectual Property Rights and Standardization: the Case of GSM. *Telecommunications Policy, 26*, 171-188.

Benkler, Yochai (1998). Communications Infrastructure Regulation and the Distribution of Control Over Content. *Telecommmunications Policy, 22,* p. 183.

Benkler, Yochai (1998). Overcoming Agoraphobia: Building the Commons of the Digitally Networked Environment. *Harvard Journal of Law and Technology, 11*, pp. 287-290.

Benkler, Yochai (2001). Coase's Penguin, or, Linux and the Nature of the Firm. *29th Telecom Policy Research Conference.* Alexandria, VA. September 24, 2001. Available at <http://www.arxiv.org/abs/cs.CY/0109077>

Benkler, Yochai (2001). Through the Looking Glass: Alice and the Constitutional Foundations of the Public Domain. *Conference on the Public Domain*, Duke Law School, Durham, NC, November 9-11, 2001.

Benkler, Yochai (2002). Intellectual Property and the Organization of Information Production. *International Review of Law & Economics, 22*(1), pp. 81-107.

Besen, Stanley (1990, December). European Telecommunications Standards Setting: A Preliminary Analysis of the European Telecommunications Standards Institute. *Telecommunications Policy, 14*(6), 521-530.

Besen, Stanley (1995). The standards process in telecommunication and information technology. In R. Hawkins, R. Mansell, & J. Skea (Eds.), *Standards, Innovation and Competitiveness: The Politics and Economics of Standards in Natural and Technical Environments.* Brookfield, VT: Edward Elgar Publishing Ltd.

Besen, Stanley, & Farrell, Joseph (1991, August). The Role of the ITU in Telecommunications Standards-Setting: Pre-Eminence, Impotence, or Rubber Stamp. *Telecommunications Policy, 15*(4), 311-321.

Besen, Stanley, & Farrell, Joseph (1994, spring). Choosing How to Compete: Strategies and Tactics in Standardization. *Journal of Economic Perspectives, 8*(2), 117-131.

Bijker, Wiebe E. (1993, winter). Do Not Despair: There Is Life After Constructivism. *Science, Technology & Human Values, 18*(1), p. 119.

Bijker, Wiebe E. (1999). *Of Bicycles, Bakelites, and Bulbs: Toward a Theory of Sociotechnical Change.* Cambridge, MA: MIT Press.

Bijker, Wiebe E.(1993, winter). Do Not Despair: There Is Life After Constructivism. *Science, Technology & Human Values, 18*(1), 119.

Bilalis, Zacharias, & Didier, Herbert (2003, January-March). (IT) Standardisation from a European Point of View. *International Journal of IT Standards & Standardization Research, 1*(1), 46-49. Hershey, PA: Idea Group Publishing.

Bizzell, Patricia, and Bruce Herzberg (Eds.) (1990). *The Rhetorical Tradition: Readings from Classical Times to Present.* Boston: St. Martin's Press.

Black, Edwin (1965). Frame of Reference in Rhetoric and Fiction. In Donald C. Bryant (ed.). *Rhetoric and Poetic.* Iowa City: University of Iowa Press. pp. 26-35.

Black, Edwin (1970, April). The Second Persona. *The Quarterly Journal of Speech, 56*(2), pp. 109-119.

Boldrin, Michelle, & Levine, David (2002, March). *Perfectly Competitive Innovation*. Federal Reserve Bank of Minneapolis, Research Department Staff Report 303.

Bourdieu, Pierre (1990). *The Logic of Practice.* trans. by R. Nice. Stanford CA: Stanford University Press.

Boyle, James (1996). *Shamans, Software and Spleens*. Cambridge: Harvard University Press. p. 226.

Boyle, James (2001). The Second Enclosure Movement and the Construction of the Public Domain. *Conference on the Public Domain.* Duke Law School, Durham, NC, November 9-11.

Bradner, Scott (2001, spring/summer). The Internet Engineering Task Force. *OnTheInternet, 7*(1), Reston, VA: Internet Society, pp. 22-26.

Brenner, Robert (1998). The Economics of Global Turbulence: A Special Report on the World Economy, 1950-98. *New Left Review.* no. 229, May/June 1998. 264 pp.

Burk, Dan L., & Lemley, Mark A. (2003). Policy Levers in Patent Law. *Virginia Law Review. 89*(7), 1575-1695.

Burke, Kenneth (1950) *(1969). Rhetoric of Motives*. Berkeley: University of California Press.

Calabrese, Andrew (1991). The Periphery in the Center: The Information Age and the 'Good Life' in Rural America. *Gazette.* vol. 48. pp. 105-128.

Carey, James W. (1989). *Communication As Culture: Essays on Media and Society.* Boston: Unwin Hyman.

Cargill, Carl (1997). *Open Systems Standardization: A Business Approach.* Upper Saddle River, NJ: Prentice-Hall. 327 pp.

Cargill, Carl F. (1989). *Information Technology Standardization: Theory, Process, and Organizations*. Bedford, MA: Digital Press.

Cargill, Carl (1997). *Open Systems Standardization: A Business Approach* (p. 327). Upper Saddle River, NJ: Prentice-Hall.

Cargill, Carl (2001). Consortia Standards: Towards a Re-Definition of a Voluntary Standards Organization. Testimony before the House of Representatives, Sub-Committee on Technology, Environment and Standards on behalf of Sun Microsystems. June 28, 2001.

Cargill, Carl (2001). *The Role of Consortia Standards in Federal Procurements in the Information Technology Sector: Towards a Re-Definition of a Voluntary Standards Organization.* Submitted to the House of Representatives, Sub-Committee On Technology, Environment and Standards. Palo Alto: Sun Microsystems. June 28, 2001. 30 pp. Available at <http://www.house.gov/science/ets/Jun28/Cargill.pdf/>

Cargill, Carl (2005). Eating Our Seed Corn: A Standards Parable For Our Time. Available at http://islandia.law.yale.edu/isp/GlobalFlow/paper/Cargill.pdf

Cargill, Carl F. (1989). *Information Technology Standardization: Theory, Process, and Organizations*. Bedford, MA: Digital Press.

Cargill, Carl F. (1999). Consortia and The Evolution of Information Technology Standardization. *SIIT '99 Proceedings - 1st IEEE Conference on Standardisation and Innovation In Information Technology*, Aachen, Germany, September 15-17, 1999. p. 37-53

Cargill, Carl, & Bolin, Sherrie (2007). Standardization: a failing paradigm. In S. Greenstein & V. Stango (Eds.), *Standards and Public Policy*, Cambridge: Cambridge University Press.

Cargill, Carl. (2001a). *The Role of Consortia Standards in Federal Procurements in the Information Technology Sector: Towards a Re-Definition of a Voluntary Standards Organization.* Submitted to the House of Representatives, Sub-Committee On Technology, Environment and

Standards. Palo Alto: Sun Microsystems. June 28, 2001. 30 pp. Availiable at <http://www.house.gov/science/ets/Jun28/Cargill.pdf/>

CEN (2002). *CEN/ISSS Survey of Standards Related Fora and Consortia*. Brussels: CEN/ISSS Secretariat. October, 2002. Available at <http://www.cenorm.be/isss/Consortia2/>

Cerni, Dorothy M. (1984, December). *Standards in Process: Foundations and Profiles of ISDN and OSI Studies*. NTIA Report 84-170, Washington, DC: U.S. Department of Commerce. 247 pp.

Chambers, Simone (1996). *Reasonable Democracy: Jürgen Habermas and the Politics of Discourse*. Ithaca: Cornell University Press. 250 pp.

Charland, Maurice (1987, May). Constitutive Rhetoric: The Case of the *Peuple Québécois*. *The Quarterly Journal of Speech 73*,(2), pp. 133-150.

Cheit, Ross E. (1990). *Setting Safety Standards: Regulation in the Public and Private Sectors*. Berkeley: University of California Press.

Choh, Kwonjoong (1999). Governance Mechanisms of Standard-Making in Information Technology. *SIIT '99 Proceedings: 1st IEEE Conference on Standardization and Innovation in Information Technology* (pp. 49-54). University of Aachen. September 15-16, 1999.

Christensen, Clayton M. (1997). *The Innovators Dilemma: When New Technologies Cause Great Firms to Fail*. Cambridge, MA: Harvard Business School Press.

Christensen, Clayton M. (2002, June). The Rules of Innovation. *Technology Review*. Cambridge: MIT, pp. 33-38.

Codding, George A., Jr., & Rutkowski, A. M. (1982). *The International Telecommunications Union in a Changing World*. Dedham, MA: Artech House, Inc. (p. 414).

Congress of the United States, Office of Technology Assessment (1992). *Global Standards: Building Blocks for the Future*. TCT-512. Washington, DC: U.S. Government Printing Office. March 1992. 114 pp.

Cowan, R. (1992). High Technology and the Economics of Standardization. Dierkes, Meinolf, and Ute Hoffmann (Eds.) *New Technology at the Ourset: Social forces in the shaping of technological innovation*. Frankfurt, New York: Campus-Verlag and Westview. Chapter 14, pp. 279-300.

Craig, Robert T. (1996). Practical Theory: A Reply to Sandelands. *Journal for the Theory of Social Behavior 26*, (1), pp. 65-79.

David, Paul A. (1985). Clio and the Economics of QWERTY, *American Economic Review, 75*, pp. 332-337

David, Paul A. (1987). Some New Standards for the Economics of Standardization in the Information Age. In P. Dasgupta & Stoneman (Eds.) *Economic Policy and Technological Performance*. Cambridge: Cambridge University Press. pp. 206-234.

David, Paul A. (1995). Standardization policies for network technologies: the flux between freedom and order revisited. In Hawkins, Richard. Robin Mansell & Jim Skea (Eds.) (1995) *Standards, innovation and competitiveness*. Brookfield, VT: Edward Elgar Publishing Ltd.

David, Paul A. (1996). Formal standard-setting for global telecommunications and information services. *Telecommunications Policy*, *20*, 789-815.

David, Paul A. (2006). Using IPR to expand the research common for Science: New moves in 'legal jujitsu', *Intellectual Property Rights for Business and Society* conference sponsored by DIME-EU Network of Excellence, held in London, September 14-15, 2006.

David, Paul A., & Greenstein, Shane. M. (1990). The Economics of Compatibility Standards: An Introduction to Recent Research, *Economics of Innovation and New Technology, 1*, pp. 3-41.

David, Paul A., & Monroe, H. K. (1994). *Telecommunications Policy Research Conference*, held 1-3 October 1994 at Solomon's Island, MD.

David, Paul A., & Shurmer, M. (1996). Formal standards-setting for global telecommunication and information services. *Telecommunications policy, 20*(10), 789-815.

David, Paul A., & Steinmueller, W. E. (1994). Economics of Compatibility Standards and Competition in Telecommunication Networks, *Information Economics and Policy, 6*(3-4), pp. 217-241.

David, Paul A., and Hunter K. Monroe (1994). *Telecommunications Policy Research Conference*, held 1-3 October 1994 at Solomon's Island, MD.

De Vries, Henk J. (1999). Doctoral dissertation: Standards for the Nation, Erasmus University Rotterdam, published as *Standardization—A Business Approach to the Role of National Standardization Organizations*. Boston: Kluwer Academic Publishers.

De Vries, Henk J. (1999). *Standards for the Nation*. Doctoral dissertation, Erasmus University Rotterdam. published as *Standardization—A Business Approach to the Role of National Standardization Organizations*. Boston: Kluwer Academic Publishers.

Demsetz, Harold (1967). Toward a Theory of Property Rights. *American. Economic Review, 57*, 347-357.

DIN (2002). *Strategie für die Standardisierung der Informations und Kommunikationstechnik (ICT) - Deutsche Positionen,* Version 1.0. Berlin: DIN Deutsches Institut für Normung, eV. Available at <http://www2.din.de/sixcms/detail.php?id=3871>

DuBoff, Richard B. (1983). The Telegraph and the Structure of Markets in the United States, 1845-1890. *Research in Economic History*, *8*, 253-277.

EC (2000). *Guide to Implementation of Directives Based on the New Aporoach and the Global Approach.* Brussels: European Commission, DG Enterprise. 118 pp. Available at <http://www.newapproach.org/>

Economides, Nicholas (1989, 5 December). Desirability of Compatibility in the Absence of Network Externalities. *American Economic Review, 79*, 1165-1181.

Economides, Nicholas (1996). Network Externalities, Complementarities, and Invitations to Enter. *European Journal of Political Economy, 12*, 211-234.

Economides, Nicholas (2000). The Microsoft Antitrust Case. Stern School of Business Working Paper 2000-09, New York University.

Economides, Nicholas, & Himmelberg, C. (1995). Critical mass and network size with application to the US Fax market. Working Paper Series // Stern School of Business, NYC EC ; 95,11, Available at <http://www.stern.nyc/networks/95-11.pdf>

*European Commission. Final Report to the European Commission.* Delft: Faculty of Technology, Policy and Management, Delft University of Technology. October 2001. 69 pp.

Egyedi, Tineke (2001, October). *Beyond Consortia, Beyond Standardization? New Case Material for the European Commission. Final Report to the European Commission* (p. 69). Delft: Faculty of Technology, Policy and Management, Delft University of Technology.

Egyedi, Tineke M. (1996). *Shaping Standardization: A Study of Standards Processes and Standards Policies in the Field of Telematic Services.* (Doctoral dissertation) Delft University of Technology. 329 pp.

Egyedi, Tineke M. (2001). *Beyond Consortia, Beyond Standardization? New Case Material for the European Commission. Final Report to the European Commission.* Delft: Faculty of Technology, Policy and Management, Delft University of Technology. October 2001. 69 pp.

Egyedi, Tineke M. (2001). Why Java Was-not-standardized Twice. *Computer Standards & Interfaces, 23*(4), 253–265. Available at <http://www.elsevier.com/locate/csi/>

Egyedi, Tineke, & Arjan, Loeffen (2002). Succession in Standardization: Grafting XML Onto SGML. *Computer Standards & Interfaces, 24* (4), 279–290. Available at <http://www.elsevier.com/locate/csi/>

EPO (2007). European Patent Office. *EPO Scenarios for the Future.* (p. 128). EPO: Munch.

Farance, Frank (2002). Interview and email correspondence with Frank Farance. October 30, 2002.

Farrell, Joseph (1990). The Economics of Standardization: A Guide for Non-Economists. In Berg, John L., and Haral Schumny (Eds.). *An Analysis of the Information Process: Proceeding of the International Symposium on Information Technology Standardization, INSITS*. Amsterdam: North-Holland, pp. 189-198.

Farrell, Joseph (1995). Arguments for weaker IPR protection in network industries. In B. Kahin& J. Abate (Eds.), *Standards Policy for Information Infrastructure*. Cambridge, MA: MIT Press. pp. 368-402.

Farrell, Joseph (1996). *Choosing the Rules for Formal Standardization.* Berkeley: UC Berkeley.

Farrell, Joseph, & Saloner, Garth (1985, spring). Standardization, Compatibility, and Innovation. *Rand Journal of Economics, 16*(1), 70-83;

Farrell, Joseph, and Garth Saloner (1988, summer). Coordination through Committees and Markets. *Rand Journal of Economics, 19*(2), pp. 235-252.

Farrell, Thomas B. (1976, February). Knowledge, Consensus, and Rhetorical Theory. *The Quarterly Journal of Speech, 62*(1), pp. 1-14.

Farrell, Thomas B. (1993). *Norms of Rhetorical Culture.* New Haven: Yale University Press. 373 pp.

Feenberg, Andrew (1991). *Critical Theory of Technology*. Oxford: Oxford University Press.

Fiske, John, and John Hartley (1978). *Reading Television*. London: Methuen.

Flood, Tony (2002). *Presentation and subsequent personal conversation*. ANSI Annual Meeting, October 15, 2002.

Fomin, Vladislav V. (2001). *The Process of Standard Making: The Case of Cellular Telepony.* Doctoral Dissertation, Univeristy of Jyväskylä, Jyväskylä, Finland.

Foucault, Michel (1973). *The Birth of the Clinic: An Archaeology of Medical Perception,* Trans. by, A.M. Sheridan Smith. New York: Random House.

Frank, Thomas (2000). *One Market Under God: Extreme Capitalism, Market Populism, and the End of Economic Democracy.* New York: Anchor Books, Random House.

Fraser, Nancy (1990). Rethinking the Public Sphere: A Contribution to the Critique of Actually Existing Democracy. *Social Text, 25-26*, pp. 56-80.

Freeman, Chris, and Luc Soete (1997). *The Economics of Industrial Innovation* (3rd Ed.), Cambridge, MA: MIT Press, 470 pp.

Froomkin, Michael A. (2002). *Habermas@discourse.net: Towards a Critical Theory of Cyberspace.* Draft paper. University of Miami School of Law.

Fukuyama, Francis (1992). *The End of History and the Last Man.* New York: Avon Books.

Gaynor, Mark, Bradner, Scott, Lansiti, Marco and H.T. King (2001). The Real Options Approach to Standards for Building Network-based Services. in Timothy Schoechle and Carson Wagner (eds.), *SIIT 2001 Proceedings: 2nd IEEE Conference on Standardization and Innovation in Information Technology*. Boulder, Colorado, October 3-6, 2001. pp. 217-228.

Geertz, Clifford (1973). *The Interpretation of Cultures: Selected Essays.* Boston: Basic Books.

Gibson, David, and Everett M. Rogers (1994). *R&D Collaboration on Trial: The Microelectronics and Computer Technology Corporation*. Boston, MA: Harvard Business School Press, 605 pp.

Gifford, Jonathan L. (1997). ITS Standardization: Assessing the Value of a Consortium Approach. in *Proceedings of the ITS Standards Review and Interoperability Workshop.* George Mason University Dec. 17-18, 1997. pp 1-7. Available at <http://www.itsdocs.fhwa.dot.gov/jpodocs/proceedn/2lz1!.pdf>

Gowan, Peter (1999). *The Global Gamble: Washington's Faustian Bid for World Dominance.* New York, London: Verso. 320 pp.

Gray, John (1998). *False Dawn: The Delusions of Global Capitalism* (p. 262). London: Granta Books.

Greenstein, Shane M. (1992, September). Invisible hands and Visible Advisors: An Economic Interpretation of Standardization, *Journal of the American Society for Information Science, 43*(8).

Greenstein, Shane M., and V. Stango (Eds.) (2005). *Standards and Public Policy.* Cambridge: Cambridge Press.

Greider, William (1997). *One World, Ready or Not: The Manic Logic of Global Capitalism.* New York: Simon & Schuster. 528 pp.

Habermas, Jürgen (1962) *(1991). The Structural Transformation of the Public Sphere: An Inquiry into a Category of Bourgeois Society* (Thomas Burger, Trans.). Cambridge, MA: MIT Press.

Habermas, Jürgen (1975). *Legitimation Crisis* (Thomas McCarthy, Trans.). Boston: Beacon Press.

Habermas, Jürgen (1984). *The Theory of Communicative Action.* vol. 1, *Reason and the Rationalization of Society.* vol. 2. *Lifeworld and System: A Critique of Functionalist Reason* (Thomas McCarthy, Trans.). Boston: Beacon Press.

Habermas, Jürgen (1995). *Justification and Application: Remarks on Discourse Ethics* (Ciaran Cronin, Trans.). Cambridge, MA: MIT Press. (portions originally published in 1990 and 1991).

Habermas, Jürgen (1996). *Between Facts and Norms: Contributions to a Discourse Theory of Law and Democracy* (William Rehg, Trans.). Cambridge, MA: MIT Press. 630 pp.

Hardin, Garrett (1968). The Tragedy of the Commons. *Science, 16*(2), pp. 1243.

Hartley, John (1994). In John Fiske (ed.). *Key Concepts in Communication and Cultural Studies.* 2nd Edition. New York: Routledge. pp. 92-95.

Haug, Thomas (2002). A Commentary on Standardization Practices: Lessons from the NMT and GSM Mobile Telephone Standards Histories, *Telecommunications Policy, 26*, pp. 101-107.

Hauser, Gerard A. (1999). *Vernacular Voices: The Rhetoric of Publics and Public Spheres.* Columbia, SC: University of South Carolina Press.

Hawkins, Richard (1999). The rise of consortia in the information and communication technology industries: emerging implications for policy. *Telecommunications policy, 23*, pp. 159-173.

Hawkins, Richard, Robin Mansell, and Jim Skea (Eds.) (1995). *Standards, Innovation and Competitiveness: The Politics and Economics of Standards in Natural and Technical Environments.* Brookfield, VT: Edward Elgar Publishing Ltd.

Hesser, Wilfried, and Axel Czaya (1999, March). Standardization as a Subject of Study in Higher Education—A Vision. Unpublished manuscript, Universität der Bundeswer, Hamburg.

Horwitz, Robert B. (1989). *The Irony of Regulatory Reform: The Deregulation of American Telecommunications.* New York: Oxford University Press. (p. 414).

Humphries, Jane (1990, March). Enclosures, Common Rights, and Women: The Proletarianization of Families in the Late Eighteenth and Early Nineteenth Centuries, *Journal of Economic History, L*(1), The Economic History Association, pp. 17–42.

Hunter, Dan (2002). Cyberspace as Place. *TPRC 2002: The 30th Annual Research Conference on Communication, Information and Internet Policy.* September 28-30, 2002, Alexandria, VA.

IEC (2003). *Introduction to the IEC.* Slide presentation. Geneva: International Electrotechnical Commission.

Innis, Harold A. (1972). *Empire and Communication.* Toronto: University of Toronto Press.

Isenberg, David S. (1997, August). "The Rise of the Stupid Network." *Computer Telephony.*

ISO (1999). [Web site for] International Organization for Standardization. January 8, 1999. Availiable at <http://www.iso.ch>

ISO/IEC (1996). *Guide 2: Standardization and Related Activities—General Vocabulary.* Seventh Edition, Geneva: ISO/IEC.

ITU-T/ISO/IEC (1996). *Guide for ITU-T and ISO/IEC JTC1 Cooperation.* Annex A to WTSC Recommendation A.23, October 1996; and Annex K to ISO/IEC JTC1 Directives. December 1996.

Iversen, Eric J., Bekkers, Rudi & Blind, Knut (2006). Emerging coordination mechanisms for multi-party IPR holders: linking research with standardization. *Intellectual Property Rights for Business and Society* conference sponsored by DIME-EU Network of Excellence, held in London, September 14-15.

Jakobs, Kai (2000). *Standardisation Processes in IT: Impact, Problems and Benefits of User Participation.* Braunschweig/Wiesbaden: Vieweg & Sohn Verlagsgesellschaft mbH.

Jakobs, Kai (2001). Broader View on Some Forces Shaping Standardization. *SIIT 2001 Proceedings: 2nd IEEE Conference on Standardization and Innovation in Information Technology.* Boulder, CO, October 5, 2001. (pp. 133-143).

Jakobs, Kai, Rob Procter, and Robin Williams (1998). Telecommunication Standardisation - Do We Really Need the User? *Proceedings of ICT '98: the 6th International Conference on Telecommunications*, IEEE Press.

Kahin, Brian, and Janet Abbate (Eds.) (1995). *Standards Policy for Information Infrastructure,* Cambridge: MIT Press. 653 pp.

Katz, Michael L., and Carl Shapiro (1985, June). Network Externalities, Competition and Compatibility. *American Economic Review 75*(3), pp. 424-440.

Katz, Michael L., and Carl Shapiro (1986, August). Technology Adoption in the Presence of Network Externalities. *Journal of Political Economy*, *94*(4), pp. 822-841.

Katz, Michael L., and Carl Shapiro (1992). Product Introduction with Network Externalities. Journal of Industrial Economics, *40*(1). pp. 55-83.

Katz, Michael L., and Carl Shapiro (1994). Systems Competition and Network Effects. *Journal of Economic Perspectives, 8,* pp. 93-115.

Katzenstein, Peter (1985). *Small States in World Markets: Industrial Policy in Europe.* Ithaca: Cornell University Press.

Keidel, Albert (2008, July). "China's Economic Rise—Fact and Fiction," *Policy Brief 61*. Washington, DC: Carnegie Endowment for International Peace.

Kennedy, George (1991). *Aristotle on Rhetoric: A Theory of Civic Discourse.* New York: Oxford University Press. pp. 190-215.

Khanna, Parag (2008). *The Second World: Empires and Influence in the New Global Order.* New York: Random House.

Koolhaas, Rem (2002). *Newshour Interview by Ray Suarez*. June 25, 2002. Available at <http://www.pbs.org/newshour/bb/entertainment/jan-june02/koolhaas_transcript.html/>

Krechmer, Ken (1998). The Principles of Open Standards. *Standards Engineering, 50*(6), November/December 1998. p. 1. Available at <http://www.crstds.com/openstds.html>

Krechmer, Ken (1998, November/December). The Principles of Open Standards. *Standards Engineering, 50*(6), 1. Available at <http://www.crstds.com/openstds.html>

Krechmer, Ken (2000, July/August). Market Driven Standardization: Everyone Can Win. *Standards Engineering, 52*(4). pp. 15-19. Available at <http://www.crstds.com/fundeco.html>

Krechmer, Ken (2000, June). The Fundamental Nature of Standards: Technical Perspective. *IEEE Communications Magazine, 38*(6), p. 70. Available at <http://www.crstds.com/fora.html>

Krechmer, Ken (2000). The Fundamental Nature of Standards: Economics Perspective. *Schumpeter 2000: Eighth International Joseph A. Schumpeter Society Conference*, Manchester, UK, June 28-July 1, 2000. p. 70. Available at <http://www.crstds.com/fundeco.html>

Krechmer, Ken, and Elaine Baskin (2001). Standards, Information and Communications: A Conceptual Basis for a Mathematical Understanding of Technical Standards. In Timothy Schoechle and Carson Wagner (eds.). *SIIT 2001 Proceedings: 2nd IEEE Conference on Standardization and Innovation in Information Technology*. Boulder, Colorado, October 3-6, 2001. pp. 106-114. Available at <http://www.crstds.com/siit2001.html>

Kuhn, Thomas S. (1962). *The Structure of Scientific Revolutions.* Chicago: University of Chicago Press.

Kupchan, Charles A. (2002, November). The End of the West. *The Atlantic Monthly*, *290*(4), 42-44.

Kupchan, Charles A. (2002). *The End of the American Era: U.S. Foreign Policy and the Geopolitics of the Twenty-first Century* (p. 391). New York: Alfred A. Knopf.

Lash, William, III (2002). Keynote speech at ANSI Annual Meeting, Washington, DC. October 15, 2002.

Lehenkari, Janne, and Reijo Miettinen (2001). Standardization in the Construction of a Large Technological System—The Case of the Nordic Mobile Telephone System. *Telecommunications Policy*, *26*, pp. 109-127.

Lehr, William (1992, September). Standardization: Understanding the Process, *Journal of the American Society for Information Science, 43*(8).

Lemley, M., & McGowan, David. (1998). Could Java Change Everything? The Competitive Propriety of a Proprietary Standard. *Antitrust Bulletin, 43*, 715

Lemley, Mark A. (1998). The Law and Economics of Internet Norms. *Kent Law Review, 73*, pp. 1257-1288.

Lemley, Mark A. (1999). Standardizing Government Standard-Setting Policy for Electronic Commerce. *Berkeley Technology Law Journal, 14*, pp. 745-752.

Lemley, Mark A. (2001). Antitrust, Intellectual Property Rights and Standard Setting Organizations. In Timothy Schoechle and Carson Wagner (eds.), *SIIT 2001 Proceedings—2nd IEEE Conference on Sandardization and Innovation in Information Technology*, Boulder, Colorado, October 3-5, 2001. pp. 157-169.

Lemley, Mark A. (2002). Intellectual Property Rights and Standard-Setting Organizations. *California Law Review, 90*(6), pp. 1889-1979.

Lemley, Mark A. (2002). Place and Cyberspace. *TPRC 2002: The 30th Annual Research Conference on Communication, Information and Internet Policy.* September 28-30, Alexandria, VA.

Lemley, Mark, and David McGowan (1998). Could Java Change Everything? The Competitive Propriety of a Proprietary Standard. *Antitrust Bulletin, 43*, p. 715

Lessig, Lawrence (1999). *Code and Other Laws of Cyberspace*. New York: Basic Books. 297 pp.

Lessig, Lawrence (2001). *The Future of Ideas: The Fate of the Commons in a Connected World.* New York: Random House. pp. 352

Lessig, Lawrence (2001, November/December). The Internet Under Siege. *Foreign Policy.*

Lessig, Lawrence (2002). Remarks at the conference, *The Regulation of Information Platforms*, Silicon Flatirons Telecommunication Program. University of Colorado, Boulder, CO. January 27-28, 2002.

Libicki, Martin (1995). *Information Technology Standards: Quest for the Common Byte*. Boston: Digital Press.

Libicki, Martin, James Schneider, Dave R. Frelinger, and Anna Slomovic (2002). *Scaffolding the New Web: Standards and Standards Policy for the Digital Economy*. Santa Monica: RAND Corporation. Available at <http://www.rand.org/publications/MR/MR1215/>

Lim, Andriew S. (2002). *Standards Setting Processes in ICT: The Negotiations Approach.* Unpublished working paper 02.19. Eindhoven Centre for Innovation Studies. Technische Universiteit Eindhoven, Faculteit Technologie Management. 21 pp. <http://www.tm.tue.nl/ecis/>.

Lipset, Seymour Martin (1996). *American Exceptionalism: A Double Edged Sword.* New York: W.W. Norton & Co. 348 pp.

List, Fredrich (1841). *(1904). The National System of Political Economy.* London: Longman.

Litman, Jessica (1990, fall). The Public Domain. *Emory Law Journal 39*, 965. Availiable at <http://www.law.wayne.edu/litman>

MacIntyre, Alasdair (1981). *After Virtue*. Notre Dame: University of Notre Dame Press.

Majoras, Deborah Platt (2005). Recognizing the Procompetitive Portential of Royalty Discussions in Standard Setting. *Standardization and the Law: Developing the Golden Mean for Global Trade* conference at Stanford University, Stanford, CA, 23 September. In S. Bolin (Ed.), *The Standards Edge: The Golden Mean* ( pp. 101-107). Chelsea, MI: Sheridan Books.

Mansell, Robin (1993). *The New Telecommunications: A Political Economy of Network Evolution.* London: Sage Publications Ltd.

Mansell, Robin (1995). Standards, Industrial Policy and Innovation. In Hawkins, Richard, Robin Mansell, and Jim Skea (Eds.). *Standards, Innovation and Competitiveness: The Politics and Economics of Standards in Natural and Technical Environments.* Brookfield, VT: Edward Elgar Publishing Ltd., pp. 213-227.

Mansell, Robin, & Richard Hawkins (1992). Old Roads and New Signposts: Trade Policy Objectives in Telecommunication Standards. In Klaver, F. & P. Slaa (Eds.). *Telecommunication, New Signposts to Old Roads*. Amsterdam: IOS Press, pp. 45-54.

Mansell, Robin, and Roger Silverstone (Eds.) (1996). *Communication by Design: The Politics of Information and Communication Technologies.* Oxford: Oxford University Press.

Mansfield, Edwin (1970). *Microeconomic Theory and Application.* New York: W.W. Norton

Marks, Roger B. and Robert E. Hebner (2001). Government Activity to Increase Benefits from the Global Standards System. *SIIT 2001 Proceedings: 2nd IEEE Conference on Standardization and Innovation in Information Technology*. Boulder, CO, October 5, 2001. pp. 183-190.

Marks, Roger B. and Robert E. Hebner (2004). Government/Industry Interactions in the Global Standards System. In Bolin, Sherrie, (Ed.) *The Standards Edge: Dynamic Tension.* Ann Arbor, MI: Sheridan Books. pp. 103-114.

McClosky, Donald M. (1972, March). The Enclosure of Open Fields: Preface to a Study of Its Impact on the Efficiency of English Agriculture in the Eighteenth Century, *Journal of Economic History, 32*, The Economic History Association, pp. 15-35.

Messerschmitt, David G. (1999). Prospects for Computing—Communications Convergence. Lecture at the University of Colorado, Boulder, CO. October 24, 1999. Available at <http://www.eccs.berkeley.edu/~messer/talks/99/Convergence.pdf>

Metcalfe, Robert M. (1995, 2 October). Metcalfe's Law. *InfoWorld.*

MIC (2004). *IT839 Strategy: The Road to $20,000 GDP/Capita*. Ministry of Information and Communication, Republic of Korea. Availiable at <http://www.mic.go.kr>

Mills, C. Wright (1940, December). Situated Action and Vocabularies of Motive. *American Sociological Review.* pp. 904-913.

Mueller, Milton (2002). Interest Groups and the Public Interest: Civil Society Action and the Milton Globalization of Communications Policy. *TPRC 2002: The 30th Annual Research Conference on Communication, Information and Internet Policy.* September 28-30, 2002. Alexandria, VA.

Murray, Alan (2002). No Longer Business As Usual for Forces of U.S. Capitalism. *Wall Street Journal* 1 October, p. A4.

Negroponte, Nicholas (1995). *Being Digital.* New York: Alfred Knopf.

NIST (1998). *National Technology Transfer and Advancement Act homepage.* Retrieved October 19, 2002 from <http://www.ts.nist.gov/ts/htdocs/210/nttaa/nttsaa.htm>

NSPAC (1979). *National Policy on Standards for The United States and a Recommended Implementation Plan.* Washington, DC: National Standards Policy Advisory Committee (reproduced in Cerni (1984), appendix c.2: pp. 223-226).

O'Reilly, Tim (1999). Open Sources: voices from the Open Source Revolution. O'Reilly and Associates. 280 pages. Available at <http://www.oreilly.com/catalog/opensources/book/toc.html>, read August 27, 2003.

Oksala, Steve (2000, 21 August). *The Changing Standards World: Government Did It, Even If They Didn't Mean To.* ANSI/NIST World Standards Day paper.

Outhwaite, William, & Bottomore, Tom (Eds.) (1993). *The Blackwell Dictionary of Twentieth-Century Social Thought.* Oxford: Blackwell Publishers Ltd.

Peters, John Durham (1989, November). John Locke, the Individual, and the Origin of Communication. *The Quarterly Journal of Speech, 75*, pp. 387-399.

Peters, John Durham (2002). The Problem of Conversation in Media. *Josephine Jones Lecture.* Department of Communication. University of Colorado, Boulder, CO. October 22, 2002.

Phillips, Kevin (1994). *Arrogant Capitalism: Washington, Wall Street and the Frustration of American Politics.* Boston: Little Brown and Co.

Popper, Karl (1945). *The Open Society and Its Enemies.* 2 vols. London: Routledge.

Post, David G. (2000). What Larry Doesn't Get: Code, Law and Liberty in Cyberspace. *Stanford Law Review, 52*, pp. 1439-1451. Available at <http://www.temple.edu/lawschool/dpost/Code.pdf/>

Random House (1967). *The Random House Dictionary of the English Language.* New York: Random House, Inc., 1960 pp.

Raymond, Eric (1998). *The Cathedral and the Bazaar.* Available at <http://www.tuxedo.org/~esr/writings/cathedral-bazaar/cathedral-bazaar/>

Rice, John, and Peter Galvin (1999). The Development of Standards in the Mobile Telephone Industry and Their Effect on Regional Industry Growth. *SIIT '99 Proceedings: 1st IEEE Conference on Standardization and Innovation in Information Technology*. University of Aachen. September 15-16, 1999. pp. 153-156.

Ritterbusch, Gerald. H. (2001). *Standards-Setting and United States Competitiveness.* Testimony before the House of Representatives, Sub-Committee On Technology, Environment and Standards on Behalf of Caterpillar Inc., June 28, 2001.

Robinson, Gary S. (1999). There are no Standards for Making Standards. *SIIT '99 Proceedings - 1st IEEE Conference on Standardisation and Innovation in Information Technology*, Aachen, Germany, September 15-17, 1999. pp. 249-250.

Rowland, Willard D., Jr. (1986, April). American Telecommunications Policy Research: Its Contradictory Origins and Influences. *Media, Culture & Society.*

Rowland, Willard D., Jr. (1997, autumn). "The Meaning of "The Public Interest" in Communications Policy—Part I: Its Origins in State and Federal Regulation." *Communication Law and Policy, 2*(4), pp. 309-328.

Samuelson, Robert J. (2002). Global Economics Under Siege. *Washington Post* 16 October. p. A25.

Saussure, Ferdinand de (1916) *(1974). Course in General Linguistics.* London: Fontana.

Scannell,Kara, & Slater, Joanna. (2008, August). SEC Moves to Pull Plug On U.S. Accounting Standards. *The Wall Street Journal, CCLII,* 28(70)August 2008, A1.

Schiller, Dan (1996). *Theorizing Communication: A History.* Oxford: Oxford University Press.

Schiller, Dan (2000). *Digital Capitalism: Networking the Global Market System.* Cambridge: MIT Press.

Schoechle, Timothy (1995, April). The Emerging Role of Standards Bodies in the Formation of Public Policy. *IEEE Standards Bearer*, *9*(2), pp. 1, 10. Available at <http://www.acm.org/pubs/articles/proceedings/cas/332186/p255-schoechle/>

Schoechle, Timothy (1998). *A Public Policy Debate: Standardsmaking Practice and the Discourse on International Privacy Standards.* Standards Policy Research Paper ICSR98-211. Boulder: International Center for Standards Research. Availiable at <http://www.standardsresearch.org>

Schoechle, Timothy (1999). Toward a Theory of Standards. *SIIT '99 Proceedings: 1st IEEE Conference on Standardization and Innovation in Information Technology.* University of Aachen, September 15-16, 1999. pp. 175-181.

Schoechle, Timothy (2001, fall). Re-examining Intellectual Property Rights in the Context of Standardization, Innovation and the Public Sphere. *Knowledge, Technology & Policy, 14*(3), 109-126.

Schoechle, Timothy (2003). Digital Enclosure: The Privatization of Standards and Standardization, *SIIT 2003 Proceedings: 3rd IEEE Conference on Standardization and Innovation in Information Technology.* Delft University of Technology. 22 October, pp. 229-240.

Schoechle, Timothy, Shapiro, Stephen, Rinow, Michael., & Richards, Barnaby (2002). Evolving Approaches to Technical Standardization: Hybrid Standards Setting. *EASST 2002: Conference of the European Association for the Study of Science and Technology*, York, UK, August 1.

Shapiro, Carl (2001). Navigating the Patent Thicket: Cross Licenses, Patent Pools, and Standard-Setting, In Jaffe, Adam, Joshua Lerner, and Scott Stern, (Eds.), *Innovation Policy and the Economy, I*, Cambridge, MA: MIT Press.

Shapiro, Carl, and Hal R. Varian (1999). *Information Rules: A Strategic Guide to the Network Economy.* Boston: Harvard Business School Press.

Shapiro, Carl, Richards, Barnaby, Rinow, Michael, & Schoechle, Timothy (2001). Hybrid Standards Setting Solutions For Today's Convergent Telecommunications Market. *SIIT 2001Proceedings: 2nd IEEE Conference on Standardization and Innovation in Information Technology* (pp. 348-351). Boulder, CO, October 5, 2001.

Sherif, Mostafa (2001, April). A Framework for Standardization in Telecommunications and Information Technology. *IEEE Communications Magazine, 39*(4), 94-100.

Sherif, Mostafa (2002). *When Is Standardization Slow?* Conference of the European Association for the Study of Science and Technology, York, UK, August 1, 2002. Revised version published in *International Journal of IT Standards and Standardization Research, 1*, (2003, spring), pp. 19-32.

Skitol, Robert A. (2005). Concerted Buying Power: It's Potential for Addressing the Patent Holdup Problem in Standard Setting. *Antitrust Law Journal, 72*, 727-744.

Smith, Adam (1776) (1976). *An Inquiry into the Nature and Causes of the Wealth of Nations*. Inn R. H. Campbell, *et al (*Eds.), Oxford & New York: Oxford University Press.

Smoot, Oliver R. (2001). *Standards-Setting and United States Competitiveness*. Statement of Oliver R. Smoot, Chairman of the Board of Directors, American National Standards Institute before the House Science Committee, Subcommittee on Technology, Environment and Standards, June 28, 2001. Available at <http://www.house.gov/science/ets/Jun28/Smoot.htm/>

Smoot, Oliver R. (2003). SIIT 2003 Address by Mr. Oliver Smoot, President of ISO. *SIIT 2003 Proceedings: 3rd IEEE Conference on Standardization and Innovation in Information Technol-*

*ogy*. Delft University of Technology. October 22, 2003. Visited November 5, 2003. Available at <http://www.siit2003.org/slides/smoot/ISO_Presentation to SIIT.pdf>

Soros, George (1998). *The Crisis of Global Capitalism [Open Society Endangered]*. New York: PublicAffairs, Perseus Books Group.

Soros, George (2002). *George Soros on Globalization*. New York: PublicAffairs, Perseus Books Group.

Spring, Michael B. *et al* (1995). Improving the Standardization Process: Working with Bulldogs an Turtles. In Kahin, Brian, and Janet Abate (eds), *Standards Policy for Information Infrastructure*. Cambridge, MA: MIT Press, pp. 220-250.

Steinmo, Sven (1994). "American Exceptionalism Reconsidered: Culture or Institutions?" In Larry Dodd and Calvin Jillson (eds.). *The Dynamics of American Politics: Approaches and Interpretations*. Boulder, CO: Westview Press.

Steinmueller, W. E. (1995). The political economy of data communication standards. In Hawkins, Richard, Robin Mansell, and Jim Skea (Eds.) *(*1995) *Standards, innovation and competitiveness*. Aldershot: Edward Elgar.

Stiglitz, Joseph E. (2003). *Globalization and its Discontents*. New York: W.W. Norton, 288 pp.

Streeter, Thomas (1986). *Technocracy and Television: Discourse, Policy, Politics and the Making of Cable Television*. unpublished doctoral dissertation. University of Illinois at Urbana-Champaign.

Streeter, Thomas (1990, spring). Beyond Freedom of Speech and the Public Interest: The Relevance of Critical Legal Studies of Communication Policy. *Journal of Communication.*

Stuttmeier, Richard P., Xiangkui, Yao, & Alex, Zixiang Tan (2006). *Standards of Power? Technologies, Institutions and Politics in the Development of China's National Standards Strategy* (p. 52). Seattle: The National Bureau of Asian Research.

Taschdjian, Martin (2001). Standards and the Velocity of Knowledge: Accelerator or Inhibitor? Paper presented at *SIIT 2001 2nd IEEE Conference on Standardization and Innovation in Information Technology*, Boulder, CO, October 5, 2001.

Techapalokul, Soontaraporn, James H. Alleman, and Yongmin Chen (2001). Economics Of Standards: A Survey And Framework, in Timothy Schoechle and Carson Wagner (eds.). *SIIT 2001 Proceedings: 2nd IEEE Conference on Standardization and Innovation in Information Technology*, Boulder, CO, October 3-6, 2001. pp 193-205.

Thomas, James (2002). Remarks of James Thomas, President, ASTM International at *Standards Activity Promotion Workshop: How to Promote Voluntary Consensus Standards Activities in Korea*. Invited panel expert at Korean Standards Association, Intercontinental Hotel, Seoul, Korea, December 4, 2002.

Updegrove, Andrew (1995). Consortia and the Role of the Government. In Kahin, Brian, and Janet Abate (eds), *Standards Policy for Information Infrastructure*. Cambridge, MA: MIT Press, pp. 321-348.

Updegrove, Andrew (1995, December). Standard Setting and Consortium Structures. *Standard View*.

*Vaidhyanathan*, Siva (2001). *Copyrights and Copywrongs: The Rise of Intellectual Property and How It Threatens Creativity.* New York: New York University Press. 241 pp.

Van Houweiling, Molly S.(2002). Cultivating Open Information Platforms: A Land Trust Model. *The Journal on Telecommunications Law and High Technology*. Vol. 1. Also presented at the Silicon Flatirons Conference. "Regulation of Information Platforms." University of Colorado, Boulder, CO. January 28-29.

Vardakas, Evangelos (1998). Director, EC Directorate for Industry (DG III), speech before the *Standardization Summit* sponsored by ANSI and NIST, Washington DC, September 23, 1998.

Vardakas, Evangelos (2002). Director, EC Directorate General Enterprise, presentation before the ANSI Annual Conference, Washington DC, October 15, 2002, and subsequent personal interview.

Veblen, Theodore (1919) *(1990)*. The Place of Science in Modern Civilization. In Veblen, T. B., *The Place of Science in Modern Civilization and Other Essays*. New Brunswick: Transaction.

Vercoulen, Frank, & van Wegberg, Marc (1998). *Standard Selection Modes in Dynamic, Complex Industries: Creating Hybrids between Market Selection and Negotiated Selection of Standards,* Maastricht: Universiteit Maastricht, (pp. 1-14).

Wagner, Douglas K. (1999). *Discourse and Policy in the Philippines: The Construction and Implementation of a Strategic IT Plan.* Unpublished doctoral dissertation, University of Colorado, Boulder, CO.

Wagner, Douglas K. (1999). Embracing Incommensurate Paradigms of Technology and Development: Strategic IT Planning for Developing States. *The 49th Annual Conference of the International Communication Association,* San Francisco, May 27-31, 1999.

Wagner, R. Polk (2002). Information Wants To Be Free: Intellectual Property and the Mythologies of Control. *TPRC 2002: The 30th Annual Research Conference on Communication, Information and Internet Policy,* Alexandria, VA, September 28-30, 2002 <http://papers.pennlaw.net/>.

Warshaw, Stanley I., & Saunders, Mary H. (1995). International Challenges in Defining the Public and Private Interest in Standards. In R. Hawkins, R. Mansell, & J. Skea (Eds.), *Standards, Innovation and Competitiveness: The Politics and Economics of Standards in Natural and Technical Environments.* Brookfield, VT: Edward Elgar Publishing Ltd. (pp. 67-74).

Weber, Max (1921-2). *Economy and Society: An Outline of Interpretive Sociology*. 2 vols. G. Roth and C. Whittich (eds.), (1978). Berkeley and New York: Bedminister Press.

Webster (1977), *Webser's New World Dictionary, College Edition*. New York: The World Publishing Company, p. 1133.

Wegberg, Marc van (2002). Positioning Strategies Of Standard Development Organizations in the Internet: The Case Of Internet Telephony. *EASST 2002: Conference of the European Association for the Study of Science and Technology*, York, UK, August 1, 2002.

Wegberg, Marc van (2002). Interview, August 2, 2002, at the EASST Conference, York, UK.

Wegberg, Marc van (2003). The Grand Coalition versus Competing Coalitions: Trade-Offs in How to Standardize. *SIIT 2003 Proceedings: 3rd IEEE Conference on Standardization and Innovation in Information Technology*. Delft University of Technology. October 22, 2003. pp. 271-284.

Wegberg, Marc van (2004). Standardization Process of Systems Technologies: Creating a Balance between Competition and Cooperation, *Technology Analysis & Strategic Management, 16*(4), pp. 457-478.

Weiser, Philip J. (2001). Internet Governance, Standard-Setting, and Self-Regulation. *Northern Kentucky Law Review, 28*(4), pp. 822-846.

Weiss, Martin B.H. (1990): The Standards Development Process: A View from Political Theory. *Standard View, 1*(2), pp. 35-41.

Weiss, Martin B.H., and Carl Cargill (1992). Consortia and the Standards Development Process. *Journal of the American Society for Information Science, 43*(8), pp. 559-565.

Weiss, Martin B.H., and Marvin Sirbu (1990). Technological choice in Voluntary Standards Committees: an Empirical Analysis. *Economics of Innovation and New Technology, 1*(1), pp. 111-134.

Weiss, Martin B.H., and Ronald T. Toyofuku (1996). Free-Ridership in the Standard Setting Process: The Case of 10baseT. *Standard view, 4*(4), pp. 205-212.

Weiss, Martin H.B., and Michael Spring (2000). Selected Intellectual Property Issues in Standardization, In Jakobs, Kai, (Ed.) *Information Technology Standards and Standardization: A Global Perspective*. Hershey, PA: Idea Group Publishing.

Williams, Raymond (1975). *Television: Technology and Cultural Form*, New York: Schocken Books.

Williams, Robin (1997). The Social Shaping of Information And Communication Technologies. In Herbert Kubicek, William H. Dutton and Robin Williams (eds.). *The Social Shaping of Infor-*

*mation Superhighways: European and American Roads to the Information Society.* Frankfurt: Campus and New York: St. Martins Press. Chapter 18, pp. 299-337.

Williams, Robin, & David Edge (1996). The Social Shaping of Technology. *Research Policy, 25*, pp. 865-899.

Winner, Langdon (1985). Do Artifacts Have Politics? In MacKenzie, Donald, and Judy Wajcman (Eds.). *The Social Shaping of Technology*. Milton Keynes: Open University Press.

Yamada, Hajime (2006, April). Patent Exploitation in the Information and Communications Sector—Using Licenses to Lead the Market. *Science and Technology Trends Quarterly Review, 19*, 11-21. [original Japanese version published October 2005]

Zhao, Houlin (2001). The IT Standardization and ITU. Speech at *SIIT 2001: 2nd IEEE Conference on Standardization and Innovation in Information Technology*. Boulder, CO, October 5.

# Glossary

**1394 TAL:** The 1394 High Performance Serial Bus Trade Association, formed to commercialize support the IEEE 1394 standard

**3G:** Third generation wireless (mobile telephony)

**AAP:** Alternative Approval Process

**ACEA:** Advisory Committee on Environmental Aspects (IEC)

**ACEC:** Advisory Committee on Electromagnetic Compatibility (IEC)

**ACET:** Advisory Committee on Electronics and Telecommunications (IEC)

**ACOS:** Advisory Committee on Safety (IEC)

**AFNOR:** *Association Française de Normalisation*, French national body

**ANSI:** American National Standards Institute, U.S. national body

**ASHRAE:** American Society of Heating Refrigeration and Air Conditioning

**ASME:** American Society of Mechanical Engineers

**ASTM:** American Society for Testing and Materials

**ATIS:** Alliance for Telecommunications Industry Solutions

**ATM Forum:** Asynchronous Transfer Mode Forum

**AT&T:** American Telephone and Telegraph

**BSI:** British Standards Institution

**CA:** Conformity Assessment

**CD:** Committee Draft

**CCITT:** *Comité Consultatif International Telegraphique et Telephonique* (ITU)

**CCIR:** *Comité Consultatif International de Radiocommunication* (ITU)

**CEBus:** Consumer Electronic Bus, the EIA-600 standard

**CENELEC:** *Comité Européen de Normalisation Electrotéchnique*

**CITI:** Columbia Institute for Tele-Information

**Committee T1:** U.S. standards committee for telephony networks, now administered by ATIS

**COPOLCO:** Consumer Policy Committee (ISO)

**COPANT:** Pan American Standards Commission

**CPE:** Customer Premises Equipment

**DIN:** *Deutsche Institute für Normung,* Germany's national body

**DIS:** Draft International Standard (ISO/IEC)

**DP:** Draft Proposal (same as CD) (ISO/IEC)

**DS:** *Dansk Standardisenngsrad,* Denmark's national body

**DSL:** Digital Subscriber Line

**DVD:** Digital Video Disk recording standard format (controlled by DVD Forum)

**E.C.:** European Commission

**E.U.:** European Union

**EASST:** European Association for the Study of Science and Technology

**EC:** European Community

**ECMA:** European Computer Manufacturers Association

**EIA:** Electronic Industries Alliance

**ELOT:** Hellenic Organization for Standardization, Greece's national body

**ESPRITL:** European Strategic Programme for Research and Development in IT

**ETSI:** European Telecommunications Standards Institute

**EURAS:** European Academy for Standardisation

**FASB:** Financial Accounting Standards Board

**FCC:** Federal Communications Commission

**GAAP:** Generally Accepted Accounting Practices

**GATT:** General Agreement on Tariffs and Trade

**GPL:** General Public License

**GSM:** Global System Mobile

**IAS:** International Accounting Standards set by the IAS Committee

**IBN:** *Institut Belge de Normalisation*, Belgium's national body

**ICSR:** International Center for Standards Research

**ICT:** Information and Communication Technology

**IEC:** International Electrotechnical Commission

**IEEE 802.3:** Ethernet Local Area Network standards committee

**IEEE:** Institute of Electrical and Electronics Engineers

**IETF:** Internet Engineering Task Force

**IMF:** International Monetary Fund

**IPR:** Intellectual Property Right

**IS:** International Standard (ISO/IEC)

**ISA:** International Federation of National Standardization Associations (ISO)

**ISOC:** Internet Society

**ISO 9000:** Management system standard for quality

**ISO 14000:** Management system standard for environment

**ISO:** International Organization for Standardization

**ISO/IEC JTC1 SC 25 WG1:** Home Electronic System working group

**IT:** Information Technology

**ITA:** Industry Technical Agreement (IEC)

**ITI Council:**Information Technology Industries Council

**ITU:** International Telecommunication Union

**ITU-T:** ITU Telecommunication Sector, formerly called the CCITT

**ITU-R:** ITU Radiocommunication Sector, formerly called the CCIR

**Java™:** Computer language developed by Sun Microsystems

**JPEG:** Joint Picture Experts Group, standard for still picture compression

**JTC1:** Joint Technical Committee 1 of ISO and IEC on Information Technology

**JVC VHS™:** Japan Victor Corporation's Video Home System recording format

**MDC:** Market-Driven Consortia

**MIT:** Massachusetts Institute of Technology

**MPEG:** Motion Picture Experts Group, standard for video compression

**MoU:** Memorandum of Understanding

**NEK:** *Norsk Elektroteknisk Komite*, Norway's national body

**NFPA:** National Fire Protection Association

**NGO:** Non Governmental Organization

**NIST:** National Institute for Standards and Technology

**NNI:** *Nederlands Normalisatie Instituut*, The Netherlands' national body

**NSPAC:** National Standards Policy Advisory Committee

**O-Member:** Observing member

**OMB:** Office of Management and Budget, U.S. Congress

**OSI:** Open Systems Interconnection

**OTA:** Office of Technology Assessment, U.S. Congress

**P-Member:** Participating member

**PACT:** President's Advisory Committee on Future Technologies (IEC)

**PAS:** Publicly Available Specification

**PASC:** Pacific Area Standards Congress

**PSTN:** Public Switched Telephone Network

**PTT:** Post, Telephone, and Telegraph

**Q:** Question (ITU)

**RAND:** Reasonable and Non Discriminatory License

**RBOC:** Regional Bell Operating Company

**RFC:** Request for Comment (IETF)

**ROA:** Recognized Operating Agency (ITU)

**SC:** Sub-Committee (ISO/IEC)

**SC25:** Subcommittee 25: Interconnection of IT Equipment

**SCTE:** Society of Cable Telecommunications Engineers

**SDO:** Standards Developing Organization

**SG:** Study Group (ITU)

**SIIT2001:** 2nd IEEE Conference on Standardization and Innovation in Information Technology

**SIO:** Scientific and Industrial Organization (ITU)

**SRO:** Standards Related Organization

**STS:** Variously as Studies of Technology and Society; or earlier, Science, Technology and Society; or most currently as Socio-historical Technology Studies

**TAG:** Technical Advisory Group

**TBT:** Technical Barriers to Trade (WTO)

**TC:** Technical Committee (ISO/IEC)

**TIA:** Telecommunications Industry Association

**TPRC:** Telecommunication Policy Research Conference

**TR:** Technical Report

**TS:** Technical Specification

**TSB:** Telecommunications Standardization Bureau (ITU)

**TTA:** Technology Trend Assessment (IEC)

**UK:** United Kingom

**UNI:** *Ente Nazionale Italiano di unificazione,* Italy's national body

**UPnP Forum:** Universal Plug 'n Play Forum

**VoIP:** Voice over Internet Protocol

**W3C:** World Wide Web Consortium

**WD:** Working Draft

**WG:** Working Group

**Wi-Fi:** Wireless local area network (standardized by IEEE 802.11 committee)

**WTO:** World Trade Organization

**WTSC:** World Telecommunications Standardization Conference (ITU)

**WP:** Working Party

**WRC:** World Radio Conference (ITU)

**XML:** Extended Markup Language

# About the Author

**Timothy Schoechle,** PhD has been in the communications and computer engineering fields since the mid-1970s and engaged in standardization since the early 1980s in a variety of both expert and leadership roles. He is active in various international standards educational and research initiatives including the biennial SIIT conference (Standardization and Innovation in IT) and an international journal on standards research. He presently serves as Secretariat of ISO/IEC JTC1 SC32 (Data Management and Interchange).

Dr. Schoechle played pioneering roles in such technologies as microprocessors, home networks, barcodes, RFID, VoIP, and energy systems. He served on the University of Colorado, Boulder, College of Engineering faculty, teaching one of the few standards courses in the world. He served a key role in founding its standards research center.

Dr. Schoechle holds a BS from Pepperdine University, and an MS and PhD from the University of Colorado. When not traveling for work, he lives in Boulder and is a consultant, advising governments, organizations, and law firms on standards policy and patent issues.

# Appendix

## APPENDIX 1: IEC TECHNICAL COMMITTEES-2008

| Committee | Title |
| --- | --- |
| **JTC 1** | Joint Technical Committee for Information Technology |
| **TC 1** | Terminology |
| **TC 2** | Rotating Machinery |
| **TC 3** | Information Structures, Documentation and Graphical Symbols |
| **TC 4** | Hydraulic Turbines |
| **TC 5** | Steam Turbines (in STAND BY) |
| **TC 7** | Overhead Electrical Conductors |
| **TC 8** | Systems Aspects For Electrical Energy Supply |
| **TC 9** | Electrical Equipment and Systems For Railways |
| **TC 10** | Fluids for Electrotechnical Applications |
| **TC 11** | Overhead Lines |
| **TC 13** | Equipment for Electrical Energy Measurement and Load Control |
| **TC 14** | Power Transformers |
| **TC 15** | Insulating Materials |
| **TC 16** | Basic and Safety Principles for Man-Machine Interface, Marking and Identification |
| **TC 17** | Switchgear and Controlgear |
| **TC 18** | Electrical Installations of Ships and of Mobile and Fixed Offshore Units |
| **TC 20** | Electric Cables |
| **TC 21** | Secondary Cells and Batteries |
| **TC 22** | Power Electronic Systems and Equipment |
| **TC 23** | Electrical Accessories |
| **TC 25** | Quantities and Units, and Their Letter Symbols |

| | |
|---|---|
| **TC 26** | Electric Welding |
| **TC 27** | Industrial Electroheating Equipment |
| **TC 28** | Insulation Co-Ordination |
| **TC 29** | Electroacoustics |
| **TC 31** | Electrical Apparatus for Explosive Atmospheres |
| **TC 32** | Fuses |
| **TC 33** | Power Capacitors |
| **TC 34** | Lamps and Related Equipment |
| **TC 35** | Primary Cells and Batteries |
| **TC 36** | Insulators |
| **TC 37** | Surge Arresters |
| **TC 38** | Instrument Transformers |
| **TC 39** | Electronic Tubes |
| **TC 40** | Capacitors and Resistors for Electronic Equipment |
| **TC 42** | High-Voltage Testing Techniques |
| **TC 44** | Safety of Machinery - Electrotechnical Aspects |
| **TC 45** | Nuclear Instrumentation |

| **Committee** | **Title** |
|---|---|
| **TC 46** | Cables, Wires, Waveguides, R.F. Connectors, R.F. and Microwave Passive Components and Accessories |
| **TC 47** | Semiconductor Devices |
| **TC 48** | Electromechanical Components and Mechanical Structures for Electronic Equipment |
| **TC 49** | Piezoelectric and Dielectric Devices for Frequency Control and Selection |
| **TC 51** | Magnetic Components and Ferrite Materials |
| **TC 55** | Winding Wires |
| **TC 56** | Dependability |
| **TC 57** | Power System Control and Associated Communications |
| **TC 59** | Performance of Household Electrical Appliances |
| **TC 61** | Safety of Household and Similar Electrical Appliances |
| **TC 62** | Electrical Equipment in Medical Practice |
| **TC 64** | Electrical Installations and Protection Against Electric Shock |
| **TC 65** | Industrial-Process Measurement and Control |
| **TC 66** | Safety of Measuring, Control and Laboratory Equipment |
| **TC 68** | Magnetic Alloys and Steels |
| **TC 69** | Electric Road Vehicles and Electric Industrial Trucks |
| **TC 70** | Degrees of Protection Provided By Enclosures |

| | |
|---|---|
| **TC 71** | Electrical Installations for Outdoor Sites Under Heavy Conditions (Including Open-Cast Mines and Quarries) (in STAND BY) |
| **TC 72** | Automatic Controls for Household Use |
| **TC 73** | Short-Circuit Currents |
| **TC 74** | (Transformed Into TC 108) |
| **TC 76** | Optical Radiation Safety and Laser Equipment |
| **TC 77** | Electromagnetic Compatibility |
| **TC 78** | Live Working |
| **TC 79** | Alarm Systems (in STAND BY) |
| **TC 80** | Maritime Navigation and Radiocommunication Equipment and Systems |
| **TC 81** | Lightning Protection |
| **TC 82** | Solar Photovoltaic Energy Systems |
| **TC 85** | Measuring Equipment for Electrical and Electromagnetic Quantities |
| **TC 86** | Fibre Optics |
| **TC 87** | Ultrasonics |
| **TC 88** | Wind Turbines |
| **TC 89** | Fire Hazard Testing |
| **TC 90** | Superconductivity |
| **TC 91** | Electronics Assembly Technology |
| **TC 92** | (Transformed Into TC 108) |
| **TC 93** | Design Automation |
| **TC 94** | All-Or-Nothing Electrical Relays |
| **TC 95** | Measuring Relays and Protection Equipment |
| **TC 96** | Small Power Transformers, Reactors, Power Supply Units and Similar Products |

| **Committee** | **Title** |
|---|---|
| **TC 97** | Electrical Installations for Lighting and Beaconing of Aerodromes |
| **TC 98** | Electrical Insulation Systems (EIS) |
| **TC 99** | System Engineering and Erection of Electrical Power Installations in Systems With Nominal Voltages Above 1kv A.C. and 1.5kv D.C., Particularly Concerning Safety Aspects |
| **TC 100** | Audio, Video and Multimedia Systems and Equipment |
| **TC 101** | Electrostatics |
| **TC 103** | Transmitting Equipment for Radiocommunication |
| **TC 104** | Environmental Conditions, Classification and Methods of Test |
| **TC 105** | Fuel Cell Technologies |
| **TC 106** | Methods for The Assessment of Electric, Magnetic and Electromagnetic Fields Associated With Human Exposure |

| | |
|---|---|
| **TC 107** | Process Management for Avionics |
| **TC 108** | Safety of Electronic Equipment Within The Field of Audio/Video, Information Technology and Communication Technology |
| **TC 109** | Insulation Co-Ordination for Low-Voltage Equipment |
| **TC 110** | Flat Panel Display Devices |
| **TC 111** | Environmental Standardization for Electrical and Electronic Products and Systems |
| **TC 112** | Evaluation and Qualification of Electrical Insulating Materials and systems |
| **TC 113** | Nanotechnology Standardizaton for Electrical and Electronics Products and Systems |
| **CISPR** | International Special Committee On Radio Interference |

## APPENDIX 2: ISO TECHNICAL COMMITTEES-2008

| Committee | Title |
|---|---|
| **JTC 1** | Information technology |
| **TC 1** | Screw threads—STAND BY |
| **TC 2** | Fasteners |
| **TC 4** | Rolling bearings |
| **TC 5** | Ferrous metal pipes and metallic fittings |
| **TC 6** | Paper, board and pulps |
| **TTC** | Ships and marine technology |
| **TC 10** | Technical product documentation |
| **TC 11** | Boilers and pressure vessels |
| **TC 12** | Quantities, units, symbols, conversion factors |
| **TC 14** | Shafts for machinery and accessories |
| **TC 17** | Steel |
| **TC 18** | Zinc and zinc alloys |
| **TC 19** | Preferred numbers—STAND BY |
| **TC 20** | Aircraft and space vehicles |
| **TC 21** | Equipment for fire protection and fire fighting |
| **TC 22** | Road vehicles |
| **TC 23** | Tractors and machinery for agriculture and forestry |
| **TC 24** | Sieves, sieving and other sizing methods |
| **TC 25** | Cast iron and pig iron |
| **TC 26** | Copper and copper alloys |
| **TC 27** | Solid mineral fuels |
| **TC 28** | Petroleum products and lubricants |
| **TC 29** | Small tools |
| **TC 30** | Measurement of fluid flow in closed conduits |
| **TC 31** | Tyres, rims and valves |
| **TC 33** | Refractories |
| **TC 34** | Food products |
| **TC 35** | Paints and varnishes |
| **TC 36** | Cinematography |
| **TC 37** | Terminology and other language resources |
| **TC 38** | Textiles |
| **TC 39** | Machine tools |
| **TC 41** | Pulleys and belts (including veebelts) |
| **TC 42** | Photography |
| **TC 43** | Acoustics |
| **TC 44** | Welding and allied processes |

| | |
|---|---|
| **TC 45** | Rubber and rubber products |
| **TC 46** | Information and documentation |
| **TC 47** | Chemistry |
| **TC 48** | Laboratory glassware and related apparatus |
| **TC 51** | Pallets for unit load method of materials handling |
| **TC 52** | Light gauge metal containers |

| **Committee** | **Title** |
|---|---|
| **TC 54** | Essential oils |
| **TC 58** | Gas cylinders |
| **TC 59** | Building construction |
| **TC 60** | Gears |
| **TC 61** | Plastics |
| **TC 63** | Glass containers |
| **TC 67** | Materials, equipment and offshore structures for petroleum, petrochemical and natural gas industries |
| **TC 68** | Banking, securities and other financial services |
| **TC 69** | Applications of statistical methods |
| **TC 70** | Internal combustion engines |
| **TC 71** | Concrete, reinforced concrete and pre-stressed concrete |
| **TC 72** | Textile machinery and machinery for dry-cleaning and industrial laundering |
| **TC 74** | Cement and lime |
| **TC 76** | Transfusion, infusion and injection equipment for medical and pharmaceutical use |
| **TC 77** | Products in fibre reinforced cement |
| **TC 79** | Light metals and their alloys |
| **TC 81** | Common names for pesticides and other agrochemicals |
| **TC 82** | Mining - STAND BY |
| **TC 83** | Sports and recreational equipment |
| **TC 84** | Devices for administration of medicinal products and intravascular catheters |
| **TC 85** | Nuclear energy |
| **TC 86** | Refrigeration and air-conditioning |
| **TC 87** | Cork |
| **TC 89** | Wood-based panels |
| **TC 91** | Surface active agents |
| **TC 92** | Fire safety |
| **TC 93** | Starch (including derivatives and by-products) |
| **TC 94** | Personal safety -- Protective clothing and equipment |

| | |
|---|---|
| **TC 96** | Cranes |
| **TC 98** | Bases for design of structures |
| **TC 100** | Chains and chain wheels for power transmission and conveyors |
| **TC 101** | Continuous mechanical handling equipment |
| **TC 102** | Iron ore and direct reduced iron |
| **TC 104** | Freight containers |
| **TC 105** | Steel wire ropes |
| **TC 106** | Dentistry |
| **TC 107** | Metallic and other inorganic coatings |
| **TC 108** | Mechanical vibration and shock |
| **TC 109** | Oil and gas burners and associated equipment |
| **TC 110** | Industrial trucks |
| **TC 111** | Round steel link chains, chain slings, components and accessories |

| Committee | Title |
|---|---|
| **TC 112** | Vacuum technology |
| **TC 113** | Hydrometry |
| **TC 114** | Horology |
| **TC 115** | Pumps |
| **TC 116** | Space heating appliances |
| **TC 117** | Industrial fans |
| **TC 118** | Compressors, pneumatic tools and pneumatic machines |
| **TC 119** | Powder metallurgy |
| **TC 120** | Leather |
| **TC 121** | Anaesthetic and respiratory equipment |
| **TC 122** | Packaging |
| **TC 123** | Plain bearings |
| **TC 126** | Tobacco and tobacco products |
| **TC 127** | Earth-moving machinery |
| **TC 128** | Glass plant, pipeline and fittings—STAND BY |
| **TC 129** | Aluminium ores—STAND BY |
| **TC 130** | Graphic technology |
| **TC 131** | Fluid power systems |
| **TC 132** | Ferroalloys |
| **TC 133** | Sizing systems and designations for clothes—STAND BY |
| **TC 134** | Fertilizers and soil conditioners—STAND BY |
| **TC 135** | Non-destructive testing |
| **TC 136** | Furniture |
| **TC 137** | Sizing system, designations and marking for boots and shoes—STAND BY |

| | |
|---|---|
| **TC 138** | Plastics pipes, fittings and valves for the transport of fluids |
| **TC 142** | Cleaning equipment for air and other gases—STAND BY |
| **TC 144** | Air distribution and air diffusion—STAND BY |
| **TC 145** | Graphical symbols |
| **TC 146** | Air quality |
| **TC 147** | Water quality |
| **TC 148** | Sewing machines |
| **TC 149** | Cycles |
| **TC 150** | Implants for surgery |
| **TC 152** | Gypsum, gypsum plasters and gypsum products—STAND BY |
| **TC 153** | Valves |
| **TC 154** | Processes, data elements and documents in commerce, industry and administration |
| **TC 155** | Nickel and nickel alloys |
| **TC 156** | Corrosion of metals and alloys |
| **TC 157** | Mechanical contraceptives |
| **TC 158** | Analysis of gases |
| **TC 159** | Ergonomics |
| **TC 160** | Glass in building |

| **Committee** | **Title** |
|---|---|
| **TC 161** | Control and protective devices for gas and oil burners and gas and oil burning appliances |
| **TC 162** | Doors and windows |
| **TC 163** | Thermal performance and energy use in the built environment |
| **TC 164** | Mechanical testing of metals |
| **TC 165** | Timber structures |
| **TC 166** | Ceramic ware, glassware and glass ceramic ware in contact with food |
| **TC 167** | Steel and aluminium structures |
| **TC 168** | Prosthetics and orthotics |
| **TC 170** | Surgical instruments |
| **TC 171** | Document management applications |
| **TC 172** | Optics and photonics |
| **TC 173** | Technical systems and aids for disabled or handicapped persons |
| **TC 174** | Jewellery |
| **TC 175** | Fluorspar |
| **TC 176** | Quality management and quality assurance |
| **TC 177** | Caravans |
| **TC 178** | Lifts, escalators, passenger conveyors |

| | |
|---|---|
| **TC 179** | Masonry - STAND BY |
| **TC 180** | Solar energy |
| **TC 181** | Safety of toys |
| **TC 182** | Geotechnics |
| **TC 183** | Copper, lead, zinc and nickel ores and concentrates |
| **TC 184** | Industrial automation systems and integration |
| **TC 185** | Safety devices for protection against excessive pressure |
| **TC 186** | Cutlery and table and decorative metal hollow-ware |
| **TC 188** | Small craft |
| **TC 189** | Ceramic tile |
| **TC 190** | Soil quality |
| **TC 191** | Animal (mammal) traps - STAND BY |
| **TC 192** | Gas turbines |
| **TC 193** | Natural gas |
| **TC 194** | Biological evaluation of medical devices |
| **TC 195** | Building construction machinery and equipment |
| **TC 196** | Natural stone - STAND BY |
| **TC 197** | Hydrogen technologies |
| **TC 198** | Sterilization of health care products |
| **TC 199** | Safety of machinery |
| **TC 201** | Surface chemical analysis |
| **TC 202** | Microbeam analysis |
| **TC 203** | Technical energy systems |
| **TC 204** | Intelligent transport systems |
| **TC 205** | Building environment design |
| **TC 206** | Fine ceramics |
| **TC 207** | Environmental management |

| **Committee** | **Title** |
|---|---|
| **TC 208** | Thermal turbines for industrial application (steam turbines, gas Expansion turbines) - STAND BY |
| **TC 209** | Cleanrooms and associated controlled environments |
| **TC 210** | Quality management and corresponding general aspects for medical devices |
| **TC 211** | Geographic information/Geomatics |
| **TC 212** | Clinical laboratory testing and in vitro diagnostic test systems |
| **TC 213** | Dimensional and geometrical product specifications and verification |
| **TC 214** | Elevating work platforms |
| **TC 215** | Health informatics |
| **TC 216** | Footwear |

| | |
|---|---|
| **TC 217** | Cosmetics |
| **TC 218** | Timber |
| **TC 219** | Floor coverings |
| **TC 220** | Cryogenic vessels |
| **TC 221** | Geosynthetics |
| **TC 222** | Personal financial planning |
| **TC 223** | Civil defence |
| **TC 224** | Service activities relating to drinking water supply systems and wastewater systems - Quality criteria of the service and performance indicators |
| **TC 225** | Market, opinion and social research – PROVISIONAL |
| **TC 226** | Materials for the Production of Primary Aluminum |
| **TC 227** | Springa |
| **TC 228** | Tourism and Related Services |
| **TC 229** | Nanotechnologies |

## APPENDIX 3: RESEARCH GATHERINGS

*SIIT2003: 3rd IEEE Conference on Standardization and Innovation in Information Technology.* (served as organizer and presented paper) Delft University of Technology, Delft, The Netherlands, October 22-24, 2003.

ISO/IEC JTC1 SC 25 WG1 Standards Meeting (Home Electronic System) (served as secretary), IBM T.J. Watson Research Center, Yorktown, N.Y., September 19-23, 2003.

INCITS (InterNational Committee for Information Technology Standards) Executive Board Meeting (observer), NIST, Boulder, Colorado, February 10-14, 2003.

*Standards Activity Promotion Workshop: How to Promote Voluntary Consensus Standards Activities in Korea.* (served as invited panel expert and invited speaker on "Theoretical Approach to Standards: Academic and Research Activities") Korean Standards Association, Seoul, Korea, December 4, 2002.

*International Workshop on Smart Home.* (served as invited speaker on "Patents and Standards: Do They Fit?") 1394 Forum and Circuit and System Study Group of IEEK, Seoul, Korea, December 5, 2002.

World Standards Day Exhibition and Banquet. U.S. Chamber of Commerce, Washington DC, October 16, 2002.

ANSI Annual meeting, (theme: "Global Competitiveness") (served as panel participant on "Academic Outreach,") and annual awards banquet. Washington DC, October 15-16, 2002.

ISO/IEC JTC1 SC 25 WG1 (Home Electronic System) (served as secretary) and SC25 (Interconnection of IT Equipment) Plenary Meeting. McLean, Virginia, September 23-27, 2002.

Special Panel: "University Education and Research on Technical Standards," organized by the International Center for Standards Research (ICSR), (served as organizer and presenter) supported by ANSI and NIST and hosted by the Columbia Institute for Tele-Information (CITI), at Columbia University, New York City, September 9, 2002.

EASST 2002 (European Association for the Study of Science and Technology), presented paper , workshops, etc., York, United Kingdom, July 30-Aug 2, 2002.

*2002 International Symposium on Advanced Radio Technologies.* "Spectrum Management in the New Millenium." National Telecommunications and Information

Administration/Institute for Telecommunications Sciences, Boulder, Colorado, March 4, 2002.

*The Regulation of Information Platforms*, Silicon Flatirons Telecommunication Program, University of Colorado. January 27-28, 2002

ISO/IEC JTC1 SC 25 WG1 Standards Meeting (Home Electronic System) (served as secretary), Massachusetts Institute of Technology, Cambridge, Massachusetts, January 21-25, 2002.

*SIIT2001: 2nd IEEE Conference on Standardization and Innovation in Information Technology*. (served as organizer and general chair) University of Colorado, Boulder, Colorado, October 3-6, 2001.

*6th EURAS Workshop: European Academy for Standardisation*, "Standards, Compatibility and Infrastructure Development," (presented paper ) Delft, The Netherlands, June 28-29, 2001.

ISO/IEC JTC1 SC 25 WG1 Standards Meeting (Home Electronic System) (served as secretary), British Standards Institution, London, UK, June 4-8, 2001.

*SIIT '99: 1st IEEE Conference on Standardization and Innovation in Information Technology*. (presented paper) Technical University of Aachen, Aachen, Germany. September 15-16, 1999.

# APPENDIX 4: THE "PAPER"

# THE ROLE OF CONSORTIA STANDARDS IN FEDERAL GOVERNMENT PROCUREMENTS IN THE INFORMATION TECHNOLOGY SECTOR: TOWARDS A RE-DEFINITION OF A VOLUNTARY CONSENSUS STANDARDS ORGANIZATION

## Submitted to the House of Representatives Sub-Committee On Technology, Environment, and Standards

Version 2.0 - 06/19/01

## EXECUTIVE SUMMARY:

I. Standardization is an essential and growing element in the success of the Information Technology industry. The success of the Internet, the World Wide Web, e-Commerce, and the incipient wireless revolution are all predicated upon successful standardization. A majority of the standards that drive these evolving areas of technology are created in consortia, a form of standardization organization that falls outside the standardization regime prescribed by the American National Standards Institute (ANSI).

II. A definition of both "Information Process" and "consortia" is provided to limit the scope of this change to a precise set of problems.

III. The laws that govern procurement for Federal agencies within the Information Technology sector are written and interpreted in a fashion such that consortia specifications are excluded from consideration unless the procuring agency requests a waiver from the OMB to permit use of a "non-standard" specification.

IV. An amendment to the Section 12(d) of Public Law 104-113, the "National Technology Transfer and Advancement Act of 1995" can be used to redefine a "Voluntary consensus standards bodies" within the IT sector in order to allow agencies to select from a more complete and realistic set of offerings than can be offered under the current law.

## THESIS OF THE PAPER

Standardization is essential to the growth of the IT industry. Within the IT industry, well-developed consensus consortia standards should be placed on an equal footing with standards developed by ANSI accredited organizations. The current Federal procurement practices - as mandated by OMB A-119 discourage the use of consortia specifications. The paper concludes with a proposal for a legislative change to permit and encourage Federal use of consortia-created standards in procurement.

## SECTION I: THE EVOLUTION AND ROLE OF STANDARDIZATION IN THE INFORMATION TECHNOLOGY INDUSTRY

Standardization is an essential element to the growth of the computer industry. Most new Information Technology (IT) industry initiatives center around the concept of interoperability, one of the fundamental goals of IT standardization (and most standardization, for that matter.) There are no more "homogeneous islands of computing" which marked the late 1980s; today's environment is worldwide, fast paced, and completely heterogeneous. The impact of this changing environment on business, society, and culture cannot be overstated. Just as the common gauge for railroads changed the face of the United States in the last half of the 1800's, the creation and growth of the standards-based digital economy will have a profound effect on the nature and future of life in the United States. Nearly a decade ago, *The Economist* published the following in its Survey of Information Technology:

*The noisiest of those competitive battles (between suppliers) will be about standards. The eyes of most sane people tend to glaze over at the very mention of technical standards. But in the computer industry, new standards can be the source of enormous wealth, or the death of corporate empires. With so much at stake, standards arouse violent passions.*[1]

This statement - echoed in one form or another in most literature on the subject of standardization - is even more applicable today in the IT industry. With the advent of the Internet and the World Wide Web (WWW), open standards[2] are becoming more and more a part of the "infratechnologies"[3], a term used by NIST to describe a superset of technologies (the technological infrastructure) which "...provide the technical basis for industry standards"[4]. As Martin Libicki of RAND notes, "(w)ith each passing month, the digital economy grows stronger and more attractive. Much, perhaps, most of this economy rests upon the Internet and its World Wide Web. They, in turn, rest upon information technology standards".[5]

This fundamental change in the focus of information technology (from one of homogeneous computing to one of interoperable information sharing) has had a significant impact on the standardization activities of the IT industry. The initial standardization organizations were those that operated under the rules and organizational constricts of the American National Standards Institute (ANSI), following in the footsteps of all the other industrial standardization activities in the United States. This was during the period that much of the fundamental hardware standardization activities were occurring - from common interconnections for the keyboard and mouse to printers and storage systems. The negotiations that created these standards which were complex and confined to a relative handful of providers - were usually under the aegis of one or two standardization committees in the United States[6]. They usually dealt with things that would stay standardized for a long time. The formal national bodies under the aegis of ANSI in the U.S., and the international bodies under the International Organization for Standardization and the International Electrotechnical Commission (ISO and IEC) were referred to as Standards Developing Organizations (SDOs) and were the source of standardization for the IT industry.

However, in the later 1980s, a different form of standardization activity appeared, beginning with an organization called "X/Open".[7] Providers began to move technology standardization away from the formal ANSI and ISO recognized SDOs to those of consortia, which did not have the intricate processes of the SDOs. The formal processes, which were both time consuming and often Byzantine, were necessary because "[m]ost delegates represent[ed] personal, professional, national, disciplinary, and industry goals..."[8], and managing this vast and sometimes contradictory set of expectations forced these groups to create intricate rules to make sure that all voices were heard. Consortia, on the other hand, because they usually consisted of groups of like minded participants (either for technical or market reasons), did not need to have the lengthy discussions over the mission and intent of the proposed standardization activity - an organization's presence was, in many cases, proof of a general agreement.[9] The archetypal consortium was the Internet Engineering Task Force (IETF), the group that manages the Internet. The success of this group in both keeping the Internet a leading-edge technical architecture leader as well as clear of greed, parochialism, and lethargy is a significant accomplishment.[10]

This shift was amplified by the introduction and ensuing popularity of the World Wide Web in the early 1990s. The establishment of the World Wide Web Consortium (W3C)[11] in October 1994 was a turning point within the IT industry; after this date, consortia were the logical place to develop joint specifications, while before they had been the "alternative place". The generation of IT practitioners who are now leading much IT development, which is largely focused on Internet technologies, do not have an awareness of ANSI and ISO as sources for standards. Their world is largely bounded by consortia such as W3C and the IETF. They see no need for ANSI or ISO standardization - a message that they carry to their companies.[12] With the maturity of the Web, an increasing number of consortia are being created to standardize Web based technology. (Nearly all e-Commerce organizations develop their specifications in arenas that are either consortia or consortia-like.)

The reason for the use of consortia lays not so much in the speed of technical development, but rather in the willingness of the consortia to use expedited processes. The IETF has been using the Internet to communicate among interested parties, post specifications, achieve rough consensus on technical features and functions, and then move forward on standardization. The specifications that the IETF adopts are usually based upon extant practice, with at least two implementations required for specifications on the standards track, and are available for widespread public review and comment. This practice - using its own technology to permit faster standardization of follow-on technology - is another step that sets the IETF apart from its contemporary organizations of the 1980s. The use of its technologies as a basis for its standardization practices ensures workable and implementable specifications, but more importantly allows the IETF to develop into a truly international organization. When the specification is complete, it is posted on the IETF web site with free access for all.

The W3C operates in a similar, though somewhat more formal, manner. W3C is a good model for the operation of many other consortia. These consortia realize the key elements are speed and accessibility accessibility to those who are concerned about their work. As *The Economist* has pointed out, "...the Internet has turned out to be a formidable promoter of open standards that actually work, for two reasons. First, the web is the ideal medium for creating standards; it allows groups to collaborate at almost no cost, and makes the decision-making more transparent. Second, the ubiquitous network ensures that standards spread much faster. Moreover, the Internet has spawned institutions, such as the Internet Engineering Task Force (IETF) and the World Wide Web Consortium (W3C), which have shown that it is possible to develop robust common technical rules."[13] These features have made the IT community turn to consortia and similar structures for their standardization needs, in both hardware and software. The creation of highly open, highly visible specifications - widespread in their adoption and use - is essential to the continuing evolution of the IT sector and IT industry.

Another aspect of consortia that separates them from the traditional SDOs is their dependence upon the market, rather than institutions, for relevance. A consortium succeeds or fails by its ability to attract members to accomplish its technical agenda. It receives little or no funding other than what its membership is willing to pay; money received from the government is rare, and is usually in return for some exact service that the consortium renders to a specific government agency in the role of a contractor.[14] While this dependence upon its members for financing can be seen as a limitation on the consortium's freedom of action, it reflects the state of the market in formal SDOs as well, except that formal SDOs do not shut down if all of the commercially important members (those who would implement the specification) walk away. There is a delicate balance between an independence that leads to an unused standard and a financial dependency that produces a constrained specification.

## SECTION II: DEFINITION OF THE INFORMATION TECHNOLOGY

Within the scope of this paper, the term Information Technology shall be the same as the definition found in "The United States Code, Title 40, Chapter 25--Information Technology Management, Section 1401.Definitions, (3) (A) and (B), to include "...any equipment or interconnected system or subsystem of equipment, that is used in the automatic acquisition, storage, manipulation, management, movement, control, display, switching, interchange, transmission, or reception of data or information by the executive agency. For purposes of the preceding sentence, equipment is used by an executive agency if the equipment is used by the executive agency directly or is used by a contractor under a contract with the executive agency which (i) requires the use of such equipment, or (ii) requires the use, to a significant extent, of such equipment in the performance of a service or the furnishing of a product.

(B) The term ``information technology'' includes computers, ancillary equipment, software, firmware and similar procedures, services (including support services), and related resources."

## SECTION III: DEFINITION OF A CONSORTIUM

The definition of a "consortium" used in this submission derives from several taxonomies developed in the previous decade, all of which were focused on the Information Technology sector. Weiss and Cargill (1992) identified three separate types that focused on implementation, application, and proof-oftechnology;[15] Updegrove (1995) identified research consortia, specification groups, and strategic consortia[16], while Ketchell (2001) identified specification creating consortia and "fora" (consortia whose function was to define user and market requirements for further technical development)[17]. The three taxonomies share enough common definitional concepts to constitute a basis for development of a model for this paper.

Of the varieties of consortia enumerated, only two general types meet the requirements of the proposal to modify the Federal procurement process. Both of these types share a common characteristic - the creation of specifications from which products can be developed and implemented in the larger industry. The first type can be identified as a group that is focused on creating a specification that acts to bridge a gap left by other standards or which fills a small niche market. These groups are "...often formed to develop a standard to fill an important niche-industry technical gap that is not large enough to merit the attention of an industry standard setting body..."[18]. These groups include consortia such as the 10 Gigabit Ethernet Alliance, Frame Relay Forum, the Small Form Factor Committee, and the WEB3D Consortium, all of which are focused on creating specifications that address a niche problem or small portion of a larger problem. These consortia are usually small and very focused in the solutions they provide - typically producing robust and implementable specifications in a short time. The players in these groups are usually organizations, which have an interest (product or service offering) that relies upon completion and wide acceptance of a specification. This type of consortia is especially widespread among providers of hardware interfaces and point software solutions. They are characterized by a relatively restricted field of application, and tend to be short lived. The work that they do is published and implemented in products relatively quickly, where it either will gain adherents and survive or will find no market and disappear.

The other type of consortia, which Updegrove labels "strategic", deal with systems, architectures, or new emerging markets where there is a need for a large number of interrelated and/or continuous specifications. These consortia, typified by W3C, the IETF, and The Object Management Group, are usually larger, concerned with a broad spectrum of specifications, and tend to be more long lived. Many of the consortia in this space are attempting to create, grow, and stabilize a market. They also have a more diverse membership, often making consensus harder to attain. As they succeed in obtaining consensus and in moving forward, however, their results can be impressive and cause a major shift, sometimes revolutionary, in the IT arena.

As noted above, both types of consortia share a common attribute - the creation of specifications from which products or services can be developed and sold. The first and primary requirement of consortia, as they are defined for the purposes of this proposal, is they must create useable specifications. This leads to a description of other attributes that a consortium must have.[19] Appendix A, Section 2, provides an overview of consortia, their rationale, and practice. However, as Updegrove notes "Effective, efficient, and representative evolution of standards by consortia is impossible without an appropriate structure of administration and technical decision making. When the authors law firm first began representing consortia, it performed a wide examination of possible forms under various jurisdictions, and settled eventually on the Delaware not-for-profit, non-stock membership corporation.... This structure has stood up extremely well in practice."[20]

This then, would appear to be a potential second criterion by which a consortium may be judged. In the case of a non-U.S. consortium, however, such a ruling would be inappropriate. What may be sought, however, is a structure that indicates some form of reality in law - something that would indicate that there is a legal basis under which the consortium operates and which subjects it to some form of governmental oversight. The intent is to ensure that the consortium is serious by its commitment to achieve legal standing.

"The heart and soul of any consortium may be found in a humble home: its bylaws and charter. Although a few important rules may come to rest in a membership application, most of the regulations and rights of the organization will be found in these legal documents. Whether or not they are carefully conceived will determine whether or not the organization is easily managed, whether it incurs needless exposure to its members under antitrust laws, whether its members feel themselves fairly represented and therefore renew their membership, and whether or not the organization is sufficiently flexible to evolve and flourish."[21] This is another important criterion - the organization must have a set of governing rules that explain how the consortium works, how its members are treated, and the rights and responsibilities of the members. Definition of how the consortium creates its technical specifications - including the methodologies of the creating committees - should also be present. While it is acceptable to have various levels of membership, the criteria for gaining these levels must be clear and unambiguous. There is also the necessity to ensure that there is no exclusivity on joining the consortium; anyone meeting the requisite entry requirements must be allowed to join and participate under the same terms and conditions as other members.

Examination of the intellectual property (IP) regime of the consortium is also necessary. The consortium must have clear IP Rules (IPR) no less rigorous than those of the ISO - since most consortia operate in the international arena. ISO patent policy[22] mandates, as a minimum, commitment to reasonable and nondiscriminatory (RAND) licensing by participants. How RAND is implemented is a matter left to the organization, as are any other rules governing IPR. However, the rules must be complete, spelling out the requirements of members, the penalties for non-compliance, and remedies available to members for such non-compliance. Basically, there must be clear assurance that the holder of IPR will not attempt to treat other consortia participants and users of the standard unfairly.

With respect to participation, ANSI-accredited SDOs cite "balance of participation" (parity between the various affected parties, usually providers, users, and others) as one of the criteria for judging whether an organization is legitimate. By definition, a consortium tends to be biased towards those who are interested enough to "pay to play", which may be enough to violate the ANSI rule of balance. What mu st be assured is that no party is denied the right to participate based upon the nature of the would-be participant, unless the participant is unwilling or unable to meet the common entrance requirements of the consortium.

The key to judging the "openness of the consortia" is one of the major differentiators between the consortia and the SDO forms of standardization. Openness has traditionally been viewed as the willingness to admit all concerned parties to the table. Consortia typically do not do this. Only consortium members may be allowed at the table to discuss specifications. This is why the members are willing to pay - they are trading money or other resources for the ability to determine the specification. This is not substantially different than the SDOs, where participants trade resources (time and travel budget) for the right to participate. Both groups traditionally charge fees - the difference is the amount of the fee charged. Therefore, it is necessary to create new criteria for "openness" among consortia.

The primary test for openness should be the outcome of the consortia – (1) the specification should provide an open (RAND minimum) reference implementation, (2) two or more competing implementations should exist, and (3) there should be, if appropriate, a testing regime to ensure interoperability among the various implementations.[23] This approach focuses on the rationale for standardization - that is, there should be a mechanism by which the users have a choice of implementations from which to choose, providing guaranteed alternative sources for critical products.

In summary, the criteria for a "good" consortium, for the purposes of this paper, includes:

1. The consortium must develop technical specifications.
2. The consortium must be some type of legal entity.
3. The consortium must have a well-defined, legally acceptable set of procedures and processes.
4. The consortium must have a clear and legitimate IPR policy that requires, at a minimum, RAND licensing of all IPR included in its specifications.
5. The membership of the consortium must not be arbitrarily restricted. The consortium must not restrict participation based on non-economic criteria (e.g. competitors, organizational origin, or purpose for joining).
6. There should be reference implementations, competing implementations, and test methods to validate conformance as appropriate.

## SECTION IV: THE ROLE OF NATIONAL POLICY WITH RESPECT TO THE IT SECTOR

In a major Congressional Office of Technology Assessment (OTA) study completed early in the 1990's, the following comment commands attention:

*Other goods, like education and standards, are impure public goods. These combine aspects of both public and private goods. Although they serve a private function, there are also public benefits associated with them. Impure public goods may be produced and distributed in the market or collectively through government. How they are produced is a societal choice of significant consequence.* [24][Emphasis added]

The major contention of this paper is that current legis lation regarding governmental procurement is weighted in favor of the SDOs and does not encourage consideration of the production of standards and specifications produced by consortia - except in special circumstances.

The basic law covering Federal Procurement with respect to standardization is Public Law 104-113, the "National Technology Transfer and Advancement Act of 1995".[25] The applicable section of PL 104-113 is "Section 12 (d) Utilization of Consensus Technical Standards by Federal Agencies; Reports", passed by the Congress in order to establish the policies of the existing OMB Circular A-119 in law. The first subsection, 12 (d) (1), states:

*In general. — Except as provided in paragraph (3) of this subsection, all Federal agencies and departments shall use technical standards that are* ***developed or adopted by voluntary consensus standards bodies*** *(emphasis added), using such technical standards as a means to carry out policy objectives or activities determined by the agencies and departments.*

This section sets the intent and establishes specific guidance to the National Institute for Standards and Technology (NIST) to ensure that the Federal agencies and departments are not creating their own standards, but are using commercially developed standards to carry out their missions. Sections (2) offers guidance on the participation in or the joining of a standards organization, and section (3) provides an exception clause, through which agencies can explain why they have chosen not to use commercial standards. Section (4) provides a definition of standards as:" the term `technical standards' means performance-based or design-specific technical specifications and related management systems".
The determination of what is a "voluntary consensus standards body" has been left to OMB. In OMB Circular A119, we find the following explication:

4. What are Voluntary, Consensus Standards?
    a. For purposes of this policy, voluntary consensus standards are standards developed or adopted by voluntary consensus standards bodies, both

domestic and international. These standards include provisions requiring that owners of relevant intellectual property have agreed to make that intellectual property available on a non-discriminatory, royalty-free or reasonable royalty basis to all interested parties. For purposes of this Circular, ``technical standards that are developed or adopted by voluntary consensus standard bodies'' is an equivalent term.

1. Voluntary consensus standards bodies are domestic or international organizations which plan, develop, establish, or coordinate voluntary consensus standards using agreed-upon procedures. For purposes of this Circular, "voluntary, private sector, consensus standards bodies,'' as cited in Act, is an equivalent term. The Act and the Circular encourage the participation of federal representatives in these bodies to increase the likelihood that the standards they develop will meet both public and private sector needs. A voluntary consensus standards body is defined by the following attributes:
   i. Openness.
   ii. Balance of interest.
   iii. Due process.
   iv. An appeals process.
   v. Consensus, which is defined as general agreement, but not necessarily unanimity, and includes a process for attempting to resolve objections by interested parties, as long as all comments have been fairly considered, each objector is advised of the disposition of his or her objection(s) and the reasons why, and the consensus body members are given an opportunity to change their votes after reviewing the comments.

b. Other types of standards, which are distinct from voluntary consensus standards, are the following:
   1. "Non-consensus standards, ''Industry standards,'' "Company standards,'' or "de facto standards,'' which are developed in the private sector but not in the full consensus process.
   2. "Government-unique standards,'' which are developed by the government for its own uses.
   3. Standards mandated by law, such as those contained in the United States Pharmacopeia and the National Formulary, as referenced in 21 U.S.C. 351.[26]

This definition - specifically with the requirement for "(ii) Balance of interest"[27] would appear to limit standards to formal (non-consortia) standardization, since, by definition, the participants in a consortium are self-selecting for a particular technology specification. At the same time, consortia standards do not fall under the conditions set forth in Section 4.b.(1), as they are developed in full consensus and then are actually implemented by the industry. Section 4.b.(1) seems to speak to "proprietary standards", which are usually implementation standards - that is, standards based upon a single vendor's implementation, and usually described as "de facto" standards.

- In section 6 g., however, we read: " Does this policy establish a preference between consensus and non-consensus standards that are developed in the private sector? This policy does not establish a preference among standards developed in the private sector. Specifically, agencies that promulgate regulations referencing non-consensus standards developed in the private sector are not required to report on these actions, and agencies that procure products or services based on non-consensus standards are not required to report on such procurements. For example, this policy allows agencies to select a non-consensus standard developed in the private sector as a means of establishing testing methods in a regulation and to choose among commercial-off-the-shelf products, regardless of whether the underlying standards are developed by voluntary consensus standards bodies or not."[28]

This section, by reading in light of the previously examined sections, seems to state that "proprietary standards" or "de facto standards" are permissible, meaning that the use of consortia based standards, which are open, consensus driven, and lack only the "balance" described in 4.a.(1)(ii) are the equivalent of proprietary or *de facto* standards, which they are not. Consortia standards represent standards that have been developed in an atmosphere that is as rigorous - if not more so - than most SDO standards, yet it is deprecated because it does not meet the five voluntary criteria.

The intent of A119 appears to be clear - standards developed in an open process are preferable to those that are not. Yet, because of the definition of a voluntary consensus standard contained in Section 4, the use of consortia developed standards is specifically disallowed, while standards developed in proprietary environments, or standards that are derived from a product (implementation standard), are permitted (Section 6.g.).

In a larger sense, however, for the IT sector the exclusion of consortia developed standards in Section 4.a. is flawed. A majority of standards that are driving the next generation of computing - specifically, those from the IETF (the standards of the Internet), those of the W3C (the standards of the Web and of e-Commerce), the wireless phone standards (those created by the WAP Forum and by ETSI), as well as the standards of the spatial industry (Open GIS Consortium), the Object

Oriented technology movement (Object Management Group), and of Linux - are all excluded.

We do not agree with those who argue that the problem is not significant. Appendix B provides background on one of these issues, while Appendix C argues that the use of proprietary standards in procurements appears to be the result of a policy that recognizes that the formal standards process has broken down and that proprietary offerings are as good as, if not better (in the eyes of the purchaser) than the currently mandated standardization regime.

We disagree with the defense that the current system addresses the problem, and that there is no real issue here. This is a serious and substantial issue to participants in the standardization process. The following quote, from a leading European standardization site, explains the issue succinctly: "To us formal ICT standardizers, sometimes consortia are a pain in the neck. We recognize they are quick, industry solutions to produce necessary specifications, which they call "standards" but we don't. These bodies don't always take full account of the real needs of end users, and it is difficult to find information on them and what exactly they are doing."[29]

While it can be argued that this is not the perception of ANSI, ANSI's strategic plan includes the following: " In successful standards processes:

- Decisions are reached through consensus among those affected.
- Participation is open to all affected interests.
- Balance is maintained among competing interests.

...

- Governments use voluntary consensus standards in regulation and procurement.
- U.S. Government should encourage more use of the principles embodied in accreditation by recognizing the ANSI process as providing sufficient evidence that American National Standards (ANS) meet federal criteria for voluntary consensus standards;
- Non-traditional standards organizations should review their objectives to determine where closer interaction with the formal system will help add value to their efforts;"

All of these assertions, if read from the perspective of a consortium, would seem to indicate that ANSI is focused on maintaining its hegemony and expanding the use of its definition of the "voluntary standards process". It does not indicate that there is an attempt to make all standards equal; rather, the above text would seem to indicate that ANSI is attempting to position its process as superior - something that consortia frequently take strong exception to.

The role of the government - within the IT sector - should be to equalize the activities of all of the standards players, so that all legitimate interests are fairly represented in the IT arena. The next section proposes legislation to achieve this end.

## SECTION V: TOWARDS AN EXPANDED DEFINITION OF A VOLUNTARY CONSENSUS S TANDARDS BODY

To unify U.S. standardization activities in the IT sector, a specific amendment to the Public Law 104-113, the ``National Technology Transfer and Advancement Act of 1995" should be proposed.

1. The proposed legislation would have to contain specific language limiting the intent of this change to only the IT community (as defined in Section II).
2. It would deal only with voluntary, market driven IT standardization, and would not impact regulatory standards (such as health, safety, or the environment).
3. It would have as criteria for a "legitimate consortia" the items listed in Section III as attributes of a "good consortium".
4. It would not exclude anyone or any organization from seeking either the ANSI or the ISO imprimatur.
5. It would make exceptions to the legislation difficult to obtain.
6. It would put in place and enforce a tracking mechanism to monitor the use of non-open standards.
7. It may be appropriate to include a directive to NIST to expand the role of the National Voluntary Laboratory Accreditation Program (NVLAP) in an effort to "train the trainers" if the private sector demands consortia accreditation.

The purpose of the legislation would be to make the formal and structured informal processes equal for the voluntary, market driven IT sector and to reunify the quarreling parts of the standardization discipline to permit the continued growth of the IT sector in the United States.

## APPENDIX A: THE EVOLUTION AND HISTORY OF STANDARDS SETTING ORGANIZATIONS (SSOs)

This section provides background on the differences between the various standardization organizations, why they evolved the way that they have, and reviews the strengths and limitations of each within the context of the Information Technology sector.

There are five basic variants of standards setting organizations within the IT sector[30]. Each variant has a place in the IT sector because there is no single optimal choice for development of standards for the entire industry. This section of the paper looks at these five organizational variants, and provides some history and background on all of them as they relate to the unique aspects of IT standardization.[31] ANSI is examined in particular detail, since it is the primary stakeholder for the U.S. in all formal organizations (national or international), that currently are the primary providers of specifications used in procurement in the United States.
The five types of organizations are:

- (1) Trade associations, (2) formal Standards Developing Organizations (SDOs)
- (3) Consortia, (4) Alliances
- (5) The Open Source software movement

## 1. Trade Associations and Standards Developing Organizations (SDOs)

These two types of organizations are linked because they both belong to the formal school of standards that is, a standards process that is heavily focused on maintaining due process, openness of participation, and a comprehensive appeals process. As will be seen, the process that these organizations have created within the U.S. is a result of legal challenges to their work, and is absolutely necessary for the regulatory or similar arenas, where there is an implied legitimacy ascribed to a specification labeled as an official standard.

The trade association activities in standardization take the place of pride for being the oldest form of standardization activity of those listed here, dating as it does from the late 1800's. Generally, the associations were gatherings of professional men who were experts in a particular field (boilers, fire prevention, mechanical engineering). Their intent in setting up these groups was to create a professional discipline and to preserve this discipline by creating specifications embodying their wisdom for the sake of their colleagues. Hence, societies like the American Society of Mechanical Engineers (ASME), the Institute of Electrical and Electronics Engineers (IEEE), and the American Society For Testing and Materials (ASTM) came into being. In most cases, the primary mission of these groups is the education of members in their professional discipline, with standards as a secondary activity to fulfill some of the training requirements[32]. These groups were directly responsible for technical practices that could impact public safety, and needed to ensure that their specifications were correct. Peer review was not only desirable, it was necessary and expected.

In many cases, the specifications developed by the trade organizations have become the basis for codes and statutes, and have acquired a regulatory patina that permits them to be used as defense in liability cases. By definition, if you follow the specifications published by the National Fire Protection Code, you are using techniques and practices that have been tested, tried and proven to be safe. This makes trade associations excellent for codifying successful past practices - things that are stable, structured, and time insensitive. Within the IT industry, in areas that do not touch upon, for example, safety issues, looking to past practices for future guidance is usually a prescription for failure.

It is necessary to note that the regulatory use of standardization has another and darker side. In two Supreme Court cases, American Society of Mechanical Engineers vs. Hydrolevel [33](1982) and in Allied Tube and Conduit vs. Indian Head[34] (1988), the standards bodies were found to have abused their ability to impact the market. While the cases varied with respect to details, the economic power of the organization was cited as a major point of contention. In both cases, there were process violations on the part of the organization. It is the necessity to have a process - and the need to adhere to that process - which makes the association a subset of the formal process, since the formal process for developing standards, in the U.S. is created, maintained, and administered by American National Standards Institute (ANSI). The U.S. government has not created a national standards body. Instead, ANSI is the "first among equals", the rule setter, the interface to ISO and the IEC, and currently the only organization that can give the *imprimatur* of an American National Standard (ANS) to the specifications produced by most U.S. standards organizations. It does not, however, create standards. It has no expertise in the subject matter of standards; it has expertise only in the maintenance of its process.

A brief examination of the history of standardization within the U.S. is necessary to put an organization like ANSI into its proper perspective. Following the First World War, there was a national standardization initiative sponsored by Herbert Hoover to make sense of the chaotic state of standards in the U.S. Voluntary cooperation between the organizations was a goal; it was initiated in the Twenties and then stopped as the Depression began. However, following the Second World War, the initiative took off again and eventually the organization that was to become the American National Standards Institute (ANSI) came into prominence.[35] While not a governmental entity, ANSI was meant to regularize standardization in the U.S. Several serendipitous legal incidents happened to strengthen ANSI's hand (an antitrust case, a Congressional investigation), and eventually ANSI came out as the first among equals in U.S. formal standardization. It alone (of the myriad of standards organizations in the United States) has the right to publish standards which bear the appellation "American National Standard"; because ANSI does itself not create standards, it acts as a publishing arm for the more than 170 organizations which have sought ANSI accreditation.[36] At the same time, other nations (especially Germany, France, the U.K., and Japan) began to strengthen their nationally chartered bodies to pursue standards as a part of their national industrial policies.

A European-style national standards body makes sense in the context of the post-World War II industrial environment. Nations were trying to strengthen their individual industrial capacity; many were rebuilding after a devastating war. The creation of "standards" allowed an industrial policy that could be controlled (to varying degrees) by the nation. The U.S. chose instead to lead by encouraging the private sector to enter into standards partnerships. This allowed the trade associations to continue to act as "standards organizations", while encouraging the formation of new organizations devoted only to standardization. Examples of this last include the Accredited Standards Committees (ASC) X3 (IT), X9 (Banking), X12 (EDI) and so on.

As national and regional economies became more interdependent, however, it was necessary to establish an international standardization authority. Following WWII, and with the growth of the internationalism, ISO was established and the IEC and ITU had more credence given them, so that there could be truly international standards. However, there is a cultural sensitivity that was overlooked at times - the concept of "international" did not necessarily mean "good" to a country, unless it was that country's specification being carried forward. And since the basis of the international formal activity was the national body, the biases of the various national bodies were brought forward. Within the IT industry, the balance of power turned to the U.S., since U.S. based IT companies were more successful than their counterparts worldwide, due in some part to the larger size and homogeneity of the U.S. market, which made economies of scale possible for U.S. firms. With the economies of scale came the ability to innovate more quickly, which in turn fed the need and use requirements of users, leading to more innovation, an increased market, and increased sales. By 1985, the U.S. dominance in IT - in market share, in intellectual property, in research and development, and in deployed base - was firmly established. Because of this market dominance, the dominance of the U.S. in formal standards was also es tablished; a majority of IT standards were those proposed or initiated by U.S. companies, either through the U.S. standardization bodies (e.g. ASC X3 or the IEEE Computer Society) or through U.S. company representatives acting in foreign standards bodies (such as the Deutsches Institute for Normung [DIN], the German national body where U.S. subsidiaries exercised heavy influence).

In the early 1990s, the European Community began to coalesce. One of the favored methods of creating a "single European market" was to require the various nations to abandon "unique" national standards in favor of "Pan-European" (or regional) standards. By eliminating a multitude of competing and conflicting standards, a British manufacturer, for example, would not have to make multiple separate products or go through national conformance test regimes. By adhering to a single "pan-European" standardization regime, it was felt that European providers could begin to realize economies of scale, similar to those of the

U.S. manufacturers. To further this purpose, the European Union recognized (or created) three Regional standards organizations - the European Committee for Standardization (CEN), the European Committee for Electrotechnical Standardization (CENELEC), and the European Telecommunications Standards Institute (ETSI).[37] The mission for all of these groups was to "... promote voluntary technical harmonization in Europe in conjunction with worldwide bodies and its partners in Europe." [38] The key to understanding the activities of the EU is to remember that European National Body standardization activities were often a barrier to the unification of European economic activity. By requiring the unification of standards (and a common acceptance of a single standard), the EU was seeking to unify its markets and to provide for economic growth as a unified Europe.

This was not, however, the way that the activity was seen in the United States. The unfortunate appearance of ISO 9000 Quality Management series of standards in 1989 gave the impression that the Europeans were creating a "Fortress Europe" by using standards and certification schemes as non-tariff trade barriers[39]. The debate was exacerbated by the use of common standards phrases with substantially different meanings, depending upon which side of the Atlantic Ocean you lived. The involvement of ANSI - at the behest of some of its members - began a long, torturous, and losing battle to stop the pan-European standardization activity. The requirement that the European national standardization bodies must accept a CEN standard, and that CEN has a "special" relationship with ISO[40] gave rise to U.S. concerns that the vote in ISO could be rigged in favor of the Europeans, since the Europeans might vote in concert with one another.

However, the accusations by ANSI that the Europeans were block voting became (and remains) shrill[41]. While this may be necessary for national positioning, it is not helpful to the IT industry, which has a substantial international market for its products. The appearance of a " National Standards Strategy for the United States" has placed IT companies with a significant presence in European standardization bodies in an awkward position - they must either accept the concept of an overriding U.S. national position or they must be willing to dismiss the statements of an organization in which many of them are members.

At the same time, the lack of clarity within the U.S. standardization regime has made many of its counterparts in ISO uneasy with ANSI[42]. Because ANSI is only the "first among equals" in the U.S., it has no absolute mandate as the sole international representative of the U.S. at ISO. ANSI sits at ISO and the IEC because it is the single "most representative" body on all standardization, and because it has the singular right to grant the title of an American National Standards (ANS) to a specification. This right is enforced by ensuring that those who wish to publish an ANS follow the ANSI procedures for creating standards. As noted above, ANSI has as its only contribution to standardization the process and coordination between groups. ANSI's mission statement reads "ANSI does not itself develop American National Standards (ANSs); rather it facilitates development by establishing consensus among qualified groups. The Institute ensures that its guiding principles -- consensus, due process and openness -- are followed by the more than 175 distinct entities currently accredited under one of the Federation's three methods of accreditation (organization, committee or canvass)".[43] The way that a group becomes "qualified" is to embrace ANSI's development rules - which are the "formal process rules".[44]

It is this "formal process" which is the value of the "formal organization", whether a trade association doing standards, ANSI, any of the ANSI accredited Committees, or the international organizations of ISO. The process is specified; variations are not allowed. The mantra of ANSI is:

- Decisions are reached through consensus among those affected.
- Participation is open to all affected interests.
- Balance is maintained among competing interests.
- The process is transparent — information on the process and progress is directly available.
- Due process assures that all views will be considered and that appeals are possible.

Absent any of these conditions, an organization cannot become accredited. And because their fundamental rationale for existence may not meet the ANSI conditions, consortia have always been outside of the pale of formally accepted standards.

## 2. Consortia and Alliances

Within the IT standardization context, consortia and alliances are collections of like-minded organizations and/or individuals who come together to act as advocates for a particular change. The desired change may be a new specification, a new way of approaching a problem, or a new research and development activity. The legal basis of the organizational style known as "consortia" or "alliance" is found in the National Cooperative Research and Production Act of 1993 (15 U.S.C. §§4301, et seq.), which has as its purpose "...to promote innovation, facilitate trade, and strengthen the competitiveness of the United States in world markets by clarifying the applicability of the rule of reason standard and establishing a procedure under which businesses may notify the Department of Justice and Federal Trade Commission of their cooperative ventures and thereby qualify for a single-damages limitation on civil antitrust liability.'"[45] The Act lists a lengthy series of activities which are prohibited if an organization wishes to take advantage of the Act; in many cases, the charter of an organization specifically writes these prohibitions into their charter to make sure that participants understand the purpose of the organization is to encourage innovation and commercialization of technology (two purposes of the act.)[46]

Consortia initially were created to deal with the "clarity and time to market" problem that was seen as a major obstacle in the formal arena. Much of the problem in the formal arena lay with its arcane rules for openness and review; several of the formal review process steps required six months and could expand to even more time. The consortia, responding to the pressure of "time is money, especially when the product life cycle was shrinking", wanted a faster system. The proponents and opponents of consortia have focused on this "speed issue", not realizing that increased speed was achieved in a consortium by changing the process. The argument has never been about speed; it has been about the process needed to achieve the speed necessary to satisfy the market needs of the members of the organization.

In most of the cases, the consortia modified the traditional standardization process by formally imposing some limitation on participation. The limitation usually took the form of dues - that is, there is a requirement to "pay to play.'"[47] The payment could be modest or significant (from approximately $3,000 per year to the $50,000 that large corporations are often taxed.) The consortia also announced their intentions - when you have like minded companies, you can announce and drive to a solution with a greater degree of freedom than can a formal SDO, which usually has no way of controlling where its efforts will lead. Finally, a consortium does not have to be broad spectrum - that is, it can focus on and solve only those problems that it wishes to solve. There is no requirement for it to create committees to solve all problems; rather it should (by definition) be working on problems that its me mbers need to have solved in order to produce products.

Finally, and perhaps most damaging to the formal standardization process, consortia specifications are usually turned into product offerings immediately by the participating companies. The rationale for playing (and paying) within a consortium is to create and then market a technology. To participate in a consortium (paying both dues and committing scarce human resources) and then to not implement the specification when it appears is definitely foolish and possibly irresponsible, and is the exception more than the rule. Additionally (depending upon the cohesiveness of the consortia), the specification usually has one or more implementations that validate the specification.

There are two schools of thought on when and what to standardize. One school believes that standardizing current practice - that is, abstracting an interface specification from existing products - is the preferred method, while another school of thought revolves around standardizing future technology in its predeployment phase. The "current practice school" rewards the innovator by allowing a time to market and market share advantage, while embracing stability in the market and rapid deployment of technology. The other (future technology) permits a group design, combining the best of breed (at times), but is usually slower and can produce a specification that is filled with compromise. Both have been used successfully within consortia, but the standardization of current practice, in which the innovator opens a proprietary specification in return for a possibly transient market advantage, is usually the most preferred.[48] The classic case used to argue for "current practice standardization" is the failure of OSI (Open Systems Interconnect), which involved of standardizing technology that was not deployed and which was being created in committee. On the other hand, there is a reluctance to take a widely deployed but non-standard technology to the formal organizations, since there have been instances where formal organizations have attempted to change the technology once it arrived in their committees. When this (the changing of a deployed technology) happens, the worst of all worlds results - a standard that does not reflect installed base usage of the specification, so that one or the other is declared invalid. With either outcome, both sides lose.

Consortia are also slightly more informal in the coordination of their efforts. Unlike the formal world, where all of the players are known to one another and tracked, the consortia/alliance arena has no central clearing house or authority to coordinate activities. There are efforts made to track consortia, but new consortia appear in the IT arena at the rate of about one every other week.[49] There is nothing to prevent multiple organizations from tackling the same general topic (i.e. wireless internet communications). This is encouraged by the organizations that fund the consortia and alliances, since having multiple solutions sometimes mitigates the impact of catastrophic technical change. What the industry does not like is two Standards Setting Organizations (SSOs) solving the same problem using the same specifications (dueling specifications) or a specification being bifurcated and modified. This is where much of the concern about standardization comes in – and the old tired rubric of "The nice thing about standards is that there are so many of them" is brought up[50]. It is duplicative standards – not duplicative standardization effo rts – that are the bane of the industry.

The consortia processes are rigorous, since they must comply with the provisions contained in the National Cooperative Research and Production Act of 1993, under which many of them are chartered. There is an area of expertise on the legal implications of the creation of consortia, and nearly every consortium that is created requires the services of at least one lawyer.[51] Consortia operate as strictly under their rules as formal SDOs operate under theirs. If they fail to keep their processes legitimate, they risk all of their members and their own existence. The emphasis that consortia place upon following their rules is illustrated by the fact that, as of this writing, there has never been a successful suit brought against a consortium for anti-trust activities.[52]

Consortia and alliances (their more short lived brethren) serve a need of the IT industry as a way to stabilize the market in a time of shortened product life cycles and rapid market change. By providing processes that are open, and by providing the market with multiple implementations of the consortia specification, they have increased competition and ensured that the standardization of the high technology industry can continue.

## 3. Open Source

Open Source is another form of standardization, and is probably the most expensive type of standardization in which an organization can engage, since participation and use of open source code may require that an organization change its fundamental licensing principles with respect to its intellectual property (IP).[53] In all of the other organizational types, the contributing organization can choose the terms and conditions of its giving, as long as the terms are reasonable and non-discriminatory. The difference is that with open source, the terms and conditions of the grant are mandated in the particular licensing agreement chosen by the group.

The reason for the allure of Open Source is contained in writings by the philosopher and activist of the Open Source movement - Eric Raymond, in the *Cathedral and the Bazaar*[54], and Jamie Zawinski (formerly of Netscape who convinced Netscape's management to make the source for Netscape's browser into open source and call it Mozilla). Linus Torvalds led the creation of the popular Operating System named Linux in the same philosophical frame - which is open for all to use without exception or restriction, other than the requirement to act as part of the community. The movement has caught mindshare and market share, and many large corporations are embracing the Linux phenomena, hoping later that they can find the method to profit.

The key to understanding the open source community understands the license. The licensing itself is complex; there are at least five variants:[55]

1. No license at all (i.e., releasing software into the public domain).
2. Licenses like the BSD License that place relatively few constraints on what a developer may do (including creating proprietary versions of open source products).

3. The GNU General Public License (GPL) and variants which attempt to constrain developers from "hoarding" code, i.e., making changes to open-source products and then not contributing those changes back to the developer community, but rather attempting to keep them proprietary for commercial purposes or other reasons.
4. The Artistic License, which modifies several of the more controversial aspects of the GPL.
5. The Mozilla Public License (MozPL) and variants (including the Netscape Public License or NPL) which go further than the BSD and similar licenses in discouraging "software hoarding" but which still allow developers to create proprietary add-ons if they wish.

The intent of these various forms of licenses is to ensure that the code remains open for all to use, validate, modify, and improve. These license forms, more than anything else, are the core of the Open Source standards movement. They encourage the community to act together, and act as a re-enforcing mechanism for "open source behavior" (which is a larger good to which all standards organizations must subscribe). By tying their unique behavior to licensing activities, they are then freed to espouse rules that re-enforce the benefits of open source licensing – including rules on how to write code, how to publish code, how to correct code, and so on.

The good aspect of open source is that there are multiple implementations of the code - anyone who wishes may take the source code and write an implementation. The difficult aspect of Open Source is that there is never a stabilized standard set of source code to specify, since by its very nature, Open Source is a constant and incremental improvement in a code base. However, the creators and purveyors of Linux are working on this, and are attempting to create a Linux standard that will solve this problem. If this problem is solved (basically, a version control problem), then the Open Source organization will also be a viable candidate for procurement.

## 4. Conclusion

All of the various forms of standardization can and do serve a purpose in the IT sector. There is the need for stability (provided by the formal arena), a need for defined and structured faster change (provided by consortia and alliances) and the need for complete community involvement (provided by open source.) The groups within each arena have not learned to work together for the good of "open systems". Rather than considering proprietary and closed systems to be the force to be changed, they have dissipated their energies arguing about which form of standardization is best, forgetting that the answer is that "Standardization is best, and non-standardization is less than optimal." ANSI is a necessary, but not sufficient, standardization component for the needs of the IT sector. Consortia are central to IT standardization success - but need the stability that the formal process can offer. And for long-term change (to both the technical and legal fabric of IT sector standardization), open source is an interesting direction and may lead to an entirely different standardization environment in the future.

Standardization is a complex discipline that is constantly changing as the industry underneath it evolves. The last decade in the IT industry have seen massive change as the very nature of information use and sharing by customers has changed. The state and changes in the IT industry in the United States reflects the state and changes of its consumers - U.S. society, both commercial and private. The IT sector has been credited with making the U.S. economy much more productive, and this has aroused admiration throughout the world.[56] Uniting the various forms of standardization by allowing equivalency - in legal as well as in economic settings - would only enhance the industry. It is one of those rare situations that has no negative consequences to the industry or society.

## APPENDIX B: AIR FORCE COMPUTER ACQUISITION CENTER RFP 251

In the mid-80's the Air Force was preparing a very large procurement for computing equipment in which it wanted to replace/upgrade its aging systems (Air Force Standard Multi-user Small Computer Requirements Contract). Specifically, it needed to get UNIX environments, but (1) there was no formal standard, and (2) there was no publicly available test suite to test that the systems procured under this contract would meet the functional requirements laid out in the RFP.

This was the time frame in which there were a multitude of UNIX variants that could not necessarily interoperate. Most were based on either BSD, developed at Cal Berkeley or Unix System III developed by Bell Labs Unix Development Laboratory. It was crucial that this procurement not result in yet more non-interoperable systems. At the time, AT&T Bell Labs had heavily invested in Unix as the steward of what was in essence a precursor open source development effort where hundreds of universities and other research facilities had helped to collaboratively evolve the Unix specification.

The Air Force, after close examination of the alternatives, decided to require that systems bid for its procurement (AFCAC 251) must conform to AT&T's SVID (System V Interface Definition), where it cited specific publicly available texts that contained the specification.

At the time, AT&T also provided a Conformance Test Suite to test conformance of an implementation to the SVID. This test suite, SVVS (System V Validation

Suite) was only available from AT&T. AFCAC 251 required passing the SVVS as a condition of the procurement.

When this RFP was released, a formal protest was filed by a number of companies objecting that this procurement was not based on a formal standard, but on a proprietary, copyrighted specification. Further, and more importantly, it was claimed that the SVVS could not be used because it was the proprietary property of a potential bidder. The resulting protest was very high profile, lengthy, and very costly for all parties involved. In addition it resulted in significant delays to a critical federal procurement.

AFCAC 251 was the impetus for the proposal, and adoption, of Federal Information Processing Standard (FIPS) 151. FIPS 151 was based on the then maturing work of the IEEE Computer Society's POSIX standards committee. POSIX was an operating system specification standard based on the Unix specification provided in the SVID. Further, the National Bureau of Standards (NBS, now NIST) prevailed in establishing a test methods working group for POSIX that developed the POSIX Test Methods standard. This standard was used, along with the SVVS donated by AT&T, as the basis for the development of FIPS 151 PCTS (POSIX Conformance Test Suite) by NIST with the assistance of experts from a number of IT companies and organizations under a Cooperative Research And Development Agreement (CRADA). NIST then established an accredited POSIX test laboratory program and required the use of the PCTS in the certification of conformance of an operating system to FIPS 151.

So, how does this support the need for clearer rules for the use of Consortia standards as equals to formal standards?

Today, the leading edge evolution of most critical IT technologies is occurring in consortia, not in formal Standards Developing Organizations (SDOs). The government will be able to obtain the best information technology by requiring conformance to these consortia specifications. In the case of AFCAC 251 this would have resulted in the savings of many millions of dollars that was spent by the government and the protesting companies in defending/pursuing the AFCAC 251 procurement protest. In addition, the systems needed by the Air Force would have been obtained in a much more timely manner.

## APPENDIX C: DISA'S USE OF FIPS CERTIFICATION

One element of DISA's Defense Information Infrastructure Common Operating Environment (DII COE) is the identification of processor and OS platforms and software that will form the foundation of the COE, also called the DII COE Kernel. DISA's customer needs (DISA's customers are the CINCs, services and agencies within DoD) has resulted in DISA maintaining three platforms as "COE compliant" including Solaris, HP-UX and Window NT/2000.

Responding to vendor assertions that they were being denied access to programs that required COE compliance, yet had no way to achieve that compliance (DISA, for cost reasons, would not undertake the effort, and there was no way for the vendor to do the work themselves), DISA established the Kernel Platform Certification (KPC) program.

The program requires four main items for acquiring a DII COE Kernel compliance certificate. These include:

- Providing a FIPS 151-2 certificate of Posix compliance
- Completing several test suites out of the UNIX98 branding suite maintained by the Open Group
- Successful porting of the COE "kernel code" to the candidate platform, and completion of a series of test suites that verify proper operation.
- Passing a security checklist that is roughly equivalent to commercial grade security (e.g. passwords for all accounts, access controls, etc.).

DISA has indicated that they expect the three platforms that they currently maintain in the COE will be kept COE compliant by their respective owners using the KPC program. The issues with the KPC are twofold. The first is that the KPC program addresses only Posix-compliant platforms. Windows NT and 2000 are not subject to its requirements, and DISA maintains that Windows OS "compliance" is essentially satisfied with a pointer to Microsoft documentation.

The second issue, which is more relevant to the standards availability theme, is that the lack of an alternative to the now withdrawn FIPS 151-2 has forced DISA to continue to use that obsoleted standard. DISA had investigated simply pointing to UNIX branding maintained by The Open Group, but this ran into an interesting problem. Other vendors successfully objected to this approach because The Open Group method of implementing Unix98 branding requires an ongoing commitment and subscription to the program with TOG. This apparently violates an acquisition law that prohibits the government from requiring vendors enter into long term, third party agreements in order to do business with the government.

The current solution, which DISA negotiated with The Open Group, is to specify a subset of the specific test suites desired by DISA, and which The Open Group will offer up to vendors on a modified basis that eliminates the long-term commitment. The FIPS 151-2 specification is still used, but is expected to be replaced by the Austin group's specification within the next 6-12 months.

## ENDNOTES

1. The Economist Newspaper, 23 February, 1993
2. An open standard is a standard which is not under the control of a single vendor and which is easily available to those who need it to make products or services.
3. While the NIST usage referred to technologies which are the basis of standards, today's Internet and web standards are becoming the infratechnology upon which e-Business, e-Commerce, and all of the other "e-" activities are being built.
4. Leech, David P.; Link, Albert N.; Scott, John T.; Reed, Leon S.; *NIST Report: 98-2 Planning Report The Economics of a Technology-Based Service Sector*, January 1998, TASC, Inc. Arlington VA., p. ES-8 (http://www.nist.gov/director/prog-ofc/report98-2.pdf)
5. Libicki, Martin C. *Scaffolding the New Web: Standards and Standards Policy for the Digital Economy*, RAND, Santa Monica, CA, p. xi (http://www.rand.org/publications/MR/MR1215/)
6. These were the ANSI accredited standards committees called Accredited Standards Committee (ASC) X3 and Accredited Organization (AO) IEEE (Computer Systems). Approximately 85% of the key standards were created in X3, including storage interconnect, languages, and so on. The IEEE dealt with physical interconnects (such as local area networks) and eventually moved in to software interfaces.
7. In 1996, X/Open was merged with the Open Software Foundation to create The Open Group. X/Open was originally created in Europe to embrace and extend UNIX ® to limit the spread of U.S. companies into the European IT arena. After ten years of existence, and before its merger, it was largely dominated by major U.S. IT providers, with Siemens as its sole surviving European member.
8. Cargill, Carl F. *Information Technology Standardization,: Theory, Process, and Organizations*, Digital Press, Bedford, MA, 1989, p. 117.
9. The reason that consortia are often more visible within a company that are formal organizations is that consortia are more directly tied to the product success of a company. A company will join a consortium to promote the creation of a specification that it needs for market reasons - there is an imperative behind the consortia's creation. The same imperative is not necessarily found in formal organizations.
10. The IETF describes itself in the following way: The Internet Engineering Task Force (IETF) is a large open international community of network designers, operators, vendors, and researchers concerned with the evolution of

the Internet architecture and the smooth operation of the Internet. It is open to any interested individual. The actual technical work of the IETF is done in its working groups, which are organized by topic into several areas (e.g., routing, transport, security, etc.). Much of the work is handled via mailing lists. The IETF holds meetings three times per year. The IETF working groups are grouped into areas, and managed by Area Directors, or ADs. The ADs are members of the Internet Engineering Steering Group (IESG). Providing architectural oversight is the Internet Architecture Board, (IAB). The IAB also adjudicates appeals when someone complains that the IESG has failed. The IAB and IESG are chartered by the Internet Society (ISOC) for these purposes. The General Area Director also serves as the chair of the IESG and of the IETF, and is an ex-officio member of the IAB. See http://www.ietf.org

11 See http://www.w3.org/Consortium/ for a detailed description of both the creation of the underlying vision of the Web by Tim Berners-Lee and the initiation of the W3C by MIT, INRIA, and Keio University.

12 In the case of HTML 3.2 (a specification developed and promulgated by W3C), ISO/IEC JTC1 SC18 (the committee charged with standardization of this technology) tried to standardize HTML 3.2 with "JTC1 improvements", but only after W3C had standardized HTML 3.2, the users had implemented it in millions of Web sites. After serious negotiations by W3C and major users and providers, SC 18 agreed not to make their standard different from the W3C standard, which was in widespread use.

13 The Economist Newspaper, "The Age Of The Cloud, Survey Of Software", Special Supplement, April420th, 2001, 111 West 57th Street, New York, NY 10019-2211

14 Spring and Weiss discuss the problems of private sector funding of the formal standards organization in their article in *Financing the Standards Development Process* pp. 289-320, in *Standards Policy for Information Infrastructure*, edited by Kahin, Brian and Abate, Janet, MIT Press, 1995.

15 Weiss, Martin and Carl Cargill. "Consortia in the Standards Development Process" *Journal of the American Society for Information Science 43(8) (1992)*:559-565

16 Updegrove, Andrew, "Consortia and the Role of the Government in Standard Setting", pp. 321-348, in *Standards Policy for Information Infrastructure*, edited by Kahin, Brian and Abate, Janet, MIT Press, 1995,

17 Ketchell, John, at The CEN/ISSS web site, http://www.cenorm.be/isss/Consortia/Surveyshort.htm

18 Updegrove, op. cit. , p. 327.

[19] The rationale for this list of attributes derives from conversations with staff members of the House of Representatives Sub-Committee On Technology, Environment, and Standards, Daniel Weitzner of W3C, Stephen Oksala (Vice President, Society of Cable Telecommunications Engineers), Oliver Smoot (Chairman of the Board, ANSI), Dr. Mark Hurwitz (President, ANSI), Dr. D. Linda Garcia (Georgetown University), and others on how to describe a "good consortium". It is based upon experience (both good and bad) of the participants in many discussions, but especially to those in the W3C Patent Policy Working Group.

[20] Updegrove, op.cit., p. 338

[21] Ibid., p. 338

[22] ISO rules state: If the proposal is accepted on technical grounds, the originator shall ask any holder of such identified patent rights for a statement that the holder would be willing to negotiate worldwide licences under h is rights with applicants throughout the world on reasonable and non-discriminatory terms and conditions. Such negotiations are left to the parties concerned and are performed outside the ISO or IEC. A record of the right holder's statement shall be placed in the registry of the ISO Central Secretariat or IEC Central Office as appropriate, and shall be referred to in the introduction to the relevant International Standard (see item *e)* below). If the right holder does not provide such a statement, the technical committee or sub -committee concerned shall not proceed with inclusion of an item covered by a patent right in the International Standard without authorization from ISO Council or IEC Council as appropriate. ISO/IEC Directives, Part 2, 1992 (as amended) [Annex A, A.2, b)] http://isotc.iso.ch/livelink/livelink/fetch/2000/2123/SDS_WEB/sds_ipr.htm

[23] The criteria here are a combination of the requirements of the IETF (running code and dual, competing implementations) and the testing regime of the UNIX ® specification, run by The Open Group. The purpose of the conformance testing regime is to ensure that organizations claiming conformance to the specification actually do conform. However, it must be noted that the requirement for testing is contentious, as providers in the IT sector tend to favor "self testing and self certification" to testing provided by third parties. Allowance should be made to allow the consortium members the right to determine what level of testing they want; at the same time, the market, which on occasion has demanded third party testing, will be the ultimate arbitrator of the decision.

[24] U.S. Congress, *Global Standards*, op.cit., p. 14, footnote 23

[25] PL 104-113 is an act to amend the Stevenson-Wydler Technology Innovation Act of 1980, Public Law 96-480.

[26] OMB Circular A-119; Federal Participation in the Development and Use of Voluntary Consensus Standards and in Conformity Assessment Activities AGENCY: Office of Management and Budget, EOP. ACTION: Final Revision of Circular A -119.

[27] "Balance of interest" is a term referring to the need to have equivalent interests (vendor, user, and others) have equal representation in an organization or in a development committee. As noted, a consortium is composed of those interested enough in a technology to commit resources (usually financial) with the hope of receiving a return on their investment, usually in the form of a specification that can be employed in some form of commerce.

[28] Ibid.

[29] Web site, http://www.cenorm.be/isss/Consortia/Surveyshort.htm

[30] The concept of sectoral approach in standardization is presented in ANSI's *National Standards Strategy for the United States*, Section V, (http://www.ansi.org/Public/nss.html)

[31] A significant difference of the IT sector with other sectors is that, within the IT industry, we are, in the main, speaking of voluntary market driven standards, which are left to the discretion of the provider to supply. It is important to note that the majority of unique IT sector standards are interface standards describing a particular systems interface. They do NOT deal with safety or environmental activities. They are optional in a product - depending upon the business model of the vendor. Standards of this type are (and will continue to be) one of the costs of doing business, just as is translation of instruction manuals into a native language.

[32] The ASTM seems to have completely morphed into a standardization organization, and, while it maintains a "Yellow page listing" of consultants and expert witnesses, it doesn't seem to be educating testing experts. The mission statement of the ASTM reads: " To be the foremost developer and provider of voluntary consensus standards, related technical information, and services having internationally recognized quality and applicability..." With a complete yearly set of ASTM standards costing nearly $7000, and with ASTM standards being cited in legislation, one can understand why the ASTM has moved entirely to standardization activities. (http://www.astm.org/NEWS/Mission2.html)

[33] http://www.antitrustcases.com/summaries/456us556.html

[34] http://www.antitrustcases.com/summaries/486us492.html

[35] From ANSI's web site, their description of themselves: The American National Standards Institute (ANSI) has served in its capacity as administrator and coordinator of the United States private sector voluntary standardization system for more than 80 years. Founded in 1918 by five engineering societies and three

government agencies, the Institute remains a private, nonprofit membership organization supported by a diverse constituency of private and public sector organizations. (http://www.ansi.org/public/ansi_info/intro.html)

36 The Institute ensures that its guiding principles -- consensus, due process and openness -- are followed by the more than 175 distinct entities currently accredited under one of the Federation's three methods of accreditation (organization, committee or canvass). (http://www.ansi.org/public/ansi_info/national.html)

37 Web sites are: www.cenom.be, www.cenelec.org, and www.etsi.org.

38 Between 1983 and 1989, the EU began to focus on its internal market and the plethora of standards available within Europe. As a result, the "Council Resolution of 7 May 1985 on a New Approach to technical harmonization and standards" was passed in 1985, establishing the principles of European standardization. The essential outcome of all of these activities was to gain a "…national commitment [that] formal adoption of European Standards is decided by a weighted majority vote of all CEN National Members and is binding on all of them" (http://www.cenorm.be/aboutcen/whatis/objectives.htm).

39 ISO 9000 is an entirely problematic standard. It was originally started as a U.S. Air Force standard in the 1960s, adopted by the British in the 1970s, and then sent to ISO in the 1980s. It is a "management standard", which means that it doesn't tell you "how to do quality", but rather "how to manage a quality program, including the necessary paperwork and records retention". The appearance of this standard and its rapid acceptance and "mandatory use" (including third party certification) in many European companies and government procurements left a bitter legacy with U.S. companies who were "forced" to comply with third party testing.

40 See the "CEN Constitution and Organization" at http://www.cenorm.be/boss/co000.htm#b1 for the complete text, recognizing the Vienna Treaty and the common European norms.

41 At a presentation at the American Academy for the Advancement of Science (17 February 2001, San Francisco), ANSI President and CEO Mark Hurwitz stated that he believed that the Europeans engaged in block voting to stop U.S. SDO initiatives. From a national point of view, this has significance; from an international point of view (that normally taken by multinational companies), the existence of a standard that is meant to satisfy a large potential market (325 million people) is of substantial interest and is worth investigating and possibly implementing.

42 See the U.S. Congress, Office of Technology Assessment, *Global Standards: Building Blocks for the Future, TCT-512* (Washington, D.C.: U.S. Government

Printing Office, March 1992), pp. 13-14 for a view of the U.S. standardization process which haunts the U.S. to this day in Europe.

43 From ANSI Online, ANSI's web site, cite: http://www.ansi.org/public/ansi_info/national.html

44 It is interesting to note that both major international standardization organizations - ISO and the IEC have, within the last four years, adopted processes to recognize "Industry Technical Agreements" (ITAs), which allow any organization as "open" to progress a common industry practice through a lightweight process to achieve the appellation of either an ISO or IEC ITA. The senior organizations have recognized the need within their primary markets for a quicker and faster way to gain widespread recognition of a specification that is widely accepted, but possibly does not need the rigor of their full process. See http://www.iec.ch/ita-e.htmfor a description of the IEC program, and http://www.iso.ch/presse/ita.htm for a description of the program at ISO.

45 http://caselaw.lp.findlaw.com/casecode/uscodes/15/chapters/69/sections/section_4301.html

46 A typical statement, taken from the proposed sponsor agreement of one consortium, is " Nothing in this Agreement shall be construed to require or permit conduct that violates any applicable Antitrust Law. A Sponsoring Member consents to the disclosure of its name as a member of the Corporation, for the purpose of permitting the Corporation to invoke the protection of the National Cooperative Research and Production Act of 1993 (15 U.S.C. §§4301, et seq.), if the Corporation decides to invoke such protection."

47 It has been argued by several members of consortia that the travel and meeting requirements of formal organizations constitute a membership limitation, as very few private citizens have the ability to travel to all of the meetings of an international technical committee where the technology is decided. Some of the consortia with Internet based processes claim that their consortia dues are less than a participant would pay in travel costs.

48 The business case behind this type of decision is usually very complex and filled with enough vagaries to make the prediction of success purely Brownian. Normally, it comes down to a senior executive being willing to take a chance and go forward with opening a technology to the market.

49 The IT sectoral organization under CEN (CEN/ISSS) undertakes to maintain a list and description of consortia. It currently lists/links to approximately 260 consortia working in the areas of IT, either publishing specifications or specifying requirements. It is available at: http://www.cenorm.be/isss/Consortia/Surveyshort.htm

50 This statement amplifies the contention that there is a lack of education about standards and standardization.

[51] See Updegrove, Andrew; "Standard Setting and Consortium Structures", StandardView (Volume 3, Number 4): December 1995 for a discussion of the nature of the rules that apply when establishing a consortium.

[52] The closest successful suit was the Addamax anti-trust suit that was lost and lost again on appeal. (United States Court of Appeals For the First Circuit No. 97-1807, Addamax Corporation, Plaintiff, Appellant, V. Open Software Foundation, Inc., Digital Equipment Corporation, and Hewlett-Packard Company, Inc, Defendants, appellees, Appeal From The United States District Court For The District Of Massachusetts).

[53] The most popular types of licenses (Mozilla, GPL, and Berkeley) do not require that the owner of IP to give up the rights to their IP> Rather, these licenses require that the owner of the intellectual property grant broad, perpetual, and non-restrictive rights to use the IP, in effect making all of the users equal. The broad nature of the grant -in which the IP owner reserves few or no rights - is what has given many the impression that open source can be equated with forfeiting IP rights.

[54] Available at http://www.tuxedo.org/~esr/writings/cathedral-bazaar/

[55] Hecker, Frank "Setting Up Shop: The Business of Open-Source Software", 6 December 1999, Revision 0.7 DRAFT, http://www.hecker.org/writings/setting-up-shop.html

[56] "Despite the relatively modest share of ICT [Information and Commu nication Technologies] manufacturing in total U.S. production - 8% of total - the remarkable acceleration of productivity in that specific sector has contributed a disproportionately high 0.6% a year to total U.S. labour productivity growth." From "Europe in the e-Economy - Challenges for Enterprises and Policy-maker", Patrick Vittet-Philippe (Expert Advisor, DG Enterprise, European Commission), p.2

## APPENDIX 5: THE "REPORT"

## BEYOND CONSORTIA, BEYOND STANDARDISATION? NEW CASE MATERIAL AND POLICY THREADS

*October 2001*

*Dr. T. M. Egyedi*

**Department of ICT**
**Faculty of Technology, Policy and Management**
**Delft University of Technology**

### Management Summary

Current standards policy appears to be caught up in a polarised discussion about what type of organisation best serves the market for democratic and timely standards: standards consortia or the traditional formal standards bodies. The general feeling is that standards consortia work more effectively, but that they have restrictive membership rules and are undemocratic. The latter is a cause of concern for the European Commission, which requires democratic accountability in the standards process if it is to refer to such standards in a regulatory context. The Commission's request for new input on how to deal with consortium standards is set against this background.

### Aim

The Standardisation Unit of DG Enterprise had two objectives when it issued this grant to the Delft University of Technology. It sought:

- **New case material:** The aim was to acquire contemporary case material that illustrates how consortia work, why sometimes consortium standardisation is preferred to formal standardisation, and whether consortia work in ways that will deliver open standards.
- **New policy threads:** The aim was to develop a perspective on consortium standardisation that clarified its significance for EU standards policy. This required re-examination of current understanding of standards consortia, and of the underlying assumptions. Does the way the problem of standards consortia is defined -i.e. that their procedures are restrictive and undemocratic, and that their standards are therefore unfit as an instrument of regulatory governance -accurately describe what is at stake?

## Methodology

Two case studies took place: Java standardisation in *ECMA, an International Industry Association for Standardising Information and Communication Systems*, and standardisation of the Extended Markup Language (XML) in the *World Wide Web Consortium* (W3C). Data was gathered, foremost, by means of participant observation, i.e. attending ECMA standards committee meetings, interviews with committee participants, face-to-face and by email, and content analysis of (electronic) documents and emails regarding the standards process.

## Structure of the Report

The report consists of three parts. The two cases are presented in part I. Dominant assumptions on consortium standardisation are confronted with the case findings in part II. The current basis for standards policy is examined, and new policy threads are developed. Conclusions are drawn and recommendations are made in part III.

## Conclusions

*Why is consortium standardisation sometimes preferred to formal standardisation?*

Consortia successfully market their feats. They are associated with timely standardisation and pragmatic standards solutions, despite some critical observations to the contrary. This, and possibly the homogeneity and suggested exclusiveness of consortium standardisation, attracts companies. The two cases further show that (a) some consortia are used as a stepping stone for formal standardisation, (b) consortia are often equally relevant with respect to market co-ordination, and (c) changing a formal standard significantly is easier if the standards work is moved to a different setting, i.e. standards consortium.

*Does the current definition of the problem of standards consortia accurately describe what is at stake?*

No, it does not. A redefinition of the problem is desirable, one which addresses the themes of democracy and compatibility.

*Democracy.* According to the dominant view, consortia lack openness and are undemocratic. This view underestimates the openness of most industry consortia and overestimates the democratic procedures of formal standardisation. The research findings indicate that formal standards bodies and standards consortia work in similar ways. Consortia, too, strive for consensus, address minority viewpoints, etc.. Although the latter more explicitly target industrial parties, both settings include and exclude the same constituencies. The framework of rivalry merely leads to new hybrid forms of organisation like the CEN workshops. Speculating somewhat, these will not lure companies away from consortia but instead lead to a shift within the CEN standards domain away from the more formal procedures. Moreover, it by-passes the more significant difference between standardising and not-standardising. The real issues lie at a higher level.

*Compatibility.* The cases further highlight that company and government policies overly emphasise the means of standardisation while largely bypassing its aim, namely technical compatibility. The latter can also be achieved by other means than standardisation. Among these are the proprietary and open source strategies to Information and Communication Technology (ICT) development. In certain circumstances, the latter strategies are more effective in achieving compatibility than standardisation. A more systematic inventory of compatibility-enhancing strategies is needed to supplement those deduced from the findings of the case studies.

## Recommendations

The report pleads, firstly, for a European standards policy that bypasses possible rivalry between standardisation settings, goes beyond the inclusion of consortium standardisation, and works towards a differentiated standards policy. The latter should, on the one hand, reflect a pragmatic view where the majority of market standards is concerned (e.g. more exclusive, multi-party committees; focus more on standards implementation and market co-ordination). On the other hand, it should give more substance to the aim of democratic accountability which is required in *de jure* contexts. Secondly, a policy is desirable that goes beyond the standards process and centres on the objective of compatibility. This vantage point puts 'the consortium problem' into a very different, and clearer perspective. The Commission is therefore recommended to focus its policy on compatibility strategies, and not to restrict itself to standardisation. It is recommended that companies and governments re-assess their standardisation policy from the *de facto* compatibility standpoint.

## Questions Raised

The report raises several questions. An important one concerns a difficult issue in the ICT field, namely that the supply-side of the market often lacks the necessary incentives to prioritise compatibility. What mechanisms does the public, i.e. the demand-side of the market, have at its disposal to advance collective compatibility interests? Would it be desirable legally to anchor compatibility interests in a way similar to that of how intellectual property interests are presently represented in regulation?

## Foreword

This report presents the findings of a study on consortium standardisation, a project funded as a spontaneous grant by the European Commission DG Enterprise/ Standardisation Unit, and with additional funding from Verdonck Holding B.V. and the Delft University of Technology. It took place in the period January 2000 -June 2001 (draft report), and was finalised in September -October 2001.

There are several people whom I want to thank personally. First of all, Jan van den Beld (ECMA) and Christine Berg (DG Enterprise Standardisation Unit) for starting me off in a very efficient way; the members of the Commission of Recommendation Prof. Theun Bruins, Wim Verdonck, and Prof. Wim Vree for providing the necessary contacts, the additional financial support, and comments on papers, respectively; Arjan Loeffen for his crucial part in the W3C case; Willem Wakker for the discussions, the cooperation, the interesting material, and for commenting on papers; Richard Hawkins, Raymund Werle, and my colleagues of the ICT department for commenting on papers and ideas; ECMA TC41 members and the interviewees (see Appendix I); and last but not least members of the European Commission's Standardisation Unit/ DG Enterprise: Mr. Vardakas and Didier Herbert for their critical comments on the intermediary product, and Christine Berg, Christopher Roberts, and Michael Kirosingh, for their valuable advice along the way. They need not agree with the contents of this report.

*Amsterdam, 17 October 2001*

## 1. INTRODUCTION

In the past, the European Commission has always been very committed to formal European and international standardisation. For the reader who is less familiar with the issue at hand, *formal standardisation* refers to the voluntary consensus standards processes that take place in technical committees under the auspices of national, regional (e.g. European), and international standards bodies. The procedures that govern these committees express democratic values, aim to be inclusive (e.g. Public Enquiry of the International Standardization Organization (ISO) allows all interested parties that did not participate in drawing up a standard to comment on the draft standard), and reflect the desirability of a technical and politically neutral standards process (e.g. in the approval stage of a standard only the negative votes which are accompanied by technical arguments are counted). At stake is what could be called a *democratic ideology* (Egyedi, 1996). Its characteristic features are, for example, decision making by consensus; voluntary application of standards; broad constituency of (national) delegations; well-balanced influence of national members in the management of international standards bodies; and impartial, politically and financially independent procedures.

Formal standards are an important point of reference for European regulation (New Approach, 1985) and public procurement. Furthermore, formal standards have been at the basis of a harmonised European market. However, in the field of information and telecommunication technologies standards have emerged with high market relevance, standards that stem from other sources than the formal standards bodies. Examples are Adobe's Portable Document Format (PDF), the Internet standards developed by the Internet Engineering Task Force (IETF), and the Extended Markup Language (XML) recommendation developed under the auspices of the World Wide Web Consortium (W3C). How should these (de facto) standards be dealt with? Should the Commission revise its exclusive focus on the formal standards bodies, or should it encourage assimilation of these de facto standards by the formal standards institutions?

Questions to this intent are also raised in the European *Council Resolution* of October 1999 (Article 14). The Council observes "(...) an increasing tendency of interested parties to elaborate technical specifications outside recognised standardisation infrastructures" (Article 7). An important source for developing such specifications -and one on which the current research was requested to focus[1] -is the standards consortium[2]. A standards consortium is defined here as "an alliance of firms and organisations, financed by membership fees, formed for the purpose of coordinating technology development and/or implementation activities (...)" (Hawkins, 1998, p.1) Its outcomes are publicly available, multi-party industry specifications or standards. Usually its members are large companies, which indicates that the resulting standards are likely to be very relevance for the market. These consortia are also referred to as 'market-driven consortia' (CRE, 2000).

The common feeling is that standards consortia work more effectively than the formal standards bodies do. But, according to the same sources, their disadvantage is that they have restrictive membership rules and are undemocratic. The latter is a cause of concern for the European Commission, which requires a minimum degree of democratic accountability if it is to refer to such standards in a regulatory context. At first sight, the Commission seems to face a policy dilemma: adhere to a principled approach, one that prioritises a democratic standards process, or pragmatically include undemocratic consortia as a source of standards. The Commission's request for new input on how to deal with consortium standards is set against this background.

## Objectives

The current research was initiated, firstly, to provide contemporary case material that illustrates how consortia work, why sometimes consortium standardisation is initiated rather than formal standardisation, and whether consortia work in ways that will deliver open standards. For, although the phenomenon has been identified and studied since the early 1990s (e.g. Bruins, 1993), few case studies of consortium standardisation exist.

The second objective was to develop a new perspective on consortium standardisation that clarifies its significance for EU standards policy. This requires re-examination of current understanding of standards consortia, and of the assumptions and beliefs that underlie it.

## Method: Case Studies

Initially, one consortium standards process was to be studied from start to end by means of participant observation (i.e. attending standards committee meetings), and interviews with committee participants. Concerned was standardisation of Java, a key network technology owned by Sun Microsystems, in the ECMA consortium, an International Industry Association for Standardising Information and Communication Systems.[3] However, after two meetings the ECMA standards committee was prematurely disbanded. Since this reduced the time needed for data gathering, there was time left for a brief examination of a second case: standardisation of the eXtensible Markup Language (XML) by the World Wide Web Consortium (W3C). XML is presently viewed as a very important standard for structured information exchange. Since the standard was already finalised, data gathering for this case was done by means of document analysis, interviews with experts, and their feedback on the resulting working paper.

## Structure of the Report

The structure of the report is as follows. In Part I, the two cases are presented. They can be read separately. In Part II current assumptions on consortium standardisation are confronted with the case findings. The current basis for standards policy is examined, and new policy threads are developed. In Part III conclusions are drawn and recommendations are made. The project has led to a number of papers and articles. These are listed in Appendix
II.

## PART I: CASES OF CONSORTIUM STANDARDISATION

Industry consortia differ. Some focus solely on the development of technical standards or specifications: standards consortia, or *specification groups* (Updegrove, 1995). As the CEN/ISSS website indicates, there are many such consortia (CEN/ISSS, 2000). They may be R&D-oriented and pre-competitive (*research consortia*, Updegrove, 1995; *proof of technology consortia,* Weiss & Cargill, 1992), or focus on heightening the usability of existing standards (*implementation and application consortia;* Weiss & Cargill, 1992). Other consortia foremost aim to promote the adoption of a certain technology and seek the support of a business community (*strategic consortia*, Updegrove, 1995). To achieve a critical mass, suppliers of primary technologies and providers of complementary products and services must be directed along defined paths (Hawkins, 1998). To this end, consortia may rally support by organising educational activities for users of standards (Hawkins, 1999) or by combining promotional activities with specification development. In sum, although there are many differences between consortia, their common emphasis is on co-ordinating a segment of the market.

In this part of the report, two contemporary cases of consortium standardisation are presented: the industry consortia of ECMA and W3C. The ECMA, which was founded in 1961, is one of the oldest standards consortia, while W3C, a consortium founded in 1994, is one of the younger ones. Both of them foremost seek the support of the business community, and focus on developing standards (-according to the above typology: they are both *specification group* and *strategic consortium).* More specifically, the cases of Java standardisation in ECMA (1999-2000), and XML standardisation in W3C (1998) will be described. The findings shed light on questions such as why consortium standardisation is initiated, how consortia work, and whether consortia work in ways that will deliver open standards. An analysis follows in Part II.

In order to present the case findings in an interesting way, questions have been formulated that draw out the main characteristics of the cases. In the case of Java standardisation (Chapter II), the red thread is the question why a company would want to standardise its technical specification if the latter is already a *de facto* standard. The account is based on extracts from an article titled "Why Java was -not standardised twice" (Egyedi, 2001b, see Appendix II). In the XML case, another puzzling element is addressed. According to its developers, the XML standard is based on the Standard Generalized Markup Language (SGML), a standard developed in the formal standards setting (ISO/IEC JTC1). Why did XML developers, most of whom were SGML experts, choose to standardise XML within the World Wide Web Consortium rather than in JTC1? To a large extent, this chapter (Chapter III) is an extract of "Succession in standardisation: Grafting XML onto SGML"

(Egyedi & Loeffen, see Appendix II). Specific case-bound literature is included at the end of both chapters. General literature in included in the main reference list of this report.

## 2. JAVA IN THE ECMA CONSORTIUM

When Sun Microsystems approached the ISO/IEC Joint Technical Committee 1 (JTC1) to standardise its Java™ Technology in 1997, Java was already well on its way to become a *de facto* standard. Sun became a recognised submitter of Publicly Available Specifications (PAS)[4] late 1997 but refrained from using its submitter status, allegedly because JTC1 had changed the PAS procedure in a way that would make the actual acceptance of the Java specs difficult. In April 1999, Sun approached the ECMA standards consortium, an international industry association for standardising information and communication systems, for the same purpose. If Java became an ECMA standard, it could be submitted to JTC1 by way of the Fast Track process[5]. However, after the first meeting of the ECMA standards committee Sun again withdrew. This time ECMA's Intellectual Property Right (IPR) rules were not elaborate enough, according to Sun. Two main questions arise. Firstly, why did Sun initiate formal and consortium standards activities in the first place? Secondly, why did Sun pull back twice?

There is a host of literature that addresses why companies partake in standardisation. Standardisation is part of the competitive product development process between producers (Weiss & Sirbu, 1990; Grindley, 1995). Companies partake in order to develop new markets and protect established markets (e.g. prevent compatibility to block competitors from their market). They use standards as change agents. They use them as strategic tools to consolidate a market position or gain advantage over competitors (Cargill, 1989; Bonino & Spring, 1991). This body of literature suggests that dominant market players, whose products have become a *de facto* standard, have few incentives to standardise. They are more likely to withhold information on interface specifications or change proprietary product interfaces at regular times to put off competitive product development. Or they may try to tie complementary products of other firms to their proprietary component technology. With an eye to long-term advantages, they may give away a technology or enter into coalitions with rivals to enlarge their user base and widen support for their proprietary standard (David & Greenstein, 1990). However, the step towards formal standardisation is seldom taken.[6] In this respect, the initiative to standardise Java™ is rather unique.

## 2.1 Sun's Java Technology™ and the User Environment

Java started as a programming language. In 1995, Sun realised that it could be used for the Internet. Its platform-independence, about which more below, allowed small Java programs to be downloaded and executed by web browsers. These moving, colourful applets triggered Java's breakthrough on the Internet.

*Java's platform-independence.* One of Sun's maxims was 'Write Once Run Anywhere' (WORA): a Java software developer should not need to rewrite his or her software program for different platforms. Java programs were to be portable and scaleable. In order to achieve cross-platform compatibility, Sun created a standardised application programming environment. Each system and browser provider was to fully implement the specifications and Application Programming Interfaces (APIs)[7] of the standardised Java environment if WORA was to be achieved. Several system providers, such as IBM and HP, did so. That is, they developed compatible Java Virtual Machines (JVMs, i.e. software that runs on proprietary operating systems and is capable of interpreting compiled Java byte code). Java is also applied in dedicated devices such as household appliances, television sets, cars, etc. in which case it is referred to as *embedded Java, or real-time Java.* The emphasis is in the case study on the Java programming environment.

*Java user environment.* Sun started by giving interested parties access to it source code. It invited developers to comment on, experiment with and improve the original source code. The source code was 'open' in the sense of being accessible and free of charge, but, for example, the decision about changes to the original code lay in Sun's hands and commercial use was bound to license restrictions. Part and parcel of Sun's licensing policy were the test suites used to certify compatible Java products, and the Java-compatible logo (the steaming cup of coffee) to brand compatible products. These instruments of control were closely tied to Sun's IPRs to trademarks (e.g. Java™ and Java Compatible logo), patents (software algorithms) and copyright on the specifications.

Pressed by its commercial licensees, Sun developed a 'Community Source' licensing model, which sought to combine the advantages of the Open Source licensing model and the Proprietary licensing model (Gabriel & Joy, 1998). It did, indeed, represent a more liberal licensing regime for commercial parties, but Sun still retained ownership of the original code, the upgrades, and the test suites.

A Java community had developed. Sun tried to institutionalise this community in December 1998 with the Java Community Process (JCP) manual. However, the document was criticised for Sun's too dominant role therein (Harold, 1999; Vizard, 1998). The second version issues in 2000 differed in many ways and answered to much of the critique (Shankland, 2000; Sun, 2000a).

The idea of WORA and Sun's strategies to involve others in developing and implementing the Java platform led to a large user base. In 1999, there were more than 1,3 million Java developers (International Data Corporation, op. cit. in Babcock, 2000). This figure consists of developers who work for companies and, for the majority, of independent developers.

## 2.3 JTC1, the First Attempt

Sun was the first private company to apply as a recognised PAS submitter. IBM strongly backed up Sun's application. This happened in March 1997. It caused a stir because although the rules allowed individual companies to apply, the criteria favoured open, consensus-oriented organisations. Within the American National Standards Institute (ANSI)[8], Sun's home base, opinions already strongly differed (Rada, 1998). In July, Sun's application was turned down with comments. The comments of the JTC1 national members roughly focused on Sun's desire to keep the Java trademark for itself and have the JTC1 standard called something else; on what body would be responsible for updating and maintaining the Java standard; and on whether Sun would be open in accepting changes to the standard (Clark, 1997). Sun addressed the comments in September 1997 and reapplied as PAS submitter (Sun, 1997a). It suggested, for example, that a JTC1 working group, which would be open to all stakeholders, would address the standards maintenance work, and it offered to supply the project editor. Two months later, Sun was accepted as a PAS submitter. But, again, there were comments (ISO/IEC JTC1, 1997). The national bodies expected their comments to be addressed in the Explanatory Report that would accompany Sun's submission of the Java specs, and they added that voting 'yes' at this stage did not automatically include approval of the specs. According to Sun, the positive outcome of the voting was to be understood as international approval of Sun's open Java development process. In the following year, Sun did not take steps to actually submit the Explanatory Report or the Java specifications to JTC1. Sun silently withdrew from the PAS process, a move that became apparent when Sun's overtures to ECMA became public.

### 2.3.1 Initiative

In the following sections, the events are examined in more detail. I thereby distinguish between Sun's explanation of the events and my interpretation of them, because they do not always coincide. I use the headings of 'stated reasons' and 'interpretation' for this purpose.

*Stated reasons.* Sun said its goal always was to "have Java, already a de facto international standard, codified as a *de jure* standard" (Sun, 1997b). From a business perspective, Sun's interest in standardisation was to increase the visibility and importance of Java and to promulgate a network-centric view on ICT developments. By approaching JTC1, Sun signalled that Java was to be a specification that people could rely on as being stable and that it would not be changed unexpectedly. It allowed people to make a commitment to it.

Sun chose the PAS procedure because this was the most effective way to get the Java technology formally accepted world-wide. It was a means to get easier access to the public procurement market, and to preserve industry's substantial investment in Java. The latter argument can be understood as a way of saying that the Java submission should not undergo serious changes during the PAS review process.

*Interpretation.* Sun did not intend to hand over the evolution of Java to JTC1 (Sun, 1997d). It expected to retain control over the standards maintenance process by safeguarding the role of the Java community during JTC1 standardisation, whose input was co-ordinated by Sun itself. ("The JTC1 working group that will address standards maintenance must be responsive to international Java community." (Sun, 1997a)) Sun upheld essential IPRs, and retained its patents (although no fees are asked), its copyright (joint-copyright ownership was suggested, no fees asked), and trademarks (e.g. control over compatibility logo). An additional benefit of the PAS procedure was that ongoing Java developments would become tightly linked to standards development. The revenues from IPRs were forfeited in exchange for enlarging and stabilising the Java market -without compromising control over cross-platform compatibility (e.g. by means of the Java compatible logo and the test suites). JTC1 's role was to codify and ratify the specification development activities supervised by Sun.

Sun's PAS initiative can therefore best be understood as a means to orchestrate the orientation of market players. There are two main reasons to think so. Firstly, because JTC1 was the pre-eminent international standards body for IT matters, it was a focal point for consensus-based standards development. The PAS procedure would appear to leave room for the influence of competitive market players, keep them oriented towards Java developments led by Sun, and dissuade competitive developments. Secondly, in the years that preceded the PAS initiative Java was becoming a hype (1995-1996). Mainly by way of Netscape Navigator, copies of Sun's Java runtime environment were downloaded to the PC systems of Windows users. Sun's network-centric vision and Java's promise of platform-independence made Microsoft nervous. Sun was challenging the basis of Microsoft's software market, the Windows platform. In 1995, Microsoft had already approached other companies to withdraw from activities that supported Java™ developments (e.g. Netscape and Intel). By late spring of 1996, senior Microsoft executives were deeply worried about the potential of Sun's Java technologies to diminish the applications barrier to entry (US, 1999).

In March 1996, Sun and Microsoft signed a Technology License and Distribution Agreement (TLDA) for the use of Java. The agreement included the incorporation of Sun's Java™ Technology in Microsoft's Internet Explorer 4.0. Late 1996, Microsoft released Internet Explorer 3.0. It was a much-improved version. Some reviewers considered it competitive to Netscape Navigator. In order maximise the usage of Internet Explorer, Microsoft decided that the next version would be more tightly integrated into Windows (US, 1999). Moreover, Microsoft was using its Java license to create its own Java development tools and its own Windows-compatible Java runtime environment. It did so in a manner that undermined Java portability and that was incompatible with Sun's Java products. In the same month that Sun started the PAS application, Microsoft distributed its own incompatible Java toolkit. When Sun applied as a PAS submitter for the second time, it was preparing a lawsuit against Microsoft for copyright infringement. For Sun, the rumours of Microsoft's previous dealings with other players and a premonition of Microsoft's strategy to develop a Windows-dependent Java browser and toolkit would have been reasons not to overestimate its own position in the market. In this market, the step towards international standardisation may well have served the purpose of rallying support for Java™. Sun most likely assessed that its footing in the Java market was not secure enough, which explains its willingness to standardise. On the other hand, it also explains why Sun could not relinquish control over Java.

## 2.3.2 Withdrawal

*Stated reasons.* Sun withdrew from the PAS process because it did not agree with changes in the PAS procedure decided on in November 1998 (ISO/IEC JTC1, 1999b). The old procedures still applied, but Sun's status as a PAS submitter would have to be reconfirmed in November 1999, at which time the new rules would apply. The new procedures, according to Sun, implied that Sun would have had to turn standards maintenance and the evolution of Java over to JTC1. Moreover, standards maintenance would not be restricted to minor adjustments such as bug fixing. JTC1, on the other hand, remarked that the changes were clarifications (ISO/IEC JTC1, 1999d). Comparing the 1999 version of the PAS procedure with the previous version (1995), in the latter version handling of standards maintenance is settled 'in accordance with the agreements made between JTC1 and the recognised PAS Submitter'. The 1999 version stipulates that the normal JTC1 rules for maintenance apply, regardless of the origin of the International Standard. JTC1 would take the lead in corrections to defects and -which will have alarmed Sun -revisions of existing standards. Reacting to Sun's objections, the JTC1 chairman writes, that "the clause addressing the topic of maintenance in the revised JTC 1 PAS procedure is consistent with the comments made by a number of JTC 1 National Bodies that voted to approve Sun as a PAS Submitter but noted the need for JTC 1 involvement in the maintenance of the resulting International Standard." (ISO/IEC JTC1, 1999c).

But much had happened behind the scenes. Sun attributed the changes made to the PAS procedure to lobbying by Microsoft, Hewlett-Packard (HP) and others from the 'Wintel world' (Shankland, 1999b). (Microsoft wanted its own Java functionalities enabled.) Sun withdrew because it felt that the change of procedures was only a next stage in the opposition. The procedural changes signalled that Sun would encounter problems when submitting the Java specification. For example, a Java Study Group had been installed in JTC1 Sub-Committee 22 (SC22) and people were discussing how they were going to change the Java specification. It was at that point that Sun seriously started considering alternatives.

*Interpretation.* Sun judged that JTC1 would probably not agree to ratify Sun's work in view of the influence of the 'Wintel-world in JTC1. But, apart from the reasons Sun gave for withdrawing, there were developments in the market that threatened Sun's position, occurrences which increased Sun's desire to keep a grip on Java developments. Firstly, Microsoft did not abide to the Java licensing agreement, and posed a threat to cross-platform compatibility. In October 1997, Sun filed a complaint against Microsoft for copyright infringement. In March 1998, the court granted Sun's request for a preliminary injunction. Microsoft was not allowed to use the Java Compatible trademark unless its products passed Sun's test suites. In May, Sun filed a complaint for unfair competition. In November 1998, the court ordered Microsoft to change its software and development tools. Microsoft appealed against the ruling (Egyedi, 2000b).

Secondly, in the same period there were disquieting developments in the area of real-time embedded Java. Hewlett-Packard (HP) announced in March 1998 that it had developed a *clean-room* version of real-time embedded Java, that is, a version that was developed without looking at Sun's source code (Concerned is a manner of reverse engineering by which Sun's IPRs on Java are circumvented.). In June, the US National Institute for Standards and Technology (NIST) started organising workshops to develop specification requirements for real-time Java. Sun participated, as did competitors such as HP and Microsoft (Jensen, 1999). In November 1998, a Real-Time Java Working Group (RTJWG) led by Microsoft and HP was formed. Sun did not participate. The RTJWG approached the US national standards channels, that is, the National Committee for Information Technology Standardisation (NCITS/ NIST), to formalise its standards work. But in January 1999 its request was turned down because NCITS feared this could lead to fragmentation of the Java market. The RTJWG subsequently founded the J Consortium. Meanwhile the Real-Time Expert Group (RTEG) was formed within the Java Community Process, a group that was led by IBM.

The RTJWG activities were disquieting to Sun, because real-time Java draws on the base specifications of Java™. According to the experts whom Sun consulted, it was not possible to write real-time specs in a useful way without making changes to the base specifications. There was therefore a risk that competitive developments in the field of real-time Java would affect the work done on Java™ within Sun's JCP.

Sun reacted to the market pressure and to changes in the PAS procedure by elaborating the procedures for Sun-led Java community participation, withdrawing from JTC1, and exploring alternative options for international standardisation. In December 1998, Sun issued its first version of the JCP and presented its Community Source licensing model (see earlier). They were designed to signal that Sun had taken the criticism of 'benevolent dictatorship' to heart and accepted more far-going influence of the community on Java development. The Community Source model, which partly sympathised with the open source movement, was to underscore Sun's new approach. It mainly served to re-orient players in the field of real-time Java. Sun's JTC1 initiative had failed to keep the real-time Java dissidents in line. The withdrawal in itself was based on Sun's assessment that it would not be able to manoeuvre the Java specification through the PAS procedure unscathed. It was a move that followed from its compatibility control strategy. To keep control of Java it needed to withdraw. This time it did not attempt to re-orientate the market, since those involved with the Java™ programming environment publicly heard about Sun's withdrawal only when Sun had already approached ECMA (May 1999). To them, Sun was still pursuing the standardisation path.

## 2.4 ECMA, the Second Attempt

In April 1999, Sun formally approached the ECMA to discuss Java standardisation (Shankland, 1999a). Sun initially proposed that ECMA would carry out 'passive maintenance' of the Java standard, meaning that Sun's JCP would still determine Java development (Sliwa, 1999). But ECMA refused to endorse this approach. The two parties ultimately agreed to the instalment of a technical committee on Platform-Independent Computing Environments (TC41) which would 'standardise the syntax and semantics of both general-purpose and domain specific platform-independent computing environments.' The committee would develop a standard for a cross-platform computing environment based upon the Java 2™ Standard Edition Version 1.2.2, a specification that consists of the Java Language Specification, the Java Virtual Machine Specification, and the Java API Core Class Library Specification. The aim was to contribute the standard to ISO/IEC JTC1 by means of the Fast Track process. The ECMA General Assembly gave its approval in June 1999.

The first TC41 meeting took place in October 1999. It was chaired by IBM. During the meeting, Sun emphasised that the TC should focus on 'edition rather than addition' of the Java specifications. Sun provided the main editor. The JTC1 SC 22 Java Study Group, with which the ECMA liaised, would be asked for input before formally invoking the Fast Track process. Three task groups were installed to tackle the work. A Microsoft representative chaired the group working on the API specifications. Sun was to distribute the Java 1.2.2 specification on CD-ROM at the meeting. However, at the end of the two-day meeting a Sun representative announced that Sun lawyers required more time to consider the IPR issues involved (ECMA, 1999a). The second meeting was set in January 2000.

In December 1999, Sun made public that it would not contribute the Java specifications to ECMA. At the January meeting, the TC41 participants debated whether it would be feasible to draft a Java standard without Sun's contribution. But some large companies objected (Fujitsu, Siemens, HP and Compaq). In March 2000 the TC was disbanded.

### 2.4.1 Initiative

*Stated reasons.* Sun chose ECMA because ECMA had close ties with the formal European and international standards bodies and an A-liaison with JTC1, which gave it access to the Fast Track procedure. Sun understood that in the past ECMA standards had been submitted to a yes/no vote in JTC1 without any modifications, and often successfully so. If Java would become an international standard, customers, partners and developers would feel more confident about investing in it (Perez, 1999). But, Sun said, it would also be pleased if Java would remain an ECMA standard (Shankland, 1999b).

From Sun's standpoint, ECMA TC41 would edit the Java version that resulted from Sun's JCP trajectory, because there were products based on it and there was a developer community working to the specification. Sun was under the impression that ECMA had agreed that Sun would retain copyright of the specifications during the standards process, and that ECMA would copyright the resulting standard. The latter was necessary to submit it to JTC1 through the Fast Track procedure. (Although Sun would not claim copyright of the standard, it would hold on to IPRs such as the Java name and the Java Compatibility logo, which had a business value to Sun.)

Furthermore, TC41's program of work was specifically limited to the Java Standard Edition version 1.2.2. Any risks which Sun was taking would be restricted to this Java version. More far-reaching changes would be part of a new Java version, a development process that would take place within the JCP environment (Sliwa, 1999).

*Interpretation.* ECMA was an open standards consortium and thus an answer to continuous pressure from licensees and real-time Java developers to open up the Java development process. Many large companies were members. So ECMA processes also promised to be relevant in respect to co-ordination of the market. Sun's move further suggested consistency in its aim towards international standardisation. But at the same time, the move was an alibi for withdrawing from the PAS procedure without gravely letting down those who were pressing Sun for open standardisation.

Sun's position in ECMA was stronger than in JTC1. Sun participated at the time in the ECMA Coordinating Committee (Mr.R. Cargill) and shortly after in its Management (Ms.V. Horsnell, treasurer); and the acting chair of the JTC1 SC22 Java Study Group, with which ECMA liaised, was a Sun representative (Mr. J. Hill). Sun further controlled the conditions under which the process would take place by means of its IPRs and by restricting the scope of the program of work. Perhaps, too, in the preparatory period of defining TC41's program of work, Sun had less reason to fear Microsoft. The judicial system was partly checking Microsoft's undermining actions with regard to Java compatibility. In the set up of this standards initiative, Sun had a more focused control strategy than during the PAS initiative. Its emphasis appears to have been on technology content-oriented standardisation.

### 2.4.2 Withdrawal

*Stated reasons.* Sun's official reason to withdraw from the ECMA process was that "(...) ECMA has formal rules governing patent protections; however, at this time there are no formal protections for copyrights or other intellectual property." ( Sun, 1999) Unofficial Sun sources indicated that problems had arisen between the ECMA GA meeting (June 1999) and the first ECMA TC41 meeting (October 1999). These concerned the timing and place of the first meeting, which was scheduled months later than Sun had intended, and procedural issues. (Certain companies insisted that the committee would not be chaired by Sun, that the editors would not be Sun people, and proposed that Microsoft co-ordinate the development of API specifications.) There were also hints, according to Sun, that the oral agreement on copyright, as Sun understood it, would not be upheld. Sun became wary.

At the first committee meeting, Sun lawyers were taken by surprise by the ECMA secretary general's explanation of IPR rules regarding contributions to standardisation. As a rule ECMA documents were not copyrighted. Regarding the copyright status of the Java specs, Sun's contribution would become an ECMA document once it was assigned a TC document submission number. When Sun representatives protested, the ECMA secretary general proposed to explore means by which Sun could maintain copyright during the standards process. ("Contributions from member companies to ECMA can be copyrighted, and can retain their copyright status if the owner of such a specification allows ECMA to freely use the contents of the contribution for the development of an ECMA Standard." (ECMA 1999c)

The problem was, firstly, that the parties (Sun and ECMA) had a different view on what was previously agreed, and in particular who was to copyright the Java specs during the standards process. But, secondly, Sun's ideas with respect to the meaning of copyright at that point appeared to differ from ECMA's. Sun differentiated between a copyrighted specification and a copyright of the contents of the specification (i.e. roughly speaking, the difference between paper and software). The problematic part was how TC41 would handle the latter copyright interpretation, which was new to all concerned. At the subsequent meeting of the ECMA Coordinating Committee (November 1999), Sun explained the distinction, and said that it intended "to provide ECMA with a derivative copyright but that this has to be treated as an IPR, under a copyright license agreement" (ECMA, 1999b). The conditions of such an agreement were not yet decided on. Early December, Sun announced its withdrawal.

George Paolini, vice president of Java community development at Sun, provided another reason for Sun's withdrawal. He said in a letter to ECMA that Sun had decided to keep control of Java within its Java Community Process. "The Java Community Process has expanded its level of activity to a point where we now believe the interests of the entire Java community will be best met by continuing to evolve the Java specifications with the open JCP process." (ECMA, 1999b) By then, a proposal for the second version of the Java Community Process had been developed.

*Interpretation.* The events that took place before the first ECMA TC41 meeting, indicated that Sun's influence on the standards process was under attack: procedural issues were discussed that would undermine Sun's position. Furthermore, according to a member of the ECMA Coordinating Committee the prior informal agreement about copyright issues was ambiguous.

The steps which Sun took in the months following its withdrawal give credence to Sun's official reason to withdraw. The industry association of European Information and Communication Technology Industry Association, founded in January 2000, installed a Standards Policy Group chaired by Sun. The policy group was to develop a position on the licensing terms of software technology embedded in standards protected by copyrights rather than patents[9]. Sun also planned to raise the issue at a meeting of the European ICT Standards Board, but refrained from doing so before the meeting (ICTSB, 2000). Lastly, Sun called together a Standards IPR Forum meeting during the Open Group Conference (April 2000, London) to address, among other things, ownership of copyright on submissions.

However, the primary issue was not that the copyright agreement was ambiguous and informally arranged -probably both ECMA and Sun initially had an interest in this arrangement. The above-mentioned procedural disputes between June (approval of the TC41 work program) and October 1999 (the first TC41 meeting) seem crucial. Moreover, in August, Sun heard that in its ongoing lawsuit against Microsoft the court had granted Microsoft's appeal against the preliminary injunction for copyright infringement. The appeal was, in brief, that the punishment did not fit the crime committed (i.e. a breach of contract should not be punished by means of an injunction). This verdict was a blow to Sun, and had consequences for Sun's stance in ECMA. If Sun would loosen its IPR claims for the purpose of ECMA standardisation, it might jeopardise its position in the next stage of the lawsuit.[10] Furthermore, possibly also the outcome of the lawsuit raised Sun's doubts about what legal protection a copyright offers (-although this was to my opinion not the issue in the August trial). This would explain Sun's introduction of a dual meaning of copyright. In sum, procedural issues and the Sun v. Microsoft lawsuit fuelled Sun's wariness. By not clearing the copyright issue beforehand, Sun could introduce a new meaning of copyright, one which would not be acceptable to the ECMA TC, to pave the way for total withdrawal. "[Sun] just does not want to give up control", as the ECMA Secretary General, Jan van den Beld, told the press (Niccolai & Rohde, 2000), and it had several reasons not to do so. Possibly Sun did not believe Java was stable enough or had achieved sufficient critical mass to relinquish control (Niccolai & Rohde, 2000). Whatever reason presided with regard to ECMA standardisation, Sun's actions focused on preserving control over the Java™ specifications.

## 2.5 Conclusion

Sun primarily initiated standardisation in JTC1 and ECMA because an international standard implied stability, would increase market confidence and would therefore encourage commitment to Java. It wanted JTC1 and ECMA to 'ratify' the existing Java™ specification and did not seek the involvement of their committee members in its development. Rather, it sought commitment from the clients of these standards bodies (i.e. implementers of JTC1 standards). It withdrew from JTC1 because it suspected standards politics behind procedural changes, because of incompatible and competing market developments, and -above all -because it expected that its Java specification would not survive the PAS procedure unscathed. Sun intensified its compatibility control strategy in subsequent negotiations with ECMA. To minimise risks, it focused its standards initiative on a specific version of the Java specifications. However, the procedural disputes that preceded the first ECMA committee meeting made Sun wary. Added to new developments in the lawsuit

*Table 2.1 Summary of the findings. (The largesse of the X indicates whether technical compatibility and market orientation are significant factors (X) or not (-), or to a lesser degree (x).*

| Sun actions > | Formal Standardisation (JTC1) | | Consortium Standardisation (ECMA) | |
|---|---|---|---|---|
| Co-ordination strategies | *Initiation* | *Withdrawal* | *Initiation* | *Withdrawal* |
| Technology-oriented compatibility control | - | X Java would not survive PAS unscathed | x Focus on Java version 1.2.2 | X Procedural disputes Copyright ambiguity Sun-Microsoft lawsuit |
| Orchestration of market orientation | X Heighten market's commitment to Java | x JCP installed to attract real-time Java developers | X Heighten market's commitment to Java | - |

with Microsoft, Sun referred to ECMA's ambiguous copyright rules to pull back from ECMA standardisation.

Table 2.1 summarises the main findings. It shows that Sun's initiative to formalise its de facto standard was primarily motivated by its aim to orchestrate the market. Whereas, a basic fear of a fragmented technical platform -and ensuing market effects -motivated Sun to withdraw.

Sun pursued a protective and defensive control strategy. Whether it should instead have followed a more offensive strategy, based on confidence in a market-co-ordinated development of platform-independent Java, is a matter for debate.

## 2.6 Case-Specific References


Babcock, Ch. (2000). 'Java: Can Sun control the flood', *Inter@ctive Week, June 12 2000* [www.zdnet.com].

Brookes, N. (2000). 'Has Sun gone off on one?' contribution to a discussion on JavaLobby Pro-Sun or Pro-Java?, 26 January 2000.

Clark, T. (1997). 'Java standard voted down--for now', *CNET News.com, July 30 1997.*

ECMA (1999a). *Minutes of the 1st meeting of ECMA TC41 held in Menlo Park (USA) on 25th -26th October 1999*, ECMA/TC41/99/16.

ECMA (1999b). *Minutes of the meeting of the coordinating Committee held in Geneva on 11th -12th November 1999*, ECMA/GA/99/127.

ECMA (1999c). ECMA CC 11-12 November 1999.

Farrell, J. & G. Saloner (1986). 'Installed base and compatibility: innovation, product preannouncements and predation', *American Economic Review 76 (1986),* 940-955.

Gabriel, R.P. & W.N. Joy (1998). *Sun Community Source License' Principles (SCSL).*

Harold, E.R. (1999). *The Java Gated Community Process* [http://metalab.unc.edu/javafaq/editorials/jcp.html].

ICTSB (2000), *Software Technology Embedded in or Becoming a Standard.*

ICTSB18 (00) 22, discussion paper. ISO/IEC JTC1 (1997). N5090, November 1997. ISO/IEC

JTC1 (1999a). *JTC1 Directives*, 1999. ISO/IEC JTC1 (1999b). *The Transposition of Publicly Available Specifications into International Standards -A Management Guide* -Revision 1 of JTC 1 N3582, N5746, January 1999.

ISO/IEC JTC1 (1999c). 'JTC 1 Chairman's Comments on Recent Press Articles Regarding Sun Microsystems' PAS Submitter Status', 1999-5-1 [www.jtc1.org].

ISO/IEC JTC1 (1999d). *PAS Transposition: JTC 1 finishes trial, improves its guidelines and moves to regular operation*, 1999-9-8.

Jensen, E.D. (1999). *Real-Time Java Status Report*, Revised Draft, January 24 1999 [http://www.real-time.org/].

Niccolai, J. & L. Rohde (2000). 'Sun confers with licensees over control of Java', *IDG News service*, 3/3/2000.

Perez, J.C. (1999). 'Sun changes strategy for Java's ISO approval', *IDG News Service*, May 6 1999.

Shankland, S. (1999a). 'Sun renews Java standards effort', *CNET News.com*, May 5 1999.

Shankland, S. (1999b). 'Sun revising plan for Java recognition', *CNET News.com*, April 29 1999;

Shankland, S. (1999c). 'Sun seeks control of Java process', *CNET News.com*, May 6 1999 [www.znet.com].

Shankland, S. (2000). 'Sun offers olive branch to Java partners', *CNET News.com*, March 1 2000.

Sliwa, C. (1999). 'Java standard switch: Will new process ease cross-platform compatibility?', *Computerworld*, May 7 1999.

Sun (1997a). 'Sun response to ISO/IEC JTC1 N4811 and N4833' [www.Sun.com].

Sun (1997b). 'Sun's ISO application progresses', July 16 1997 [www.sun.com].

Sun (1997c). *Amended Complaint*, October 14 1997.

Sun (1997d). Sun press conference, Question and Answering session, November 1997 [www.sun.com].

Sun (1998). *The Java Community Process (sm) Program Manual: The formal procedures for using Java Specification development process* (version 1.0, December 1998).

Sun (1999). Sun press release, December 7 1999.

Sun (2000a). *The Java Community Process (sm) Program Manual: The formal procedures for using Java Specification development process*, version 2.0, review draft, April 2000.

Sun (2000b). *Sun Microsystems v. Microsoft Corporation*, United States District Court For The Northern District Of California, January 24 2000, No. C 97-20884 Rmw (Pvt), Order Re Sun's Motion To Reinstate November 17, 1998 Preliminary Injunction Under 17 U.S.C. § 502; Preliminary Injunction Under Cal.Bus. & Prof. Code §§ 17200 Et Seq.

US (1999). *United States of America v. Microsoft Corporation. Findings of Fact*, United States District Court For The District Of Columbia, Civil Action No. 98-1232 (TPJ), 1999.

Vizard, M. (1998). 'Java Lobby president calls for reform', *InfoWorld Electric, November 19 1998*. [www.infoworld.com]

## 3. XML IN THE WORLD WIDE WEB CONSORTIUM

The XML standard (1998) was well received by the information technology practitioner community. While the trade press mostly hailed it as a functionally rich sequel to the HyperText Markup Language (HTML), it sometimes described it as a welcome leaner version of the Standard Generalized Markup Language (SGML). According to XML developers themselves, XML (1998) was a compatible successor[11] to SGML (1986/1988), a standard developed in the formal standards setting of ISO/IEC JTC1. Was XML, indeed, compatible? If so, why did XML developers, most of whom were SGML experts, standardise XML within the World Wide Web Consortium rather than in ISO/IEC JTC1? To provide the background necessary to answer these questions, a brief sketch of the technologies concerned follows.

## 3.1 SGML

Work on SGML started in 1969 with the development of a language called the Generalized Markup Language (GML) at IBM (Goldfarb, 1990). It was used to manage the large amount of complex industrial documents at IBM. GML was designed to record document structures independent of how these structures would subsequently be processed. For example, GML documents recorded headings, paragraphs, lists and figures – that is, information that is useful for editorial applications– but no formatting instructions. In this manner, GML separated the document description from the formatting languages (IBM used several such languages for printing). Also, because GML identified document structures, fragments of documents could be addressed and reused in different contexts.

In 1978, the ANSI took an interest in IBM's work on GML. By the efforts of Charles Goldfarb, one of the three inventors of the language, work started on a more generic version: SGML. A major addition to the original design was made. In order to determine the validity of the document structure, and to support a wide variety of lexically different languages (e.g. different signs for start-tag), a formal description, or *grammar,* would accompany each document. Firstly, this grammar identified the type of components (*elements*) and their interrelations (*content model*).

*Box 3.1. Aims of the SGML and XML standardisers*

| SGML Objectives (Source: ISO 8879:1986, Clause 0.2) | Design goals for XML (Source: XML 1.0, 2nd ed., 2000) |
|---|---|
| 1. Documents "marked up" with the language must be processable by a wide range of text processing and word processing systems.<br>2. The millions of existing text entry devices must be supported.<br>3. There must be no character set dependency, as documents might be keyed on a variety of devices.<br>4. There must be no processing, system, or device dependencies.<br>5. There must be no national language bias.<br>6. The language must accommodate familiar typewriter and word processor conventions.<br>7. The language must not depend on a particular data stream or physical file organization.<br>8. "Marked up" text must coexist with other data.<br>9. The markup must be usable by both humans and programs. | 1. XML shall be straightforwardly usable over the Internet.<br>2. XML shall support a wide variety of applications.<br>3. XML shall be compatible with SGML.<br>4. It shall be easy to write programs which process XML documents.<br>5. The number of optional features in XML is to be kept to the absolute minimum, ideally zero.<br>6. XML documents should be human-legible and reasonably clear.<br>7. The XML design should be prepared quickly.<br>8. The design of XML shall be formal and concise.<br>9. XML documents shall be easy to create.<br>10. Terseness in XML markup is of minimal importance. |

It was defined separately in what was called a *Document Type Definition* (DTD). Secondly, the DTD included a descriptive lexical and syntactical model that defined how the data was to be recorded, archived and distributed.

Working drafts were published between 1980 and 1983. In 1983, the Graphic Communications Association (GCA) produced the first SGML recommendation. It was adopted by the US International Revenue Services and the US Department of Defense (DoD). The International Organization for Standardisation (ISO), too, became interested. It started a working group on SGML (ISO/IEC JTC1/SC18/WG8, now equivalent to ISO/IEC JTC1/SC34). This led to an international standard in 1986 (ISO 8879: 1986). An amendment was issued in 1988 (ISO 8879: 1988).

The 1988 version remained stable for eight years. In that period, ISO also published a number of SGML-related, supplementary standards. We mention two important ones. The first is the *Hypermedia/Time-based structuring language* (HyTime, ISO/IEC 10744:1992), a standard that addresses hypermedia relations. It offers a rich model for addressing and linking SGML documents as well as other type of information objects. Another important standard, called the *Document Style Semantics and Specification Language* (DSSSL, ISO/IEC 10179.2:1996) addresses styling. It specifies rules for transforming and formatting SGML documents. Furthermore, various tools and applications were created. Because the SGML concept was based on process-independent document structures, the same data in SGML documents could be understood by, for example, database and text processing tools. The range of SGML supporting tools included word processors, parsers, transformers, publishing engines, browsers, document management systems, and even dedicated programming languages and libraries. Areas of application included publishing (e.g. so used by the American Association of Publishers, IBM, and the US Department of Defense in the CALS initiative), text research (Text Encoding Initiative), and the exchange of product information (Society of Automotive Engineering J2008).

One of the important uses made of SGML was the HyperText Markup Language (HTML). It was developed by Tim Berners-Lee (CERN, and founder of W3C) for the World Wide Web, and first standardised by the IETF in 1995 (Berners-Lee & Connolly, 1995). HTML did not start out as a fully SGML-compliant application. It complied from the second version onwards. Many of the rules imposed on SGML documents were not –and are still not– enforced by browsers for HTML documents. Most browsers even accept and process invalid HTML documents.

## 3.2 XML

The W3C installed the SGML Editorial Review Board (ERB) in 1996 to develop XML (Connolly, 1997). Its members all had SGML expertise. Many also participated in SGML(-allied) ISO working groups. Apart from bringing the power of SGML to the web (XML), the ERB aimed to develop specifications for 'XML hypertext link types' and for DSSSL use in an Internet context.[12]

The review board became a regular working group (XML WG) the year after. Microsoft, one of the three active members of the XML WG, was an early adopter of XML for Internet Explorer. Netscape, likewise an active member, supported XML at a later stage. Together, these two companies covered a large share of the HTML market, which is of interest because at the time the web-browser was the main platform for XML document exchange. The W3C recommendation for XML 1.0 was published in February 1998 (Bray, Paoli & Sperberg-McQueen, 1998).

A wide range of XML applications, tools and standards has emerged since. Presently, the number of public applications exceeds 250. They address very different areas: publishing, electronic data interchange (XML/EDI), data modelling (UML/XMI), workflow management (WfMC), software engineering (SOAP), and so on. The functionality offered by XML-based software tools is equivalent to those for SGML. But, firstly, the advent of web content delivery, and the emergence of XML servers and middleware has led to additional XML functionality. Secondly, many libraries and XML extensions to existing programming environments have become available. Thirdly, the number of W3C XML-based specifications and standards by far exceeds those for SGML. W3C has produced additional recommendations on naming (namespaces), normalisation (XML information set), transformation (XSLT), publication (XSL, Associating style sheets), implementation (DOM), addressing (Xpath) and linking (Xlink) of XML documents.

### 3.3 Efforts to Create a Compatible Successor

The participants in XML development were SGML experts. They partly were or had been active in SGML or SGML-allied standards developments (e.g. DSSSL-O), and often knew each other from, for example, GCA conferences. The constituency of W3C's working group and JTC1's WG8 overlapped. Because of overlapping membership there was reciprocal influence. However, there was also a degree of group identification (we-them)[13] and standards politics (e.g. personal differences and the Not-Invented-Here syndrome)[14] .

When the W3C's working group started, it was clear that "(...) the ultimate goal of this effort is the creation of a form of SGML that can be used to transmit documents (or document fragments) to a future generation of Web browsers and similar Internet client applications."[15] But whether this XML would be an SGML subset, a derivative, a conformance level, or an application profile was not yet decided and, as the chair of the working group writes, "our uncertainty has two levels: we're not sure where the optimum balance is between SGML compatibility and ease of implementation as a general goal, and we're not sure which specific features of

SGML should be retained in XML. (...)"[16] The starting point was, that XML would be compatible with SGML. That is, existing SGML tools should be able to read and write XML data, and XML instances were to be SGML documents without changes to the instance.[17]

Overlap between the constituents of the W3C and the JTC1 working groups kept the compatibility intent alive. In September 1996, soon after the electronic discussion list of the W3C working group started, Eve Maler posted a contribution which illustrates some of the compatibility concerns and dilemmas that were at stake[18]. For example,

*Who is the customer/audience for XML --existing robust-SGML users, existing Web/HTML users who are not SGML-aware, or both? (...) I'd rather think of XML as an effort to define a cohesive SGML 'application profile' that benefits both tool creators and document creators, rather than a set of unrelated cool hacks that make it easier to write parsers. (...) What should happen when existing SGML documents (including valid HTML) are processed by XML tools? Should a 'round trip' between the two forms be possible, or is only XML->SGML or SGML->XML okay?*

Partly these were resolved. Some were impossible to resolve satisfactorily.[19] The outcome was a largely but not fully aligned XML specification in respect to SGML (1988). That is, the XML Recommendation included a non-normative part which, if implemented, "increased the chances" of XML-SGML interoperability. But it provided no guarantee. JTC1, on the other hand, developed a new version of SGML (SGML 1999) to re-established compatibility. (For a technical discussion of the SGML-XML relationship and efforts to re-forge compatibility between them, see Egyedi & Loeffen, Appendix II.)

## 3.4 Analysis of Standardisation Outcome

Why was XML not fully compatible with SGML (1988), as had been the aim? Did XML represent a paradigm shift? What sort of causes led to discontinuity in standardisation?

### 3.4.1 Paradigmatic Elements

To briefly refresh the reader's memory, a *paradigm* is a set of shared views, heuristics, exemplars, etc. that guide and structure the way a practitioners community normally solves its problems. Which paradigmatic features structure the way SGML practitioners (i.e. standards developers and implementers) work? That which characterises the SGML problem and started things off was the need to reuse information fragments and share documents across publication systems in a future-proof way. IBM addressed the problem by separating the syntactical and the logical document structure. It determined – as it were – the sort of answers with which to solve the puzzle and laid the fundaments for the SGML approach. XML developers were raised with the principles of SGML. SGML was a technical exemplar and an exemplar in respect to the functionality it could offer: the identification, exchange and reuse of information fragments in different contexts. XML, too, was initially document-oriented.

Furthermore, in discussions XML was called a "lean-and-mean dialect of SGML"[20]. It was to become a simpler version of SGML. It could become so, for example, because different from SGML it could refer for character sets to Unicode and the ISO 10646 standard. Simplifications like these emphasise continuity rather than deviation from SGML features. Except for the DTD-less document, which we would typify as a shift within the SGML paradigm, the general SGML mindset and strategies also apply to XML.

### 3.4.2 Causes for Discontinuity

*Context of Use.* What explains the discontinuity between the standards successors? Firstly, the web-based context of use had little in common with the context of SGML in the 1980s *(technological anachronism)*. The technology of the 1990s offered new opportunities and posed different constraints. The domains of use shifted together with the practitioners involved *(migration of use)*. See Figure 3.1. Although the information modelling approaches of SGML and XML were in principle identical, the SGML problem was foremost how to manage the *company-internal*, complex flow of documents. XML, on the other hand, developed as a solution to the limitations of HTML in respect to *company-external, web-oriented* document exchange. See Table 5.1.

*Context of Standardisation.* Some thought the XML market would only be of interest to SGML users[21] . Others hoped to target the huge, well-funded, energetic Web population.[22] There would be an important marketing advantage in being able to say "XML processors can read HTML".[23] Therefore compromises to XML

*Figure 3.1. A schema of the relative importance of domains of use in SGML and XML*

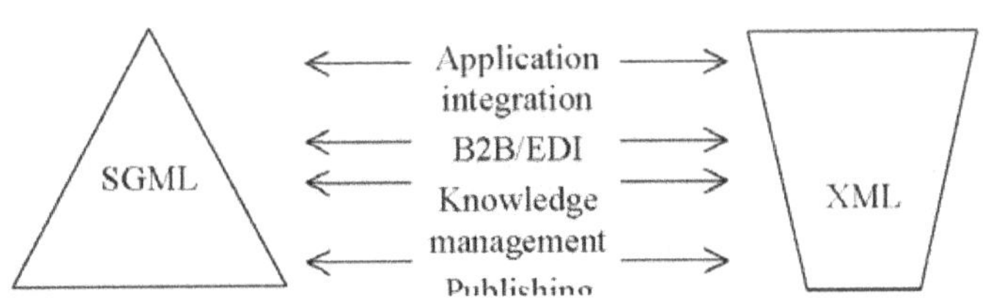

*Table 3.1. Causes for discontinuity: Differences between the problems, context of use, and context of standardisation of SGML and XML.*

| Causes for discontinuity ↓ | SGML | XML |
|---|---|---|
| Information problem | | |
| • Orientation | Company-internal | Company-external |
| Context of use | | |
| • Technology | 1980s (mainframes etc.) | 1990s (Internet, chips etc.) |
| • Domains | Publishing | B2B, application integration |
| Standardisation | | |
| • Frame of reference | GML | SGML, HTML |
| • Standards body: culture | ISO: stability, accountability | W3C: pragmatism, speed |
| • Problem, emphasis on | Ubiquitous applicability | Simplicity, implementability |

compatibility with SGML were considered that left the installed base of HTML untouched. The deliberations are well illustrated by the following quotation, and explains part of the discontinuity in the SGML trajectory.

*(...) For the 99% of the world that doesn't care a bit about SGML (...). They know HTML, so we must make things look like HTML. But when it comes to adding the important things that HTML doesn't have, we should make them as attractive as possible. (...) The SGML folks need a standard, as well as capability so they will continue to need SGML. But for the rest of the world, clean extendible markup is the biggest need, not SGML compatibility.*[22]

However, more influential was the successful spread of HTML use. It was an exemplary achievement in the eyes of XML developers. Its implication for standardisation was: aim for simplicity. If we compare the SGML aims with the design goals of XML, the latter's emphasis on ease of implementation and usability is salient (See Box 3.1). Simplicity was difficult to align with compliance to SGML.

## 3.5 Conclusion

This case study focused on XML as a successor of SGML (1988). It shows that a discrepancy existed between the claimed compatibility between XML and SGML, and the actual outcome of XML standardisation. XML deviated from the decade-old stable SGML trajectory. This did not occur because of any paradigmatic change in the way SGML principles were used in XML -although the abandonment of DTDs was a revolutionary step. There were other reasons. Firstly, XML's context of use differed from SGML's. It was *company-external and web-oriented,* whereas SGML foremost had a company-internal focus. Secondly, the HTML context of standardisation played a role. The spread of HTML use was exemplary for XML developers. The message was: aim for simplicity in standardisation, which at times conflicted with aim of compatibility with SGML. In other words, two exemplars guided XML development and coloured its heritage relation with SGML: SGML, which was a technical exemplar and an exemplar in respect to the functionality it could offer, and HTML, which was a standardisation exemplar in terms of its simplicity and widespread diffusion. In respect to the type and setting of standardisation, therefore, the W3C consortium headed by the pragmatic developer of HTML (Tim Berners-Lee) was a more likely choice for XML standardisation than the more formal ISO. This change of institutional setting made it easier for XML developers to deviate for practical or other purposes (e.g. Not-Invented-Here) from standard SGML solutions developed in JTC1. Incompatible succession is then more likely to occur.

## 3.6 Case-Specific References

Berners-Lee, T., & D. Connolly (1995). *Hypertext Markup Language -2.0*, RFC 1866, IETF. Bray, T., J. Paoli & M. Sperberg-McQueen (1998). *Extensible Markup Language (XML) 1.0. W3C Recommendation,* 10 February 1998. Connolly, D. (ed.): 'XML. Principles, Tools and Techniques'. *WWW Journal 2 (4)*, fall 1997. Goldfarb, C. (1990).

*The SGML handbook.* Oxford: Oxford University Press. ISO/IEC 10744:1992. Hypermedia/Time-based structuring language (HyTime).

Geneva, 1992. Revised edition (ed. 2), 1997. ISO/IEC 10179.2:1996. Document style semantics and specification language (DSSSL). International Standard, Geneva, 1996. Raggett, D., A. Le Hors & I. Jacobs (1999). *HTML 4.01 Specification.* W3C

Recommendation 24 December 1999. W3C (1996 -). W3c-sgml-wg@w3.org Discussion List. XHTML™ 1.0: *The Extensible HyperText Markup Language. A Reformulation of HTML 4 in XML 1.0.* W3C Recommendation 26 January 2000.

## ISO/IEC JTC1/SC18 Documents

[N1855] Document Processing and Related Communication— Document Description and Processing Languages, *Third Interim Report on the Project Editor's Review of ISO 8879,* SGML Review Group, WG8, 24 May 1996. [N1893] *Fourth Interim Report on the Project Editor's Review of ISO 8879*, SGML Review Group, 23 December 1996. [N1925] Report of the SGML Rapporteur Group (Barcelona Meeting), WG8, 9 May 1997. [N1929] Draft SGML TC2, WG8, *Proposed TC for*

*WebSGML Adaptations for SGML*, 1 June 1997, Annex K (normative), WebSGML Adaptations. [N1955] *Final Text of SGML Technical Corrigendum*, WG4, 'ISO 8879 TC 2', 4 December 1997)

## PART II: ANALYSIS

EU policy on consortium standardisation, or the lack of it, is partly based on some unchecked -but widely shared -assumptions and beliefs. In this part of the report these are confronted with the findings of the case studies and other, additional sources. Examples are discussed to underscore why 'the consortium problem' needs to be redefined. The current policy framework is examined, and a new one is developed. Specifically, Chapter IV addresses consortium standardisation, compares it to formal standardisation, and lays the basis for recommendations on this point. But the insight gained by the cases reaches beyond consortium standardisation. In Chapter V, consortium standardisation is discussed within this wider context.

## 4. CONSORTIUM STANDARDISATION

In so far as consortia focus on specification development, they are seen as rivals of the formal standards bodies (e.g. CEN/ISSS, 2000). They compete on different dimensions. For example, they compete for committee participants. Standards consortia are thought to drain away the scarce technical expertise needed in the formal standards committees, and some of the large industrial companies who tend to volunteer for the required committee secretariats. A certain degree of competition would therefore seem plausible. However, there is little research on the matter. What information there is, for example, in respect to whether or not the rise of standards consortia has led to a decline of industry participation in formal standardisation, is inconclusive (Hawkins, 1999; Cargill, 2000).

Lack of hard data also exists in respect to the two dimensions on which this chapter focuses, and on which standards consortia and the formal standards bodies are most often compared: democracy and time. Standards rhetoric refers to them in terms of the 'consensus versus speed' dilemma. In the past, the formal standards processes have often been criticised for being slow and bureaucratic. This was later seen to explain the rise of consortia in the early1990s, which were significantly more effective in developing standards. In response, the formal standards bodies introduced many new measures and procedures to speed up the process (Egyedi, 1996). However, the criticism held on. Indeed, the rationale behind the age old criticism still seemed to apply: consortia produce specifications quicker than the

*Table 4.1. Schema of the rhetorical basis of the framework for current standards policy: The formal standards bodies are slow but operate democratically, while industry consortia are undemocratic but effective in developing standards.*

| Speed > Degree of democracy | Slow | Fast |
|---|---|---|
| Democratic | Formal standards bodies | |
| Undemocratic | | Standards Consortia |

formal bodies do because they need not bother with the lengthy democratic and open process, with consensus decisions, with a well-balanced representation of interest groups, etc. (Meek, 1990; Rada, 2000). Because for consortia consensus is not a main issue, the reasoning goes. Indeed, the CRE report (2000) states that consortia deliver *non*-consensus specifications. Therefore, among other advantages[24], consortia need not compromise on standards content as much as the formal standards bodies do; they can operate in a more timely manner; and they can therefore better cater to the needs of time-sensitive technologies. Such is currently the dominant view. See Table 4.1.

This view is not self-evident. It contains several values and assumptions that are a matter for debate. For example, 'democracy' is not a concept well-understood in relation to the standards setting. Its meaning and significance for standardisation has not been held to closer scrutiny.[25] Its implications for standardisation have therefore largely remained unspecified.

The way 'democracy' is defined is of immediate relevance to the 'consensus versus speed' assumptions that lie at the basis of -among others -EU standards policy, which are

1. Formal standardisation proceeds much slower than consortium standardisation.
2. Formal standardisation is democratic in theory (i.e. standards procedures).
3 Formal standardisation is democratic in practice.
4 Consortium standardisation is undemocratic in theory (i.e. standards procedures).
5. Consortium standardisation is undemocratic in practice.
6. A generic standards policy well-serves both public and market interests.

*Speed dimension.* The first assumption addresses the speed dimension. Although very little quantitative data exists on the matter, at present doubts are being raised about the timeliness of consortium standardisation (Hawkins, 1999; Krechmer, 2000). The two cases discussed in Part I are not very helpful in this respect. In a sense, the Java/ECMA case confirms these doubts, since Java standardisation never really took off. Moreover, as was noted in Chapter II, in order to allow ECMA standards to be fast-tracked through the formal JTC1 process, ECMA procedures need to be compatible with JTC1 procedures. This comes with a certain degree of bureaucracy. Therefore, had Java standardisation proceeded, it would have been submitted to a certain degree of bureaucracy too. More in general, no significant difference in standardisation pace is to be expected between formal standards bodies and consortia that have close ties with the formal standards bodies. In the W3C case, the initial draft XML standard took 11 weeks to develop (September-November 1996)[26], which is very fast. (Of course, this high pace was made easier because the XML working group built upon ISO's decade old SGML technology, and had gained ample experience with SGML.) The standard was published in February 1998, 15 months later.

The two cases cannot confirm or disaffirm recent doubts about the high speed of consortium standardisation. Quantitative research is needed to provide a definitive answer. What about formal standardisation? Do the formal standards bodies still merit criticism for their slow pace? Since the mid-1980s many different measures have been taken at the European and international level to speed up standardisation[27]. To give an impression of the pace at that time and the changes that followed:

- According to the ISO/IEC Directives, 7 years were needed to take a standard from start to Draft International Standard in 1989, while in the Directives of 1992 this period had been reduced to 3 years.
- The approval time of recommendations in the ITU-T has been reduced from 4 years in 1988 to a maximum of 9 months in 2000 (ITU-T, 2000).

Other important innovations were the Fast-Track process and the PAS procedure of ISO/IEC (see Chapter II). These last years have seen, for example, an increasing amount of web-based information exchange among committee members; pre-standardisation initiatives for the timely development of industry specifications (e.g. CEN workshop agreements, CWAs; ISO-organised Industry Technical Agreements, ITAs); and pre-publication of standards on the website of the standards sector of the International Telecommunications Union (ITU-T). [28] Therefore, already if one only takes the changes applied by the formal bodies into account, the difference in pace of standardisation with standards consortia has been fast diminishing. See Table 4.3. The arrows in the 'speed dimension' indicate that consortia take longer than would be expected according to the dominant rhetoric, while the formal standards bodies operate faster than would be expected.

Below, the democratic dimension of the 'consensus versus speed' rationale is examined. First supposed lack of democracy in standards consortia is addressed and cross-checked with the two consortium cases (section 4.1). Next a comparison is made with formal standardisation. Again, this is to determine whether both standards environments are as different as standards literature and standards policy would have it (section 4.2). Having compared both settings on the dimensions of speed and democracy, in the final section of this chapter the more fundamental question is posed whether this comparison has actual relevance for EU standards policy. It is related to the sixth assumption listed above, which will be introduced in more detail in section 4.3.

## 4.1 Consortia: Democracy and Openness

Lack of "openness" and "democracy" sum up the main problem of governments with consortia. I take the terms to refer to a want in membership procedures (i.e. lack of diversity among committee participants and the exclusion of certain groups) and in decision making procedures (i.e. not consensus-driven and no provisions for the inclusion of minority standpoints), respectively. Is this perception of consortia correct? In the following, first the standards procedures (i.e. theory) are examined, and then the way they are applied (i.e. praxis).

### 4.1.1 Undemocratic Procedures?

The focus is, again, on the two cases of ECMA and W3C. As an illustration, their membership and decision procedures of two consortia are discussed in more detail.

*Membership.* ECMA and W3C have several membership categories. Apart from the usual industry members, small and medium-sized Enterprises (SMEs), and governmental and educational institutions can participate.[29] The consortia demand membership fees. Members pay a fee according to the membership category they are in. Only Ordinary ECMA members (i.e. full membership fee is 70,000 Swiss Franc) have a vote in the General Assembly. W3C bestows no extra voting benefits on full members ($US 50,000 per year).

*Decision procedures.* The decision structures of ECMA and W3C differ strongly. In ECMA the ultimate power lies in the hands of the General Assembly (GA). But only ordinary members have a vote in the GA. In other words, there is full democracy among members that pay the ordinary membership fee (large companies). In W3C, the ultimate power lies in the hands of the director, who formally has the role of a benevolent dictator. In other words, there is no democracy at this level. Indeed, where the chairs of W3C technical committees are concerned, 'if it is necessary to move

*Table 4.2. Characterisation of the two consortia in respect to their (un)democratic approach at the level of the standards committee and the overall organisation based on their respective procedures.*

| Consortia Degree of democratic decision making within ... | ECMA | W3C |
|---|---|---|
| Standards consortium | democratic (General Assembly) | undemocratic ('benevolent dictator') |
| Standards committee | consensus-oriented | consensus-oriented |

on' this person also has far-going powers. However, the W3C standards process is consensus-oriented, and strives for a vendor-neutral solution and an open process, according to the W3C rules. Some procedures give room to minority standpoints. W3C's combination of these democratic ideological features with 'dictatorship' leads to an interesting mix of procedural directions (W3C, 2001).

ECMA procedures are internally -ideologically -consistent. The process should be consensus-oriented and give room to minority standpoints as well. Where W3C's director appoints committee chairs, in the ECMA committee members elect the chair. If there is a deadlock in an ECMA committee, voting takes place by simple majority of the members present at the meeting. In both W3C and ECMA, each member may assign several representatives to participate in the standards work, but the company has only one vote. See Table 4.2.

The two cases do not simply confirm the widely shared bleak picture. On the contrary, although there may be practical and organisational exclusion mechanisms (e.g. membership fee[30]), in principle consortium membership is open. Indeed, in certain respects they appear more open than the formal bodies. For example, while the latter usually keep access to committee drafts restricted to participants -and, thus, seek consensus within a limited group -consortia more often post their drafts on the web and actively seek comments from outside (Rada, 2000).

The W3C case indicates that the procedures of -some -consortia allow them to be reigned in an autocratic manner. However, at committee level their procedures embody the same values as those of the formal standards bodies (i.e. strive for consensus, address minority viewpoints, etc.). There is one exception: consortia do not explicitly include in their policy the aim to involve diverse participants.

### 4.1.2 Undemocratic Practice?

The inclusion or exclusion of individuals or groups can differ during and after the standards process. Unlike the formal standards bodies, participation in the consortium standards process is not automatically linked to standards use. Often it may not be in the interest of early members of a consortium committee to seek additional committee members from other interest groups since this often makes the standards process more complex. Therefore, depending on the market, an exclusive standards process might be preferred during this phase. However, once the standard is finished -to satisfaction -committee members have an interest in the diffusion of its use. The inclusive phase begins. The attention of other market players must be directed towards the new standard in order to heighten interoperability among multi-vendor products.

An exclusive phase followed by an inclusive phase is normal practice where proprietary specifications become de facto standards (e.g. Java). However, with regard to consortium standardisation, pragmatism and instrumentalism rather than principle may sometimes exclude parties during standardisation. Indeed, exclusion does not typify the standards process in the two cases addressed in this report.

- The W3C/ XML working group primarily used an easily accessible discussion list for standards development, where technical as well as strategic and pragmatic questions were raised. Contributions to the discussion list give the impression that individuals rather than company representatives are participating. Openness and inclusion seem to characterise the standards process as well as the diffusion phase.
- Only two plenary face-to-face meetings had been held when Java standardisation in ECMA prematurely stopped[31]. Those who attended partook as company or organisational representatives. Interestingly, a large group of business users (SHARE) was represented in the technical committee. The woman in question was chair to one of the three task groups, a very inclusive move towards users.

In sum, consortia do not automatically link more and diverse participation in the standards process to wider standards use. Therefore, they do not expressly aim to be inclusive at committee level. However, in practice they may nevertheless show a high degree of inclusion (e.g. open, publicly accessible discussion lists, user participation, and user representation in key functions).

## 4.2 Formal Standardisation: Democratic Praxis?

The second assumption in the 'consensus versus speed' dilemma is that the procedures of the formal standards bodies embed democratic ideals. Content analysis of official documents confirm this assumption (Egyedi, 1996). However, are these ideals also evident in formal standards practice?

*Table 4.3. Convergence in the standards process between formal standards bodies and standards consortia along the dimensions of 'speed' and 'democracy'*

| Speed dimension -> Degree of democracy | Slow | Fast |
|---|---|---|
| Democratic | Formal standards bodies | |
| Undemocratic | 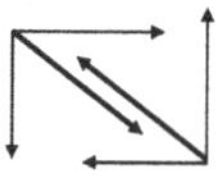 | Standards consortia |

Standards consortia typically have an informal manner of handling procedures (Hawkins, 2000). That is, theory and praxis is likely to diverge lightly. The significance of the procedures of formal standards bodies, too, must not be overemphasised. As Schoechle's (2001) tale about the Emperor and the Professor illustrates, politically correct procedures are at times leniently applied or downright misused.

If we look at the inclusive procedures of the formal standards bodies, do they lead to a greater amount of or higher quality consensus decisions, or to a higher diversity of participant categories? The formal standards bodies have not systematically monitored aspects of democracy and openness characteristics of their technical committees. To my knowledge, regrettably, no other quantitative data exists on the matter. Therefore, in this respect the experiences of individuals are an important main source. According to such sources, the democratic aims of the formal process are not met (Cargill, 2000). Often the formal standards process is an exclusive one. Generally, the participation of end-users and SMEs is very low (Jakobs, 1999); formal procedures are often exploited in 'undemocratic' ways (e.g. staging a voting during Christmas holiday; Egyedi, 2000b), or redefined (as when e.g. US multinationals participate in European national delegations; Cargill, 1999). The regional governments and formal standards bodies are well aware that in formal standardisation the objectives of democracy, diversity and openness are often not met. Under the present conditions, it would be as difficult for the formal bodies to amend this situation as it would be for consortia to meet such objectives.

In fact, the formal standards bodies and standards consortia are rather similar in other respects as well. For example, in respect to Java, JTC1 and ECMA dealt with Sun in much the same way. The same parties participated and the same issues arose. In both fora, Sun insisted on 'edition, not addition'. This illustrates that 'rubberstamping' - which is a crude generalisation of the work which 'edition' would nevertheless entail -is a practice which takes place in standards consortia[32] as well as formal standards bodies.

In sum, although consortia more explicitly target industrial parties, in practice the formal standards bodies and standards consortia include and exclude the same constituencies. Like the formal bodies, they also strive for consensus, address minority viewpoints, etc, that is, they share the values of the formal ideology at committee level. However, unlike the organisational set-up of the formal bodies, there is much variation in who is the ultimate gatekeeper, an individual director, like in W3C, or a collective, like the General Assembly in ECMA.

## 4.3 European Union: Single Policy - Two Areas of Interest

The conclusion is that by and large committee standardisation in consortia is (a) not undemocratic in theory (striving for consensus & accounting for minority standpoints), and (b) as undemocratic as formal committee standardisation in practice. Moreover, the alleged difference in pace between the two types of standards fora seems to be based on dated information about the formal standards bodies, and is overrated. Since both standards development environments increasingly seem to show similar characteristics in theory (procedures) and in practice (committee process), has the distinction made by standards policy become arbitrary? Before answering, let us return to the central issue.

In Chapter I, the following reasoning was introduced: formal standards are an important point of reference for European regulation in particular. The European Commission requires a degree of democratic accountability if it is to refer to such standards in a regulatory context. However, in the field of ICT standards have emerged with a high market relevance, standards that stem from standards consortia. How should these standards be dealt with? According to the previous discussion, one could reason, the Commission can take the pragmatic route and include consortia as a source of standards. For in democratic respect the consortium standards process does not differ notably from the formal one.

This answer, however, does not take into account that, firstly, in the field of ICT, on which this report focuses, standards will seldom be part of regulation. Overall, the regulator's need for democratic accountability of ICT standards processes - whether it takes place in a formal standards body or in a standards consortium -is absent. Secondly, where such a need does exist, should not the regulator's concern for democratic accountability be given more substance? As a first step, more clarity should be created by the European regulator about what type of democracy is needed and for what purpose. The second step would be to monitor systematically if 'democratic' standards developing organisations follow up the democratic requirements, be they formal standards bodies or consortia.

The two previous points already more or less argue the case that a comparison between and a polarisation of formal and consortium standardisation has little actual relevance for European standards policy. The next and third point takes these arguments one step further. It is related to the sixth assumption that underlies European standards policy: "A generic standards policy well-serves both public and market interests." Much of what happens in standards committees is of little interest to the Commission[33]. The standards have no relation with issues of health, safety, environment, privacy etc. The latter issues concern the general public, are subjects likely to be addressed by regulation, and therefore typically need monitoring by the European Commission. A two-fold standards policy for regulation-related, and other, mostly market-related standards would seem appropriate. Thereby the requirement of democratic accountability would only apply to standards processes that include public interest issues addressed by regulation. These processes would be closely monitored. For all other standards, no changes would be need to be made. Both formal standards bodies and consortia could apply for the status of 'accredited, monitored democratic standards development environment'. In this respect, the ITU-T has recently introduced an Alternative Approval Process for all its standards, except for those which have policy or regulatory implications. The latter still need to undergo the Traditional Approval Process (ITU, 2000).[34] The reverse would be at stake for public interest standards. A specified concept of democracy would need to lay the basis for a well-monitored type of democratically accountable standardisation.

## 5. STANDARDISATION AND OTHER COMPATIBILITY STRATEGIES

Instead of fuelling rivalry and pitting standards environments against each other, it would be more appropriate to underscore the common ground between them[35]: the belief that in certain instances standardisation is the preferred mechanism for coordination of the ICT market. In areas where coordinative action is needed, any multi-party standard -whether of consortium or formal origin -will be preferred to no standard at all. Or are there other, more effective means of coordination apart from standardisation which also enhance compatibility among products and services?

With regard to compatibility standards[36], policy developers are prone to lose sight of the aim of compatibility[37] and focus more narrowly on the means instead, namely on committee standardisation[38]. The latter is a means to co-ordinate the activities of parties that compete in the market (Schmidt & Werle, 1998; Weiss & Sirbu, 1990). Ideally, the resulting standards become the shared basis for compatible implementations. However, standards do not guarantee compatibility. Whether compatibility is achieved de facto, depends on the scale and manner in which standards are implemented. Little information exists on this aspect of standardisation[39].

Ultimately policy interest should focus on technical compatibility and on compatible implementations. These outcomes can be contributed to and achieved by other means than standardisation as well. Compatibility often results as a by-product of market dominance. In the field of computer software, for example, de facto standards such as PDF and UNIX have emerged that sometimes more forcibly induce widespread compatibility than some committee standards do. The origin of these de facto software specifications differs. Some result from in-company R&D efforts, others from various forms of co-operation between multiple parties. The type of specification development process need have no bearing on how the ownership of the specification is handled. A company may keep the proprietary technology for itself. It may, for example, monopolise the production of a key component, and define an interface which ties complementary products of other firms to the proprietary component technology (David & Greenstein, 1990). However, it may also give away its technology with an eye to expected long-term advantages, or enter into coalitions with rivals to enlarge its user base and increase support for its technology. At present, the practice of giving free access to the software source code is gaining interest. Well-known examples are Linux, TCP/IP, SMTP, DNS and C. An 'open source-mindedness' seems to be developing, also among commercial companies (O'Gara, 2000).

Whatever ownership approach is used -proprietary, open source or other approach -a sizeable market share may result. If a software specification acquires market dominance, we retrospectively speak of 'de facto standardisation'. This would seem to refer to a standards process[40], whereas it actually refers to a process of software diffusion. The term 'de facto compatibility' would better express the relevance of the compatibility outcome. Compatibility is in this situation a by-product of a successful product development and diffusion -which *may* include compatibility aims -rather than the outcome of a proprietary or multi-party standards trajectory. See Table 5.1. This table compares committee standardisation with software development. Both are specification processes. How the specifications are developed -multi-party or by a company -need have no direct effect on their use. If a committee standard is implemented widely the consequence is similar to that of a dominant product specification: de facto compatibility results. For compatibility can only be measured as an aspect of the market, and not as an aspect of the specification process.

*Table 5.1. The two specification processes that may lead to de facto standards in software (i.e. compatibility)*

| Stages > Type of Specification Process | Specification Process | | Market Process | |
|---|---|---|---|---|
| | Participation | Outcome | | |
| Committee Standardisation | Multi-party | Standard | Implemented widely? | Yes > de facto compatibility No > local or no compatibility |
| Software Development | Multi-party | Specification | Market dominance? | |
| | In-company | | | |

## 5.1 Other Compatibility Enhancing Measures: Sun's Java Strategies

There is a wide diversity of strategies which companies may use to enhance software compatibility. Sun Microsystems illustrates this in respect to its de facto Java standard. To briefly refresh the memory, Java is a middleware technology for which compatibility, and in particular the integrity of the Java platform, is crucial. Sun has therefore introduced several complementary, compatibility-enhancing measures such as the Java programmer certificate and the Java Compatibility Logo (Egyedi, 2001a).

Firstly, Sun started by giving interested parties access to its source code. It invited developers to comment on, experiment with and improve the original source code. Since the developer community worked with the same source code, this open source approach had a light coordinating effect. Likewise, the instructional books which Sun authors and others wrote on Java, and the training programs that led to 'certified' Java developers added to a coordinated Java trajectory. These activities fortified the development of a Java practitioner community, but they did not guarantee compatible Java implementations. More effective in this respect was the free distribution of the Java Software Development Kit (SDK, formerly JDK). The SDK contained a full suite of development tools to write, compile, debug, and run applications and applets. Use of the SDK narrowed down the number of possible programs to those that would run on 'standard' Java platforms.

Secondly, another means to enhance compatibility is IPR licensing. Sun's licensing policy was based on its intellectual property, which it protected by means of trademarks (Java name, Java-Compatible logo), patents (software algorithms) and copyrights (specifications). Part and parcel of Sun's licensing policy were its test suites. These were used to certify compatible Java products. They gave licensees access to the 'Java compatible' logo (the steaming cup of coffee). The logo had much goodwill among Java programmers. Different from e.g. the strategy of encouraging the use of the SDK, which is a means to coerce developers towards compatible implementations (compatibility push), the logo created an incentive towards developing compatible commercial products (compatibility pull).

The Java Community Process (JCP) was a third means to co-ordinate the development of Java. Together with a new licensing model (Community Source License), Sun issued *The Java Community Process (sm) Program Manual: The formal procedures for using the Java Specification development process* (1998, 2000). The last and most evident compatibility strategy to mention here are Sun's two failed standardisation attempts in JTC1 and ECMA (see Chapter II).

Sun applied these compatibility strategies sometimes successively, sometimes simultaneously, sometimes in combination. Some strategies were focused (e.g. the

*Table 5.2. Overview of Sun's compatibility-enhancing measures regarding Java*

| Sun initiatives that contribute towards Java compatibility | |
|---|---|
| Input Control | Output Control |
| • JTC1 Standardisation<br>• ECMA Standardisation<br>• Instructional books, conferences, certified training programs, etc.<br>• Open source code<br>• Distribution of the Java Software Development Kit | • Java Community Process (incl. Participation Agreement) •Reference Implementations<br>• IPRs<br>• Licenses<br>• Technology License and Distribution Agreement<br>• Sun Community Source Licensing model<br>• Test suites & Compatibility logo |

Java Software Development Kit), others were more diffuse (e.g. the Java Community Process); some were forceful (e.g. licensing combined with IPRs), while others were more coercive (e.g. compatibility logo); some strategies were based on proprietary control (e.g. IPRs), while others primarily aimed to broaden the Java user base (standardization); some strategies primarily aimed to direct the attention of market players towards Java developments (e.g. standardization), while others specifically targeted technical compatibility (IPRs). Its strategies are listed in Table 5.2.[41] The table distinguishes between compatibility strategies that control the initial phase of a specification process (what goes in: input control) and those that control its outcome (output control). The specification process at stake can either be a software development process or a standards process. For example, Sun's problem with JTC1 and ECMA standardization was, partly, that while it could control its own standards input (i.e. the Java specification), it could not control the outcome of the standards process. Therefore its standardisation attempts are listed in the left column. A more output-oriented means to control Java compatibility (right column) was the use of Sun test suites to determine Java compatibility.

## 5.2 Why Choose the Standards Strategy?

Why would a company want to initiate standardisation if there are other, possibly easier means to achieve compatibility, or if, as in the case of Java, de facto compatibility already exists? Firstly, the straightforward rationale for approaching a standards body is to foster technical compatibility among existing or future products and services. The resulting standard then serves as a concrete means to co-ordinate product and service development of different producers and providers. Secondly, companies also profit from another level of co-ordination. Economic studies have pointed out the relevance of compatible product pre-announcements (Farrell & Saloner, 1986) and 'embrace-and-extend' strategies regarding standards (Vercoulen & van Wegberg, 1999). These aim to direct the actions and orientations of other market players. They address the strategic level of market co-ordination and complement the operational level of technical compatibility.[42] In sum, the standardisation process unites two complementary co-ordinative functions: technical compatibility and market co-ordination. Both aspects to committee standardisation serve as an *ex ante* mechanism for structuring the market.

Turning to our example, at a very early stage of Java development Sun had announced that it intended to standardise Java. If Java were to become an International Standard, this would signal stability in Java development, increase market confidence and encourage commitment to Java. It needed the commitment of industry to counteract possible competing Java developments and prevent fragmentation of the Java market. This early promise was, similar to the effect of product pre-announcements, a means to keep Java programmers and competitor companies focused towards Sun-driven Java initiatives. The step towards formal JTC1 standardisation in 1997 was in line with the promises made. Since Java was already a de facto standard, Sun wanted JTC1 -and later ECMA -to ratify its Java specification, not change it ('edition, not addition'). It did not seek committee standardisation in the usual sense. In other words, Sun's actions mainly targeted market co-ordination. This aim was better served by the status of formal International Standard than consortium standard or de facto industry standard, according to Sun.

Standardisation remains a well-used option. Although in Sun's case, it was not a very successful strategy, Sun's competitors in the embedded Java area, the J Consortium, have applied it more successfully. The consortium has become a recognised PAS submitter, which allows it to submit its embedded Java specifications to ISO/IEC JTC1. Its efforts triggered Sun's recent decision to adapt Java in order to better suit the requirements of embedded Java users.

## 5.3 Contribution Towards Compatibility

The list of compatibility enhancing strategies in Table 5.2 puts standardisation into perspective. Of interest is then, of course, how effective standardisation is in securing technical compatibility relative to other strategies. The question is difficult to answer because little empirical data exists. For example, little is known about the actual impact of standardisation. But some preliminary observations can be made.

The problem of all voluntary standards, formal standards and consortium standards alike, lies in not being able to enforce -partial or full -compliance to standards (i.e. little output control). Even submission to conformance testing is mostly a voluntary matter. Apart from regulatory requirements, conformance to standards depends almost solely on market-pull mechanisms. That is, demand-driven conformance to standards is needed if standardisation is to lead to actual compatibility. The demand-side of the market is reasonably developed where individuals are concerned (e.g. consumer organisations). However, mechanisms to coordinate shared consumer interests in compatibility are lacking in the post-standardisation phase. This contrasts strongly with the diversity of coordination mechanisms used by the supply side of the market. In other words, because of the voluntary nature of standards implementation (voluntary technical base) the degree of compatibility that is ultimately achieved is uncertain and non-transparent. See Table 5.3 (second column). Are other types of specification processes, possibly in combination with compatibility-enhancing measures, more effective?

Several types of specification processes are imaginable. The previous chapters mentioned four of them: pure proprietary specification development, the open source approach to software development, and the two middle-of-the-road approaches used by Sun: proprietary-led multi-party specification development and the community source approach. Let us focus the comparison with standardisation on the two most extreme strategies: the proprietary and the open source strategy. They come with different licenses (Cargill, 2000; O'Mahoney, 2000). A well-known open source license is the General Public License (GPL) used for the operating system Linux. It allows one to download Linux, and use, change and distribute adapted source code without charge. These adaptations should in turn be made available in source code. This license thus removes the incentive to turn a program into a proprietary product.[43]

The open source approach faces many of the same compatibility problems as standardisation does. Sharing the same source code need not imply a compatible technical base. The license does not diminish the incentive the change source code,[44] and -if software compatibility would be prioritised -there are no easy means to control the outcome of the open source process. Therefore, with the open source strategy compatibility is uncertain and the outcome may be diffuse -albeit transparent (see Table 5.3 column 4). Market pull is needed to maintain compatibility among software products. In contrast, the proprietary approach to specification development, where IPRs are usually kept under tight control, prescribes other players to how to deal with the specifications by means of licensing agreements. The proprietary specs start out as being compatible (controlled technical base; compatibility push) and compatibility is imposed on licensees (controlled outcome), a strategy that is usually very effective (Table 5.3, column 3).

In sum, apart from standardisation, there are other, sometimes more effective compatibility strategies which lead to de facto compatibility. Whether or not the European Commission should involve itself in compatibility strategies other than

*Table 5.3. Assessment of the type of and degree of compatibility achieved in different specification processes*

| Aspects of Compatibility | Specification Processes | | |
|---|---|---|---|
| | Standardization | Software Development | |
| | | Proprietary Approach | Open Source Approach |
| Conformance Mechanism | market pull, demand-driven | compatibility push | market pull, demand-driven |
| Ultimate Technical Compatibility of Software | uncertain voluntary techn. base non-transparent outcome | high controlled techn. base controlled outcome | uncertain shared techn. base diffuse but transparent outcome |

standardisation -or even as little as possible in standardisation -is a matter for debate. For the moment, this debate should be kept open. It should not be closed beforehand with reference to the danger of reinforcing monopolies. The Java case study suggests that in certain circumstances, the public's primary interest is in solutions that prevent unnecessary fragmentation of the market. Since there are few legal means to safeguard the public interest in compatibility, if the latter coincides with a company interest, proprietary solutions deserve consideration. At the same time, we should not overemphasise the idea of compatibility control: the case also shows that even tightly controlled compatibility strategies (e.g. licensing combined with IPRs) cannot prevent incompatible developments (Microsoft's use of Java).

## PART III: CONCLUSION AND RECOMMENDATIONS

The research on which this report is based aimed, firstly, to provide contemporary case material that illustrates how consortia work. The main material of the two case studies was presented in Part I of the report. The material partly served to answer the questions: Why is sometimes consortium standardisation initiated rather than formal standardisation? Do standards consortia work in ways that will deliver open standards? These questions were answered in Part II. A summary of the answers is given in section 6.1 and 6.2. The second objective was to examine

a. The assumptions and beliefs that are part of current understanding of standards consortia (see Chapter IV, introduction), and
b. To develop a new perspective on the significance of consortium standardisation for EU standards policy (see Chapters IV and V).

In this part of the report, the consortium problem is readdressed, and core-arguments for the proposed new policy direction is summarised (section 6.3). Overall conclusions are drawn, and recommendations are made (Chapter VII).

## 6. REVIEWING CONSORTIUM STANDARDISATION

### 6.1 Why is Consortium Standardisation Sometimes Preferred?

The case studies and literature show that standards consortia successfully market their feats. They are associated with timely standardisation and pragmatic standards solutions, despite incidental critical observations to the contrary. This, and possibly the homogeneity and suggested exclusiveness of consortium standardisation attracts companies. The two case studies further show that:

- **The consortium can be a stepping stone for and offer easier access to formal recognition of technical specifications:** Sun chose ECMA because ECMA had an A-liaison with JTC1, which gave it access to the Fast Track procedure. In the past ECMA standards had been submitted to a yes/no vote in JTC1 without any modifications, and often successfully so.
- **A consortium may be seen as equally relevant to market co-ordination as a formal standards body is:** ECMA was an open standards consortium and thus an answer to continuous pressure from licensees and real-time Java developers to open up the Java development process. Many large companies were members. So ECMA processes also promised to be relevant in respect to co-ordination of the market.
- **A consortium may represent an exemplary style of standardisation (simple and widely used standards):** For XML developers, HTML was a standardisation exemplar in terms of its simplicity and widespread diffusion. In respect to the type and setting of standardisation, therefore, the W3C consortium headed by the pragmatic developer of HTML (Tim Berners-Lee) was a more likely choice for XML standardisation than the ISO.
- **Improving a standards is easier if this takes place in a new institutional setting (consortium):** The change from SGML's JTC1 setting to the setting of W3C made it easier for XML developers to deviate for practical or other purposes (e.g. Not-Invented-Here) from standard SGML solutions.

## 6.2 Do Consortia Deliver Open Standards?

The two cases do not simply confirm the widely shared assumption that consortia are undemocratic. To the contrary, although there may be some practical exclusion mechanisms, in principle consortium membership is open. Indeed, in certain respects consortia appear more open than the formal bodies. For example, while the latter usually keep access to committee drafts restricted to participants -and, thus, seek consensus within a limited group -consortia more often post their drafts on the web and actively seek comments from outside.

The W3C case indicates that -some -consortia are reigned in an autocratic manner. However, at committee level their procedures embody the same values as those of the formal standards bodies (i.e. strive for consensus, address minority viewpoints, etc.). There is one exception: consortia do not explicitly aim to involve diverse participants.

Consortia do not automatically link more and diverse participation in the standards process to wider standards use. Therefore, they do not expressly aim to be inclusive at committee level. However, in practice they may nevertheless show a high degree of inclusion (e.g. open, publicly accessible discussion lists, user participation, and user representation in key functions). In fact, the formal standards bodies include and exclude the same constituencies. Therefore, in so far as compatibility standards (i.e. market standards) are concerned, standards consortia will probably do the job as well as formal standards bodies do.[45] (In other situations, see below point 2.)

## 6.3 Redefining the Consortium Problem

Standards consortia are defined as a problem. Their procedures are held to be undemocratic and therefore unfit as an instrument of regulatory governance. Does this accurately describe what is at stake? As was discussed in Part II, there are several questionable aspects to the way the consortium problem is defined. They are listed in Table 6.1 and briefly summarised below.

**Democratic standardisation?** [1, 2, 3]. Consortia procedures are held to be unfit for use in a regulatory governance context because they are undemocratic. However, should the standards setting primarily be seen as an extension of regulation? A sharp distinction should be made between *de jure* uses of standards that touch on issues of health, safety, privacy, environment, etc., on the one hand, and compatibility standards, on the other. Consortia usually address the latter. This type of standardisation is more aptly characterised as market coordination among competitors. How this takes place, that is, whether consortia operate in true or quasi-democratic way (e.g. multi-party standardisation), the level of democracy is here foremost a marketable asset of the standard. The consumer decides its value.

*Table 6.1. The shift in perspective needed in standards policy based on an analysis of the consortium problem.*

| | Aspect of the Consortium Problem | as defined | as redefined |
|---|---|---|---|
| 1 | primary characterisation of the standards setting | regulatory governance | market coordination among competitors |
| 2 | democratic standards process •compatibility standards | consensus, well-balanced participation of interest | multi-party |
| | •health/safety/enviro nm./ etc. standards | groups | (discussion needed to further refine 'democratic standards needs' ) |
| 3 | policy | single | twofold |
| 4 | aim of technical co-ordination | standardisation | compatibility |
| 5 | stage of standardisation emphasised | standards development | standards implementation |

For *de jure* uses, where the public interest is involved, the democratic requirements of the European government need to be re-examined and defined more sharply. Once these are clarified and operationalised in measurable terms, 'democratic standardisation' need not be restricted to the formal standards bodies. Because of this difference between the use of standards in a market and in a *de jure* public interest context, in this respect a more specific, twofold standards policy for each of these areas seems appropriate.

**Co-ordination of Technology: Compatibility** [4]. The 'consortium problem' is defined as a standards development problem. Standardisation, whether by means of formal standards committees, hybrid workshops (e.g. CEN/ISSS workshops), or specification consortia, it is one means to co-ordinate technology development. However, the compatibility objective can also be achieved by other means. Other compatibility strategies are, for example, proprietary-led multi-party specification development, and the open source approach to software development.

Many issues that seem very important from a standardisation standpoint take on a different meaning in the light of the compatibility objective. For example, the distinction between specifications and standards becomes unimportant; and, instead of concerns about non-consensus consortium standards, there should be relief about the fact that companies prefer standardisation above proprietary strategies. De facto compatibility should be the central issue.

**Market Co-ordination by Specification and Strategic Consortia** [5]. Most specification consortia are implementation-oriented. They aim at coordinating technology development in a multi-vendor environment. They succeed if companies implement the specifications. The result is, ideally, a co-ordinated segment of the market. This outcome requires the support of a business community. Strategic consortia focus on developing such communities. However, the 'problem as defined' almost exclusively emphasises standards development issues. Instead issues concerning standards implementation should acquire more emphasis. A re-definition of this aspect of the consortium problem is required.

## 7. CONCLUSIONS AND RECOMMENDATIONS

**Beyond Consortia.** Current standards policy appears to be caught up in a polarised discussion about which category of organisations best serves the market for democratic and timely standards: standards consortia or the traditional formal standards bodies. It is an unhelpful discussion, this framework of rivalry.

Let us, for a moment, go along with it. In the discussion, consortia are seen as a problem. They lack an open and democratic standards process. However, neither recent literature nor the case studies in this report can confirm this. The findings show that in theory the standards committees of both settings strive for consensus and address minority viewpoints, while in practice both largely include and exclude the same constituencies. The framework of rivalry merely leads to new hybrid forms of organisation like the CEN workshops, which, speculating somewhat, will not lure companies away from consortia but instead lead to a shift within the CEN standards domain. Moreover, it by-passes the more significant difference between standardising and not-standardising. The real issues lie elsewhere. These are discussed below as main threads for standards policy. (For more detail, I refer to the previous chapter.)

*Democracy: beyond rhetoric.* European standards policy shows too little interest in whether democratic standards procedures provide the desired democratic accountability or not. Where a *de jure* need for standards exists, that is, where standardisation touches on aspects of health, safety, environment, privacy, security, etc., should not the regulator's concern for democratic accountability be given more substance? Firstly, more clarity would need to be created about what type of democracy is needed and for what purpose – in practice. This is a political decision. Secondly, the Commission should monitor systematically if 'democratic' standards developing organisations follow-up the democratic requirements, be they formal standards bodies or consortia. For where democratic accountability is still important, insight into the factual democratic course of the standards process is needed. For market coordination, on the other hand, the democratic requirement of 'balanced representation of interest groups' could be simplified to 'multi-party participation'.

A differentiated standards policy is recommended to better cater to the significance of standardisation as a means to coordinate the market and as an instrument of regulatory governance. Differentiation prevents a situation where democratic (or other political) ideals are diluted in order to be able to apply a market-oriented standards policy to *de jure* situations -or, as presently happens, vice versa. In this scenario, the assignment of a work item to either the multi-party or to the more demanding 'democratic' standards environment becomes an important decision. An interesting case for gaining experience about problems of assignment would be the ITU-T's recent introduction of a two-fold track of the Traditional and the Alternative Approval Process for *de jure* and non-*de jure* standards.

**Beyond Standardisation.** Standardisation -whether by means of formal standards bodies, hybrid workshops or specification consortia -is one means to achieve technical compatibility. There are other means to this end as well. In standards policy the vantage point of compatibility should take priority. It puts into perspective the distinction between (a) standards and specifications (de facto standards), and (b) formal and consortium standards. These distinctions, which may seem very important from the standardisation standpoint, take on a different meaning in the light of compatibility aims. In this light, both standardisation and software development are specification processes, and ownership and property issues (open source/ proprietary) are in principle irrelevant. The primacy of standards' implementations and compatible technologies then comes back into focus. For example, a proprietary multi-party specification (e.g. Java) and a standard stemming from a consortium led by a 'benevolent dictator' (XML from W3C) can be at least as effective in fostering compatibility as a formal standard. A gap appears to exist between outcome-oriented market practices and process-oriented governance ideals, which standards policy will need to address. Compatibility can only be measured as an aspect of the market, and not as an aspect of the specification process.

In addressing standardisation, current policy should not overemphasise standards development activities; it should focus more on standards implementation and market co-ordination. Furthermore, it is recommended that companies and governments re-assess their standardisation policy from the *de facto* compatibility standpoint.

Whether or not the European Commission should involve itself in other compatibility strategies than standardisation (e.g. licensed software specification processes) -or even as little as possible in standardisation -is a matter for debate. For the moment, this debate should be kept open. It should not be closed beforehand with reference to the danger of reinforcing monopolies. The Java case study suggests that in certain circumstances, the public's primary interest is in solutions that prevent unnecessary fragmentation of the market. Since there are few legal means to safeguard the public interest in compatibility, if the latter coincides with a company interest, proprietary solutions deserve consideration. At the same time, we should not overemphasise the idea of compatibility control: the case also shows that even tightly controlled compatibility strategies (e.g. licensing combined with IPRs) cannot prevent incompatible developments (Microsoft's use of Java).

*Institutionalisation of public compatibility interests: coordination of demand and legislation.* In particular in the current immature state of the ICT field, the supply-side of the market often lacks the necessary incentives to prioritise compatibility. What mechanisms does the public, the demand-side of the market, have at its disposal to advance collective compatibility interests? This question has arisen in two different contexts: (a) while comparing the effectiveness of different compatibility strategies, and (b) in relation to Microsoft's attempts to fragment the Java platform.

a. *Comparing the effectiveness of different compatibility strategies.* The standardisation and open source software strategies suffer from the same problem: whether they lead to compatible products is not clear. Taking the example of standardisation, compliance to (voluntary) standards cannot be enforced. Demand-driven conformance to standards is needed if standardisation is to lead to actual compatibility. However, in the post-standardisation phase there are no mechanisms that coordinate shared consumer interests in compatibility. This contrasts strongly with the diversity of coordination mechanisms used on the supply-side of the market. Discussions about supporting users and user coordination *during* the standards process should be extended to include the *post-standardisation phase* with the aim of fortifying demand-side interests in compatibility.
b. *Maintaining the integrity of a platform.* A regulatory asymmetry exists between IPR interests and compatibility interests. Current regulation anchors the primacy of IPR ownership and market competition in law, but it hardly recognizes the societal significance of compatibility interests (i.e. technical interoperability). Would it be desirable to legally anchor compatibility interests in a way similar to intellectual property interests?

Much research remains to be done. Among other things, the actual compatibility effects of -formal and consortium -standardisation and the open source approach remain uncertain. For example, little is known about (a) the participatory specification process of Open Source Software, which could contain leads for improving or diversifying standards development; and (b) whether the process and outcome of developing open source software involve problems of compatibility (e.g. upward compatible or stable source code). Case studies are needed to throw light on these issues.

A second area of research which this study points to, is the effectiveness of different compatibility strategies. A more systematic inventory of the compatibility implications of market strategies is needed to supplement the findings of the Java case. Of interest is which other means exist to enhance compatibility and how their contribution can be measured (i.e. quantify effectiveness).

## REFERENCES

Abbate, J. (1994) *From ARPANET to INTERNET: A history of ARPA-sponsored computer networks, 1966-1988*. Dissertation. University of Pennsylvania.

Besen, S.M. & Farrell (1990). The European Telecommunications Standards Institute, A preliminary analysis. *Telecommunications policy*, December 1990, pp.521-530.

Bonino, M.J. & M.B. Spring (1991). Standards as change agents in the information technology market, *Computer Standards & Interfaces 12*, pp. 97-107.

Bruins, Th. (1993). *Open systemen*. PTT Research, the Netherlands.

Cargill, C.F. (1989). Information Technology Standardisation. Theory, Process and Organisations. Digital Press, Digital Equipment Corporation.

Cargill, C. (1999). 'Consortia and the Evolution of Information Technology Standardisation', in: K. Jakobs & R. Williams (Eds.), *SIIT '99 Proceedings,* IEEE, pp.37-42.

Cargill, C. (2000). 'Evolutionary pressures in standardisation: Considerations on Ansi's national standards strategy', presented at *The role of standards in today's society and in the future*, Subcommittee on Technology of the Committee on Science, U.S. House of Representatives, September 13, 2000. [www.house.gov/science/cargill_091300.htm]

CEN/ISSS (2000). *CEN/ISSS survey of standards-related fora and consortia,* 4th edition, Brussels, June 2000. [www.cenorm.be/isss]

CRE, Center for Regulatory Effectiveness (2000). *Market-Driven Consortia, Implications for the FCC's Cable Access proceeding.* Working Draft 7/20/00. [www. the CRE.com]

Council of the European Union (2000). 'Council Resolution of 28 October 1999 on the role of standardisation in Europe', *Official Journal of the European Communities, 2000/C 141/1-4*, 19.5.2000.

David, P.A. & S. Greenstein (1990). The economics of compatibility standards: an introduction to recent research. *Economics of Innovation and New Technologies, Vol. 1,* pp.3-41.

Dosi, G. (1982). Technological paradigms and technological trajectories. Research Policy, 11, pp.147-162.

Egyedi, T.M. (1994). *Grey fora of standardisation: a comparison of JTC 1 and Internet* .Leidschendam: KPN, KPN Research, R&D-RA-94-1235.

Egyedi, T.M. (1996). *Shaping standardisation: A study of standards processes and standards policies in the field of telematic services.* Dissertation. Delft, the Netherlands: Delft University Press.

Farrell, J. & G. Saloner (1986). Installed base and compatibility: innovation, product preannouncements and predation. *American Economic Review, 76,* pp.940-955. O'Gara, M. (2000). 'IBM Threatens Unremitting Sun with Counter-Java'. *CSN 346-01.* Grindley, P. (1995). *Standards, Strategy and Policy: Cases and Stories.* Oxford, New

York: Oxford University Press. Hawkins, R. (1998). *Standardisation and Industrial Consortia: Implications for European Firms and Policy.* ACTS/SPRU, Working Paper 38, March 1998.

Hawkins, R. (1999). 'The rise of consortia in the information and communication technology industries: emerging implications for policy', *TelecommunicationsPolicy, 23,* pp.159-173.

Hawkins, R. (2000). *Study of the Standards-related Information Requirements of Users in the Information Society.* Final report to CEN/ISSS, 14 February 2000. [www.cenorm.be/isss]

ISO/IEC (1991). *Guide 2: General terms and their definitions concerning standardisation and related activities.* Geneva.

ITU (2000). *Alternative approval process for new and revised Recommendations. ITU/T Recommendation A.8*, August 2000.

Krechmer, K. (2000). 'Market Driven Standardisation: Everyone Can Win.' *Standards Engineering, 52 (4)*, pp.15-19.

O'Mahoney, B. (2000). "Software issues", [www.benedict.com/news/resume].

Rada, R. (1998). Corporate Shortcut to Standardisation, *Communications of the ACM 41(1),* pp.11-15.

Rada, R. (2000). 'Consensus versus Speed', in: K. Jakobs (Ed.), *IT Standards and Standardisation: A Global Perspective*, London: Idea Group Publishing, 2000, pp. 19-34.

Schmidt, S.K., & R. Werle (1998). *Co-ordinating Technology. Studies in the International Standardisation of Telecommunications.* Cambridge, Mass.: MIT Press.

Updegrove, Andrew (1995). 'Consortia and the Role of the Government in Standard Setting', in Brian Kahin and Janet Standards Policy for Information Infrastructure, Cambridge: MIT Press, pp. 321-348.

Vercoulen, F. , & M. van Wegberg (1999). Het spel van samenwerking en concurrentie: Hybride standaard-selectieprocessen in de ICT sector, *I&I 17*, pp.19-

27. Weiss, M., & C. Cargill (1992). 'Consortia in the standards Development Process', *Journal of the American Society for Information Science, 43(8)*, pp. 559-565.

Weiss, M.B.H. & M. Sirbu (1990). Technological choice in voluntary standards committees: an empirical analysis. *Economics of Innovation and New Technology, 1*, pp.111-133.

W3C (2001).*World Wide Web Consortium Process Document,* 8 February 2001. [http://www.w3.org/Consortium/Process-20010208/ ]

## APPENDIX I: INTERVIEWEES

**Java case:**
Jan van den Beld (ECMA, Secretary General)Carl Cargill (Sun, Director of Standardisation)Roger Martin (Sun, Standardisation Strategy Manager)Wim Vree (Delft University of Technology)Informal interviews were held with participants to the ECMA Technical Committee41 meetings.

**XML case:**
Arjan Loeffen (Salience B.V.)Pim van der Eijk (eXcelon Nederland B.V.)Diederik Gerth van Wijk (Kluwer B.V.)Charles Goldfarb (SGML inventor/ XML expert)

**General:**
Willem Wakker (ACE Consulting)

## APPENDIX II: LIST OF PUBLICATIONS ON THE PROJECT

Egyedi, T.M. (2000a). Compatibility strategies in licensing, open source software and formal standards: Externalities of Java 'standardisation', *Contribution to the 5th Helsinki Workshop on Standardisation and Networks, 13-14 August 2000.* Revised versions published as: Egyedi, T.M. (2000b). 'Compatibility Strategies in Licensing, Open Sourcing and Standardisation: The case of Java$^{TM}$'. In: H. Coenen, M.J. Holler & E. Niskanen (Eds.), *5th Helsinki Workshop on Standardisation and Networks, 13-14 August 2000.* Government Institute for Economic Research, Helsinki, VATT-Discussion Papers 243, pp. 5-34.

Egyedi, T. (2001a). 'Strategies for *de facto* Compatibility: Standardization, proprietary and open source approaches to Java', *Knowledge, Technology and Policy,* 14 (2), pp.113-128.

Egyedi, T.M. (2000c). Judicio-Standardisation regime: Exploring the grounds for IPR paralysis, *Contribution to the European ITS conference, Lausanne, 11-12 September 2000.*

Egyedi,T.M. (2001b). Why Java™ was -not -standardised twice, *Computer Standards & Interfaces, 23/4*, pp. 253-265. [Reprinted with permission from:

R.H. Sprague (Ed.), *Proceedings of the 34th Annual Hawaii International Conference on System Sciences 2001.* Abstracts and CD-ROM of Full Papers. California, Los Alamitos: IEEE Computer Society.

Egyedi, T.M. (2001c). 'IPR Paralysis in Standardisation: Is Regulatory Symmetry Desirable?', *IEEE Communications Magazine, 39(4), April 2001,* pp.108-114.

Egyedi, T.M. (2001d), 'The Problem of Standards Consortia: Analysis and Redefinition', in: W. Hesser (Ed.). *Proceedings of the Third Interdisciplinary Workshop on Standardization Research, University of the Federal Armed Forces Hamburg, September 23-25, 2001,* pp.75-103.

Egyedi, T.M & J. Hudson (2001), 'Maintaining the Integrity of Standards: The Java Case', in: K. Dittrich & T.M. Egyedi (Eds.), *Standards Compatibility and Infrastructure Development: Proceedings of the 6th EURAS Workshop, 28-29 June 2001, Delft, The Netherlands,* pp. 83-98.

Egyedi, T.M. & A.G.A.J. Loeffen (2002). 'Succession in Standardization: Grafting XML onto SGML', *Computer Standards & Interfaces, 24(4),* pp. 5-16 . [IEEE 2001, Reprinted with permission from T.D. Schoechle, C.B. Wagner (Eds.), *Proceedings of Standardization and Innovation in Information Technology (SIIT2001)*, October 3¯5, 2001, University of Colorado, Boulder, CO, USA, pp. 38¯49.]

Egyedi, T.M. & W.G. Vree (August, 2000). *Consortium standardisation and restructuring the ICT market..* Intermediate report for the European Commission.

Egyedi, T.M. & W.F. Wakker (2000). Java™, een geplaagde *de facto* standaard, *i&i, 18(5), pp.8-16.* [Information and Information Policy, Dutch magazine].

## ENDNOTES

1. The scope of this research does not include some interesting, recent phenomena such as the significance of the CEN workshops and the meaning of the Open Source phenomenon for standardisation. These settings deserve separate attention.
2. Other sources of what Bruins (1993) calls grey standards, are trade-or profession-oriented organisations (e.g. IEEE, ASE), and organisations like the IETF, a non-commercial multi-party forum, that work towards a specific environment (see e.g. Abbate, 1994; CRE, 2000; Egyedi, 1994, 1996).
3. The standards process was to be concluded within the initial period of the EU research grant (15 January 2000 -15 March 2001).
4. The procedure for the Transposition of Publicly Available Specifications into International Standards is based on the Fast Track process (see next note). It also allows an external organization to submit its specification as a draft International Standard, which means that according to JTC1's aims, the transposition can be completed within 11 months (ISO/IEC JTC1, 1999b). But the criteria for becoming a recognized PAS submitter are less restrictive than those for an A-liaison membership.
5. The Fast Track process is an option for consortia and other multi-party fora that have an A-liaison membership status in JTC1. The A-liaison status is meant for organizations that contribute actively to JTC1 standards committees (e.g. ECMA and IEEE). It gives access to the Fast Track procedure: an A-liaison member can submit its specification as a final Draft International Standard -and thus skip the prior phases of the JTC1 standards process. This procedure strongly reduces the time needed for standardization. ("The duration of the final ballot, to become an IS ballot is six months." (ISO/IEC JTC1, 1999a)
6. Two other significant exceptions are Adobe's PDF-format (ISO/DIS 15929; 15930) and HTML (ISO/IEC 15445:2000) which was offered first to the IETF and later to ISO.
7. APIs comprise the standard packages, classes, methods and fields made available to software developers to write programs (Sun, 1997c).
8. Of interest is that Sun had taken first steps to formalize Java through ANSI, the first most obvious step for a U.S.-headquartered company. According to Sun's head of standardization, Sun did not pursue this route because of "arcane and potentially obstructionist processes" in ANSI (Cargill, 2000).
9. The EICTA position paper is referred to in CSN 337-03 Sun's Up To Something.
10. Informal communication with ECMA TC41 participants. The ruling was confirmed in January 2000. Sun's compliant against Microsoft for unfair competition was granted. (Sun, 2000b).

[11] Compatible succession -or 'grafting' as it is called in Egyedi & Loeffen, 2002 - refers to a situation where software products that comply to the successor standard also interoperate with products based on its predecessor. Such is typically the aim when the successor is a new edition or a minor revision of a standard. Concerned are incremental innovations, where the problem addressed by the old standard has not changed and -in essence -neither has the means to solve it. Both standards are part of the same technological paradigm (Dosi, 1982).

[12] Jon Bosak: "Re: Welcome to w3c-sgml-wg@w3.org!", one of the first submissions to the discussionlist of the W3C SGML Working Group, contribution to w3c-sgml-wg@w3.org discussion list, Aug 281996.

[13] Tim Bray quoted in Charles F. Goldfarb, 'Re: Compliance with 8879, a moving target', 12 Sep 1996.

[14] Private communication.

[15] Jon Bosak, 'W3C SGML WG: The work begins', contribution to w3c-sgml-wg@w3.org discussion list, 5 Sep 1996

[16] Jon Bosak, 'W3C SGML WG: The work begins', 5 Sep 1996.

[17] Two other aims are "For any XML document, a DTD can be generated such that SGML will produce "the same parse" as would an XML processor.", and "XML should have essentially the same expressive power as SGML."

[18] Eve L. Maler, 'Compatibility issues and principle #3', contribution to w3c-sgml-wg@w3.org discussion list, Sep 16 1996.

[19] David G. Durand, 'Last unstructured discussion: SGML compatibility', 9 Oct 1996, contribution to w3c-sgml-wg@w3.org discussion list.

[20] "Re: Capitalizing on HTML (was Re: equivalent power in SGML and XML)," C. F. Goldfarb, contribution to w3c-sgml-wg@w3.org discussion list, Sept. 19 1996.

[21] Charles F. Goldfarb, 'Re: Make DTDs optional?', contribution to w3c-sgml-wg@w3.org discussionlist, 30 Sep 1996)

[22] Tim Bray, 'XML, HTML, SGML, life, the universe, and everything', contribution to w3c-sgml-wg@w3.org discussion list, 08 Nov 1996.

[23] Tim Bray, 'Recent ERB votes', 06 Nov 1996. In: Reports From the W3C SGML ERB to the SGMLWG And from the W3C XML ERB to the XML SIG, Compiled for the use of the WG and SIG by C. M. Sperberg-McQueen, 4 December 1997.

[24] Cargill (1999) notes a number of advantages. Consortia are generally more under control of their members than the formal standards bodies; they have more financial room to manoeuvre; and they need not forgo the public procurement market of the US government, which in practice also accepts consortium specifications (In the OMB circular of the United States it says that no prefer-

ence is given to formal (consensus) standards in respect to (non-consensus) consortium standards. CRE, 2000).

25 The democratic rationale is: A democratic standards process best serves the aim of widespread, international use of standards and specifications (e.g. compatible products, consumer protection, etc.). This rationale has not been checked. See Egyedi (1996).

26 Tim Bray, "Boston, for those who weren't there', 22 Nov 2006 (contribuition to w3c-sgml-wg@w3.org discussion list)."

27 Egyedi 1996, section 4.3.2

28 Placing on the website may occur from a few days to 4 weeks after approval of the text. Approval Time runs from determination/consent to final approval. (Source: "The IT Standardization and ITU", presentation of Houlin Zhao, Director of the Telecommunication Standardization Bureau of the ITU, at the IEEE SIIT 2001 conference in Boulder US, 3-5 October 2001

29 ECMA procedures are less explicit on this account than W3C procedures

30 High membership fees are often viewed as a stumbling block for participation by small-and medium-sized enterprises (SMEs) and users. However, criticisers are usually not aware, firstly, of special lower fees for non-profit organisations; secondly, of the costs incurred by participating in formal standardisation (travel, hotel and subsistence costs of going to committee meetings, the membership fees (e.g. ITU annual fee 20.000 $; associate fee 6.000 $); thirdly, of the -often publicly accessible -discussion lists of consortium standards committees, and invitations to participate in external reviews.

31 Although an electronic mailing list was installed, the committee was disbanded before it had been put to use for technical discussion.

32 W3C has been accused of rubberstamping the products of major vendors because of the 'member submission process' (Rada, 2000 p.22). This process makes it possible to consider proposals developed outside of W3C. The W3C rules explicitly state that this process is "not a means by which Members ask for 'ratification' of these documents as W3C Recommendations."

33 In exceptional cases, consortium standards are likely to be relevant as indirect means for market governance, that is, in situations of complex market coordination problems where government support is needed to break dead-locks in the market. As such, consortium standards are relevant as part of the European Union's public procurement policy.

34 In the alternative process the approval time is 2-9 months and publication time is 3-9 months Approval Time runs from determination/consent to final approval. (Source: "The IT Standardization and ITU", presentation of Houlin Zhao, Director of the Telecommunication Standardization Bureau of the ITU, at the IEEE SIIT 2001 conference in Boulder US, 3-5 October 2001.

[35] Private discussion with Carl Cargill.

[36] The term 'compatibility standard' is used in this chapter to distinguish this category of standards fromsafety and health standards (e.g. Grindley, 1995).

[37] The ISO defines compatibility as the "suitability of products, processes or services for use togetherunder specific conditions to fulfil relevant requirements without causing unacceptable interactions."(ISO/IEC, 1991) ICT practitioners also use the term 'interoperability'.

[38] The term 'standardisation' refers here to activities that are exclusively set up to lead to standards andthat take place within formal standards bodies such as ISO or in standards consortia such as W3C (i.e.multi-party industry standards fora).

[39] Presently, an EU project on the impact of standardisation is taking place, which promises to throwlight on the subject (see www.standardsimpact.org).

[40] See for an example Grindley, 1995, p.140 a.f.

[41] NB: Other case studies would probably show up additional strategies.

[42] The two types of co-ordination are complementary explanatory frameworks-although specific strategies may sometimes be explained by both.

[43] The license is 'viral': all changes to the source code automatically become part of the software commons (Op cit. from 'The world is taking to open source', Apr 12th 2001, From The Economist print edition, Opinion, Economist.com, Out in the open.)

[44] Although, when the source code diverges at least with open source software the differences are retraceable.

[45] NB: Quantitative research is needed to confirm this.

# APPENDIX 6: THE "DIN PAPER"

# STRATEGY FOR STANDARDISING INFORMATION AND COMMUNICATION TECHNOLOGY (ICT)

## 4. GENERAL STRATEGY

In conceptual terms, standardisation in information and communications technology primarily means safeguarding the cooperation of heterogeneous, distributed systems. In other words: interoperability is the main standardisation objective for ICT. This interoperability refers to all levels of the OSI reference model as well, increasingly, to semantic aspects. However, the latter is causing considerable problems since existing systems and processes (cf. product classification in e-business, for example) are typically involved bringing standardisation to its limits if 100% solutions are aimed at.

Interfaces (in hardware and software), communications protocols and data objects are standardised to safeguard interoperability. This means that interfaces can remain stable in the long term in complex distributed systems and investments are thus secured and migration made possible. By contrast, the standardisation of product-related aspects plays a very minor role in ICT.

However, the standardisation objective of ICT is not limited to interoperability. Much rather, the penetration of ICT into all areas of society, especially the consumer area, means that issues of relevance to society are also gaining in importance. Security including biometry and data protection are in great need of standardisation, as are ergonomic aspects, especially with regard to allowing the disabled barrier-free access to new technology (*'disability access', 'design for all'*). These issues require the participation of the general public and are consequently predestined for treatment in standardisation organisations instead of other bodies.

### 4.1 Standardisation Policy Aspects

#### Standards

The development of specifications and standardisation are methods for technical harmonisation called into being and supported by industry – they all have their own justification. This leads to industry's interest in combining these various methods and their products in accordance with their specific strengths and making the competent organisations and bodies cooperate purposefully, where this is necessary.

Development processes and voting methods based on consensus are the optimum precondition for the general acceptance of the standards and fair competition between rival companies. In this connection, to achieve consensus it may be appropriate to aim for a pragmatic "80% solution" instead of a perfect "100% solution" and to treat the remaining heterogeneity with other means (cf. A 4.2.1). However, the search for consensus comes up against its limits where market opportunities are jeopardised by the course of time.

In addition to the development process and voting method, revision and updating as well as consistency among themselves are also important characteristics of standards. Revision, updating and consistency are particularly marked in standards. In the case of standards for short-lived products, revision and updating do not play a major role – specifications that lose their significance when the product comes to the end of its life can be used here. The situation is completely different with standards that serve the purpose of building up the most varied structures (infrastructures) to be maintained in the long term – here the precautions for revision, updating and consistency must satisfy the most rigorous demands.

If specifications prove their value on the market and thus become part of an infrastructure, their transposition into a standard and the transfer of responsibility for revision and updating to a standardisation organisation is a desirable step.

**Statement A 0.1 Need for Standards of Different Consensus Levels**

SICT acknowledges that in the field of ICT there is a need for standards at different consensus levels, depending on the planned application. Consequently, there is a need for standardisation organisations as well as consortia/fora. These can complement each other. However, it lies in the interests of all concerned – the developer and the user of standards – if duplication of effort and conflicting solutions are avoided with cooperation between specification developers and standardisation organisations.

Depending on the planned application, in each case the most suitable form of standard should be developed and used.

**Statement A 0.2 Graduated System of Products of the Standardisation Organisations**

The proliferation of consortia in the ICT field that can be seen has considerable cost consequences for the companies affected: participation in many of these organisations requires high expenditure of personnel and financial resources, irrespective of the duplication of effort resulting from the lack of coordination between the consortia. It is therefore in the interests of the industry concerned to limit the number of consortia.

SICT therefore expressly welcomes the fact that the international (ISO, IEC) and European (CEN, ETSI; CENELEC under consideration) standardisation organisations, as well as DIN, make the establishment of fora and consortia superfluous to a certain extent by meeting the needs of industry by introducing alternative standardisation processes (workshops, etc.) and products (*Workshop Agreements,* etc.) and by a graduated system of standardisation products. SICT calls upon the standardisation organisations to introduce further improvements to the methods in a dialogue with the potential users of such workshops.

**Statement A 0.3 General Main Issues of ICT Standardisation**
SICT recommends that ICT standardisation should primarily focus on those subjects where it can bring its specific strengths into play. In particular there are

- Subjects that concern infrastructures with a long-term effect;
- Subjects where competing technologies do not encourage competition and, in fact, tend to impede it (cf. the VHS and Betamax video systems);
- Subjects with great heterogeneity in the groups concerned with respect to economic region, market share and role in the market;
- Subjects with great public interest (authorities, large group of users);
- Subject with a reference to the cultural peculiarities of individual countries and regions.

**Statement A 0.4 Upgrading Other Standards**
SICT recommends that the standardisation organisations establish methods, if they have not already done so, that allow other standards to be upgraded quickly.

## Standardisation as an Instrument of Government and Policy

Standardisation is an important instrument of (economic) policy. Standards serve

- The technical implementation of the so-called key requirements of European directives in line with the so-called New Concept of the European Union in the regulated area;
- The harmonisation of markets and, thus, the dismantling of barriers to trade or – in a negative case – the isolation of markets by means of the deliberate choice of deviating standards;
- The creation of new markets by achieving advantages of scale;
- The rationalisation of administrative processes (authorities as large-scale users).

**Statement A 0.5 Standards and the Statutory Regulated Area**
SICT supports the application of the New Concept for technical regulation in the European Union in the ICT area, too. Although the co-regulation approach is to be

welcomed in principle, from the point of view of standardisation it has not convincingly demonstrated its usefulness.

SICT calls for the reference to standards for regulation pursuant to the New Concept. The use of other specifications in this connection is rejected.

SICT recommends that the standards organisations offer their services for developing the specifications needed for self-regulation to the main parties involved in co-regulation. These can also be *Workshop Agreements* (CWAs, IWAs) if this corresponds to the wishes of the main parties involved. A binding application of *Workshop Agreements* is rejected. Financial support for workshops that is not oriented to the interests of the main parties involved is also rejected by the European Commission.

## International, European and National Standardisation

**Statement A 0.6 Primacy of International Standardisation**

SICT firmly emphasises the primacy of international standardisation because the ICT market is a global market. In this connection, persistent efforts are being made for strong impetus for international standards to come from Germany.

SICT firmly emphasises that a splintering of international standardisation into further organisations other than ISO, IEC and ITU-T would run counter to the goal of uniform international standards and vehemently rejects this.

In addition to this strong international orientation, European integration continues to play an important role. However, European standardisation only makes sense in conjunction with top international developments or pilot applications or with regionally restricted solutions. Alternative or rival developments to international standardisation should be rejected. This means that SICT recommends that European ICT standardisation should always be assessed from the aspect of the ability to transfer and apply the results at an international level. German participation in European projects should be primarily based on this consideration.

By contrast, purely German standardisation in information and communications technology is not supported unless it expressly and exclusively serves to provide a strong impetus for international standardisation.

**Statement A 0.7 National Delegation Principle and the Role of National Standardisation Organisations**

The national delegation principle for participation in international or European standardisation allows all concerned appropriate participation in the development and decision-making processes. SICT therefore recommends that this principle also be retained in ICT standardisation.

With the increasing supranational character of ICT standardisation, a priority task for national standardisation organisations is to give all groups concerned, especially SMEs and consumers, a platform for participation at international or European level.

**Statement A 0.8 Language in ICT Standardisation**
In information and communications technology, standards are preferably drawn up atinternational level and applied in the original version and language.The time-consuming development and updating of standards at three levels (international,regional and national) does not bring any value added for the groups concerned in the ICT field.In the field of ICT, standards should therefore be developed at international level and applied inthe original version.It is then necessary or sensible to transfer them into regional or national standards

- If matters relating to safety technology are dealt with,
- If the users cannot use the standard in the original version,
- If technical harmonisation/interoperability cannot be achieved in any other way.

In these cases, too, the transfer and update should be done with as little effort as possible.

## Market Orientation

Standards that do not comply with market needs do not develop an effect; developing them is a waste of resources. The difficulty is recognising market needs when initiating a project with a certain degree of reliability or recognising declining market interest on time. If the players in a standardisation project are only economically interested partied, it can be assumed that a project meets the market needs. However, the situation is different in standardisation with a potentially broad spectrum of players: with the exception of standardisation in the context of the New Concept, the market orientation of standardisation projects should be strictly reviewed and ensured.

**Statement A 0.9 Ensuring Market Orientation**
The market orientation of standardisation projects shall be ensured by: appropriate criteria for evaluating projects and their strict application, respecting minimum participation, consistent project management and halting projects if the timeframe is exceeded, appropriate involvement of all participants, even in the state sector, in the project costs. .

## 4.2 Standardisation and Remaining Heterogeneity

The previous sections contained clear indications for the fact that the traditional methods of standardisation are coming up against their limits in ever more sectors (such as virtual specialist libraries) – in spite of all the improvements to the details of the methods. On the one hand, they appear to be indispensable and clearly increase quality and efficiency in some areas. On the other hand, they can only be implemented to a certain extent within the context of global provider structures and changed framework conditions, against a background of rising costs. That is why the previous standardisation concepts should be reconsidered. The remaining and unavoidable heterogeneity must be countered by various intellectual and automatic methods for subsequent conceptual integration. A new way of looking at the remaining call for maintaining consistency and interoperability is needed. It can be described with the following premise: standardisation should be thought of from the point of view of the remaining heterogeneity. Only with the joint interaction of intellectual and automatic methods for dealing with heterogeneity and standardisation is there a strategy for a solution that does justice to the current technical, political and social framework conditions.

Only at first glance does it seem to be a contradiction that considerations about deregulation, such as the acceptance of the standards that can only be implemented incompletely, will simultaneously lead to a strengthening of efforts for integration. However, if standardisation is not viewed as an independent or primary process to which all others have to subject themselves, but as a method that is thought of from the point of view of the remaining heterogeneity and is modelled and implemented in terms of this, conditions for consistency and interoperability that can be used under current conditions.

**Statement A 0.1 Remaining Heterogeneity**
If there is no success in some areas to implement standardisation, or to do so only partially, with reasonable time being spent, the consequences of the lack of standardisation and how the remaining heterogeneity can be at least approximately countered by automatic or intellectual methods should be specifically analysed for every remaining detail. The costs and concessions to quality that may result should be compared to the effort and the prospects of success of further strengthened attempts at standardisation.

## 4.3. Operative/Organisational Aspects

**Statement A 0.1 Interdisciplinary Work**
ICT standardisation frequently calls for interaction between experts of several specialist areas. SICT therefore recommends that the existing fixed structures in

the standardisation organisations should be supplemented with flexible structures and ways of working that not only support this interdisciplinary work, but also encourage it in the long term. The establishment of strategically oriented bodies that set targets for cooperation should also be considered. For example, one solution could be project-related teams that are called by a cross-body board and report to it. (ICT SB)

**Statement A 0.2 Cooperation between ISO, IEC and ITU-T in the Field of ICT**

In line with the convergence character of ICT, relevant standardisation work affects the traditional field of work of all three international standardisation organisations, ISO, IEC and ITU-T. SICT therefore firmly calls upon ISO, IEC and ITU-T to intensify their cooperation in the field of ICT standardisation and to cooperate in a pioneering fashion.

**Statement A 0.3 Cooperation with other Specification Developers**

SICT emphasises (cf. 4.1.1) the need to cooperate with other specification developers in order to largely avoid duplication of effort and conflicting solutions. An important aid is the use of the world wide web to inform third parties about existing standards as well as projects currently being worked on.

The future structure of international ICT standardisation should be one of a cooperative network of conventional standardisation organisations and standard developers of a new type. SICT recommends that the standardisation organisations in this network should take on the active role of a catalyst in initiating standardisation work by identifying need and expressing requirements. The standardisation organisations should aid for moderation, especially with apparent conflicting recommendations and those with legal consequences.

An infrastructure that supports interaction and documentation beyond language and national borders is needed for the assumption of an active role in the desired network of standardisation networks. SICT recommends that the standardisation organisations build up the required structures on the internet. The establishment of a cooperation platform structurally based on the ICT Standards Board in Europe (ICTSB), for example, should be considered.

SICT recommends that the standardisation organisations review their processes for adopting other suitable standards into the set of standards together with other standard developers and to design them in such a way that they encourage this adoption and do not hinder it.

**Statement A 0.4 Standardisation Alongside Development**

SICT views standardisation alongside development as an important instrument for introducing trial developments to standardisation. Successful projects in standardisation alongside development show that standardisation and innovation are not a contradiction in terms. Moreover, standardisation alongside development can be used to practise other ways of working within a standardisation organisation with a level of consensus deviating from standardisation. However, in this connection, attention must be paid to sufficient coordination within the meaning of A 4.3.3.

**Statement A 0.5 Criteria for the Evaluation of Standardisation Areas and/or Individual Issues**

SICT recommends that objective criteria be developed, according to which an evaluation of the standardisation areas or individual issues with regard to their compliance with the market needs can be made at any point in the process. This evaluation should aim at providing decision-making aids for strategic decisions that need to be made. In particular, these include decisions about setting up new bodies and running their secretariats, initiating new projects, choosing an appropriate level of consensus and/or the appropriate form for publication as well as target data (cf. Statement 4.1.1). In this connection, the conclusions in 3.7.6 should also be considered. The criteria, the evaluation and the decisions derived from them strictly oriented to market relevance must be made transparently at all levels in order to ensure active support for the goals.

**Statement A 0.6 Initiating New Issues**

The initiation of new issues relevant to the market especially requires that the imminent needs within the context of research and development be recognised at an early stage. The support and incorporation of research and development in standardisation must be strengthened beyond the existing levels of the previous extent within the context of standardisation alongside development.

The standardisation organisations should observe technological developments (technology watch) on their own initiative in order to be able to recognise trends at an early stage. They should take the initiative themselves to review new issues for their need for standardisation.

The new way that the standardisation organisations see themselves as a platform for various levels of consensus for standardisation must be made transparent to the consortia and fora in particular that are to be involved to a greater extent, with the framework conditions applicable to all forms of publication being emphasised: strict project management with time and resource planning, milestones, reviews, criteria for abandonment.

As the initiation of new issues after evaluation of the market relevance on the basis of objective criteria (cf. Statement 4.3.5) also presupposes active perception of the interests, those involved must undertake to provide the necessary resources.

**Statement A 0.7 Public Relations/Information Policy/Transparency**
The transparency of the work of standardisation organisations must be permanently improved. In this connection, the instrument of the internet/WWW/email should be used more intensively. This transparency starts with initiating new projects and includes the development of draft standards. SICT recommends that interested members of the public be given the opportunity, via suitable internet fora, to follow and comment on a development project; the power to make decisions, however, will remain with the competent standardisation body.

**Statement A 0.8 Marketing**
Standardisation organisations are in competition with other standardisation organisations in the field of ICT. However, they cannot assume that the need for new standards will be brought to them. Much rather, the standardisation organisations must actively try to introduce new interested parties to standardisation and to meet the need for standardisation with all of the instruments available (cf. A 4.1.2). To do this, marketing should be intensified in the long term.

**Aussage A 0.9 Project-Related Work**
SICT recognises that the standardisation organisations have already made great efforts to meet industry's demands for targeted standard development that is on time. However, more stages are needed. Professional project management using appropriate tools should be introduced throughout and the staff should be trained accordingly. However, for the effectiveness of the project management methods it is also necessary for all involved in the standardisation to be prepared to enter into effective obligations to provide resources.

**Statement A 0.10 Other Products of the Standardisation Organisations**
Standards and other standards are addressed to a specialist audience, e.g. product developers or testers, who are usually familiar with the state of affairs described in the standard in questions and its technical environment. In addition, however, there is a large audience that does not meet these requirements and that has a great need for introductory literature, maybe in the form of tutorials or recommendations for use. Standardisation organisations, especially those at national level, have specialist staff suitable for meeting this need. This and other services by the standardisation institutes, such as specialist representation in international and European bodies with their own staff, can also help to reduce the great dependence of financing the standardisation infrastructure on the sale of standards (cf. A 4.3.11 and A 4.3.12).

SICT firmly recommends that new paths should be taken in this respect.

**Statement A 0.11 Electronic Provision of Standards**

Experience shows that the electronic provision of standards can help their distribution in the long term, especially for small and medium-sized companies (SMEs) as well as in the academic field. SICT also points out that ICT standardisation differs from standardisation in other fields due to the 'competition situation' of the bodies described (cf. Error! Reference source not found.) as well as the fact that those involved and interested in standardisation definitely prefer electronic ways of working. Standardisation must take account of this. SICT recognises that income from the sale of standards are indispensable for financing the standardisation infrastructure (cf. Statement A 4.3.12), but demands that other innovative financing models be used in addition for the field of ICT, e.g.:

- The possibility of acquiring the non-exclusive copyright for a standard so that it can subsequently be made freely available. Pricing in line with the economic benefit of the standard for the relevant buyer, e.g. lower price for the academic, by contrast higher price for the manufacturer of a relevant product.

**Statement A 0.12 Financing Standardisation**

SICT emphasises the need for standardisation to finance itself. Consequently, all involved in standardisation should also be involved in the costs appropriately. This also applies to participants from the state area, with a distinction being made between regulatory and application-oriented tasks. SICT recommends that the proportion of member-related financing be increased in order to be able to reduce the prices of the standards (cf. A 4.3.11).

# Index

## D



## E



## F



## G

## H



## I



## J



## M



## N



## O



## P

## R



## S



## T



## U



## V



## W



## X